Roman Nantwich: A Salt-Making Settlement

Excavations at Kingsley Fields 2002

Peter Arrowsmith
David Power

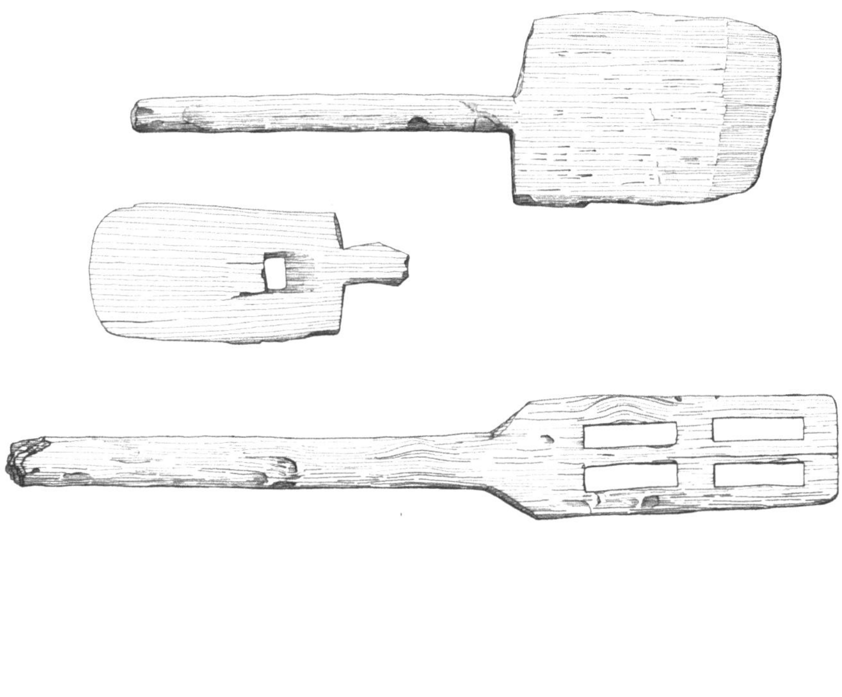

BAR British Series 557
2012

Published in 2016 by
BAR Publishing, Oxford

BAR British Series 557

Roman Nantwich: A Salt-Making Settlement

ISBN 978 1 40730959 0

BAR Publishing is the trading name of British Archaeological Reports (Oxford) Ltd. British Archaeological Reports was first incorporated in 1974 to publish the BAR Series, International and British. In 1992 Hadrian Books Ltd became part of the BAR group. This volume was originally published by Archaeopress in conjunction with British Archaeological Reports (Oxford) Ltd / Hadrian Books Ltd, the Series principal publisher, in 2012. This present volume is published by BAR Publishing, 2016.

Printed in England

BAR
PUBLISHING

BAR titles are available from:

BAR Publishing
122 Banbury Rd, Oxford, OX2 7BP, UK
EMAIL info@barpublishing.com
PHONE +44 (0)1865 310431
FAX +44 (0)1865 316916
www.barpublishing.com

Contents

Please note that the CD referred to above has now been replaced with a download available at www.barpublishing.com/additional-downloads.html

Acknowledgements

The excavation at Kingsley Fields, Nantwich, reported in this volume was carried out by the University of Manchester Archaeological Unit in 2002 and was the subject of an interim report in 2004. The Unit came to an end in 2009, at which time a final account of the excavation, the preparation of a number of analysis reports for the site and the compilation of the results in a single narrative remained outstanding. The present volume, edited by the former UMAU Deputy Directors Peter Arrowsmith and David Power, represents the completion of that process.

The programme of archaeological work at Kingsley Fields was generously funded by Bellway Homes Limited (West Lancashire). The fieldwork was monitored by Mark Leah, Development Control Archaeologist, Cheshire Archaeology Planning Advisory Service. Jill Collens, Project Manager, Cheshire Archaeology Planning Advisory Service, also provided support in the production of the present volume, and the patience and understanding of both has been appreciated.

The archaeological works were managed for UMAU by David Power and were carried out by UMAU field officers Simon Askew, Peter Connelly, Sarah Craig, Graham Mottershead and Philip Wilson, with David Aspden, Steven Bellshaw, Angela Brennan, Victoria Clements, Andrew Dickin, Maria Duggan, Robert Golding, Christopher Heyes, Michael Higgins, Robert Isherwood, John Maudsley, Matthew Ratcliffe and Adam Thompson. Additional help was provided by Liam Bowden, Laura Broughton, Philip Cooke, Ruth Garratt, Nils Mason, Sarah Newton, Malcolm Reid and Adele Shaw. All who participated in the excavation carried out the demanding and frequently difficult work with equal measures of ability, enthusiasm and care, during what at times were very wet conditions. Thanks are due to Kevin Guest, Colin Sharratt and other members of the Crewe and Nantwich Metal Detecting Society for their numerous visits to the site and support of the project. Thanks are also due to the people of Nantwich who supplied water to maintain the stability of the structural Roman timbers during the excavation and exhibited an interest and enthusiasm for the project which was an encouragement to all the professional staff involved. Technical advice during the excavation was provided by Sue Stallibrass, English Heritage. Staff of the Trent and Peak Archaeological Trust assisted with the off-site recording of the structural timbers.

Acknowledgement is gratefully made to the external specialists who have contributed reports included within the present volume: Marks Adams, Museum of Liverpool Field Archaeological Unit; Catherine Barnett, Wessex Archaeology; Nigel Cameron, University College London; Hilary Cool; John Cruse, Yorkshire Archaeological Society; Steve Davis, University College Dublin; Lorne Elliott, Archaeological Services Durham University; Jeremy Evans; Louisa Gidney, Archaeological Services Durham University; Jacqueline I McKinley, Wessex Archaeology; Philip Mills; Quita Mould; Ruth Pelling; Charlotte O'Brien, Archaeological Services Durham University; Cynthia Poole; David Shotter; Jeff Speakman, Museum of Liverpool Field Archaeological Unit; Ian Tyers; and Felicity C Wild. Also to other former UMAU members who have contributed to the volume: Simon Askew, Steven Bellshaw, Peter Connelly and Michael Nevell.

Thanks also for specialist advice and information on the finds to Steve Allen, York Archaeological Trust; Brenda Dickinson; Andrew Fielding; Susanne Jülich, LWL-Museum für Archäologie Westfälisches Landesmuseum; Graeme Lawson, McDonald Institute for Archaeological Research, Cambridge; Susan Sportun, The Manchester Museum; and J P Wild, University of Manchester. Rachel Newman and Christine Howard-Davis, Oxford Archaeology North, kindly provided information on comparative material. Dan Garner and Malcolm Reid kindly made available their forthcoming synthesis of the archaeology of Roman Middlewich.

Drawings in the present volume are by Peter Arrowsmith (Chapters 1 & 2), David Power (Chapters 3 & 17), Michael Hawkes (Chapter 4), Hilary Cool (Chapter 7), Susan Winterbottom (Chapter 9) and Steven Bellshaw (Chapters 10, 11 & 12). The samian ware rubbings (Chapter 5) are by Felicity C Wild. The X-radiographs included in Chapter 7 are by Sonia O'Connor, University of Bradford. Plates 18.1 and 18.2 are by Archaeological Services Durham University, and Appendix B, Plate B.1 by Gifford.

The finds and site archive from the archaeological investigations at Kingsley Fields have been deposited with Cheshire West Museums.

Summary

In 2002 the fullest evidence so far recovered for the Roman settlement at Nantwich, a historic salt-producing centre in Cheshire, was revealed by an excavation carried out at Kingsley Fields, on the west side of the town, ahead of a housing development. This uncovered a previously unknown Roman road, linking the settlement at Nantwich to the main road network, and, positioned along this, evidence for the collection and storage of brine and the production of salt, together with buildings, enclosures, a well and a small number of cremation burials. Waterlogged conditions meant that organic remains, including structural timbers, were well preserved on the site. These included the two finest examples of timber-built brine tanks excavated from Roman Britain. Their fills contained an exceptional assemblage of finds, including metalwork, wooden objects and animal bones, which appear to indicate ritual deposition.

Development along the Roman road began in the Hadrianic period and intensified in the Antonine, before this first phase of occupation came to an end in about the AD 180s. A second main phase of occupation began in the early 3rd century. There is strong evidence, particularly in the 2nd century, of a military presence on the site. Occupation largely ceased in about the mid 3rd century, although brine collection was still taking place on a reduced scale in the late 3rd-4th centuries.

1. Introduction

Peter Connelly & David Power

This volume describes the results of an excavation carried out between January and September 2002 at Kingsley Fields, Nantwich, Cheshire (Fig 1.1). The excavation was undertaken by the University of Manchester Archaeological Unit and was funded by Bellway Homes Limited (West Lancashire) ahead of development. It revealed for the first time evidence of a substantial Roman site located on the north side of Welsh Row, west of the River Weaver. This site was principally industrial in character, with salt production predominating.

Nantwich: A Cheshire Salt Town

Nantwich is one of Cheshire's best-known historic centres and is famous for its impressive collection of late 16th-century timber-framed buildings. These dominate the town core on the east bank of the River Weaver, particularly the area around the parish church, Hospital Street and Pillory Street. They are also found on the west bank of the river along Welsh Row, the road which leads westward from the town towards Chester and Wales (Fig 1.2). These timber-framed buildings reflect the contemporary wealth of Nantwich, which was rebuilt after a disastrous fire in 1583 had reduced much of the earlier town to ashes. The medieval town was equally wealthy, as indicated by the massive size of the 14th-century parish church of St Mary in the centre of Nantwich, originally built as a chapel of ease. Nantwich also had a castle; the line of its defences appears to be preserved today in the curve of High Street and Mill Street, between which lies Castle Street. Excavations at the Crown car park in this area in 1978 and on the north side of Mill Street in 2003 indicated the presence of ditches probably associated with the castle (McNeil Sale 1978; Earthworks 2003).

The cause of Nantwich's prosperity was primarily its role as one of Cheshire's main salt-producing centres. This industry exploited the county's extensive natural reserves of the mineral and involved a process whose basic elements remained unchanged into the modern era. Brine was collected from local brine springs and wells, heated on open pans to evaporate the water, and the salt was recovered as it crystallised out (Fielding 2005).

The Domesday Book provides documentary evidence for the salt industry in Nantwich before the Norman Conquest and shows that in 1066 the town was worth £21 a year, in contrast to the other main salt-making centres or 'wiches' of the county, Middlewich and Northwich, which were each worth only £8 (Sawyer & Thacker 1987, 328). The salt industry continued to thrive in Nantwich into the post-medieval period and appears to have been centred on both banks of the Weaver to the north of the town bridge. The town's brine well was situated on the east bank of the river in this area. Wich houses, premises engaged in salt production, have been located on the west bank around Wood Street. Excavation here in 1979-80 revealed two wich houses, with evidence of activity beginning in the late 12th century (McNeil 1983). More recent excavations in the Wood Street/Welsh Row area, around the former Cheshire Cat public house, have uncovered further wich house remains, dating to the late medieval and early post-medieval periods (Earthworks 2001 & 2005; Dodd forthcoming). These sites retained the remains of timber structures and artefacts associated with salt production, including barrels and 'salt ships', hollowed-out trunks used for the storage of brine. The survival of such well-preserved evidence is due to the extensive waterlogging in parts of Nantwich. This has proved to be as vital to the preservation of Roman deposits as it has been to those of later periods.

The Evidence for Roman Activity in Nantwich

For some time, Roman activity has been reasonably well known at the two other main Cheshire 'wiches', Northwich, situated on the Roman road between Manchester and Chester, and Middlewich, on the north-south Roman road known as King Street (Figs 1.1 & 1.3), and both places have produced evidence of Roman salt production, utilising the local brine springs. Summaries of discoveries made at these sites during the 19th and 20th centuries can be found in the *Victoria County History* (Petch 1987, 198-208).

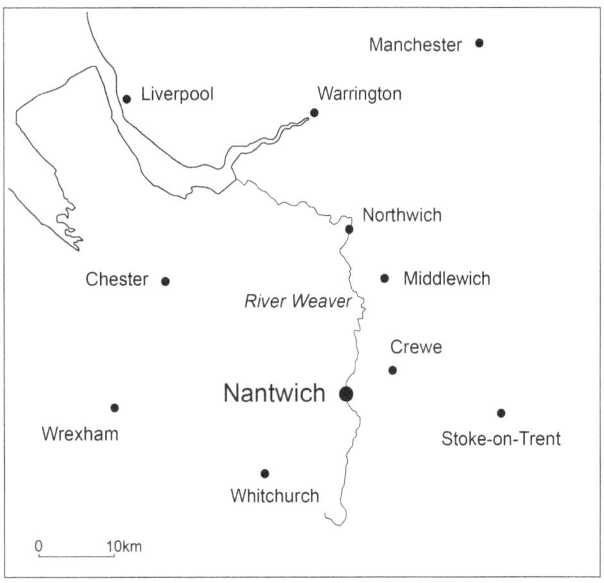

Figure 1.1: Location map of Nantwich.

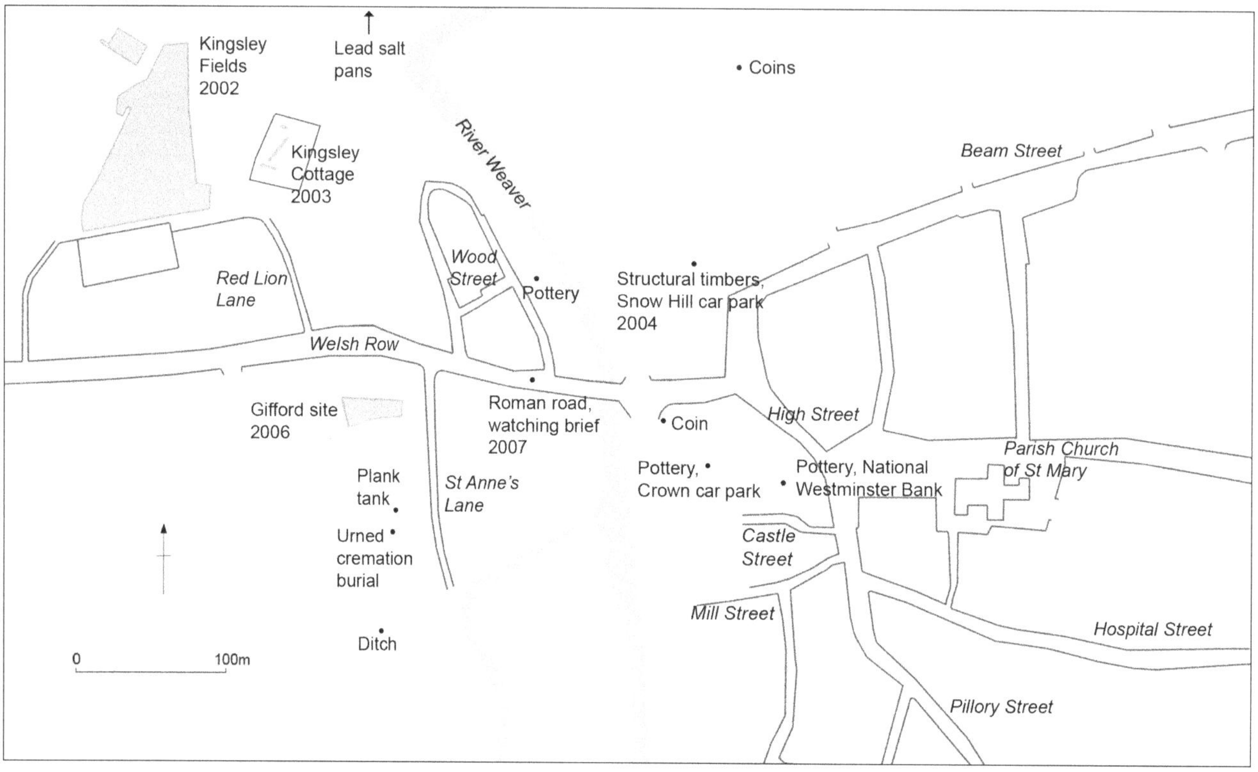

Figure 1.2: Nantwich town centre and Welsh Row, showing the location of the main Kingsley Fields excavation and the sites of other discoveries of Roman features and finds. Based on OS mapping. © Crown copyright. All rights reserved. Licence number 100022765.

Whilst no new information concerning Roman salt making at Northwich has emerged since 1987, archaeological work at Middlewich since the early 1990s has greatly expanded our knowledge and understanding of the extent and development of the Roman settlement, with the discovery of an auxiliary fort and the excavation of several salt-making areas (Strickland 2001; Dodd 2005; Garner 2005; Williams & Reid 2008; Garner & Reid forthcoming).

Outside Cheshire the main evidence for inland salt production in Roman Britain has been provided by excavations carried out between the 1960s and 1980s in the Worcestershire town of Droitwich (Woodiwiss 1992; Hurst 1997). Coastal salt making in the Roman period is mainly known from sites in the Fenland of eastern England (Lane & Morris 2001).

Until the beginning of the 21st century the evidence for Roman Nantwich was fragmentary. There were hints of an important settlement, probably connected with salt making, but no clear picture of the focus of activity or its longevity (Petch 1987, 208-11; Shaw & Clark 2003a, 3, 14) (Fig 1.2).

The Crown car park excavations in 1978 produced around 200 sherds of Roman pottery along with Roman roof tile but these were from residual contexts. This evidence did, however, suggest significant activity on the eastern bank of the River Weaver in the immediate vicinity of the medieval castle. None of the pottery seemed to predate the 2nd century and the bulk was dated between the early 3rd century and the mid 4th (McNeil Sale 1978). A smaller assemblage of *c* 29 sherds was recovered from this same general area during rescue excavations in 1974-6 at the site of the National Westminster Bank (Petch 1987, 209). Other scattered chance finds, recorded in the Cheshire Historic Environment Record, had also been made on the east side of the river and included a number of coins.

On the western bank of the river most of the Roman finds known before 2001-2 came from the area of St Anne's Lane to the south of Welsh Row. These included the

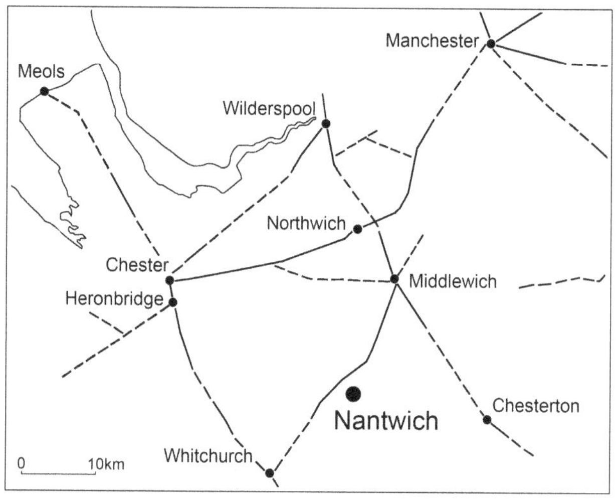

Figure 1.3: Roman settlements and roads in the Cheshire region (after Petch 1987, 186).

discovery in 1985 of a plank-built tank, *c* 1.4m square, which may have originated as a brine well but was reused as a cess pit. Dendrochronological analysis showed that timbers for this structure were felled in the summer of AD 132 (McNeil & Roberts 1987). Other finds from St Anne's Lane included a cremation burial in a 2nd- or 3rd-century urn, recovered during the digging of a modern foundation trench. Sherds of Roman pottery and fragments of a quernstone were found during the same development. In addition a Roman ditch, running roughly east-west and containing 2nd-century pottery, was identified during investigations in this area in the 1990s. It has been suggested that this ditch may mark the southern limits of Roman activity on the western bank of the river (Earthworks 1997).

The area to the north of Welsh Row had not produced the same quantity of evidence, prior to the investigations at Kingsley Fields. A number of coins had been discovered but the excavation at Wood Street in 1979-80 revealed only a few residual sherds of Roman pottery (McNeil Sale 1980, 24). Particular mention, however, should be made of the discovery in the early 1980s of two Roman lead salt pans. Both were found *c* 500m to the north of the town bridge and were inscribed with the personal name 'Cunitus' (Petch 1987, 209). They form part of a small number of inscribed Roman lead pans uncovered in Cheshire, which also include examples from Northwich, Middlewich and three from Shavington, *c* 5km to the east of Nantwich (Shotter 2005).

While the discovery of the Nantwich pans effectively confirmed that the place was a Roman salt-making site, the *Victoria County History* could still conclude that 'Little is known about the site at Nantwich, but it seems that production was modest in scale, and perhaps short-lived' (Petch 1987, 223). This paucity of knowledge about Roman Nantwich extended to its place within the local road system (Fig 1.3). It was recognised that Nantwich was bypassed by the Roman road between Middlewich and Whitchurch, which ran to the north and west of the town (Petch 1987, 221), but no clear evidence had emerged of a link between the two.

It was against this background that the investigations at Kingsley Fields were undertaken.

The Present Volume

The following chapters first outline the circumstances of the excavation and then describe the results of the excavation zone by zone. This is followed by chapters which analyse and discuss the finds and other material and data produced by the site. The concluding chapter discusses the phasing, development and usage of the site as a whole, and places it within the wider picture of Roman Nantwich.

2. Background to the Excavation

Peter Connelly & David Power

The site of the excavation at Kingsley Fields lies on the north side of Welsh Row in Nantwich, on the west bank of the River Weaver and close to the town's historic and commercial centre on the east bank of the river (NGR SJ 646 525). The area of Kingsley Fields had been the subject of successive development proposals for a number of years prior to 2002. In 1995 an archaeological desk-based assessment was carried out for a proposed residential development concentrated to the west of the River Weaver (Gifford & Partners 1995). This work documented the high potential of the area for the survival of multi-period archaeological remains, especially in the vicinity of the banks of the river where it is known that early salt production took place. In 2000, based on this assessment and in response to new development proposals by Jennings Holdings Limited, Cheshire County Council Environmental Planning prepared a brief for archaeological works at Kingsley Fields.

Figure 2.1: The area of the archaeological investigations at Kingsley Fields, showing the location of the Phase 1 geophysical survey, evaluation trenches (1-XXXVIII) and test pits (TPI-II), and the Phase 2 trenches (40-47) and excavation areas A-G. Also shown are the evaluation trenches (1-3) at Kingsley Cottage in 2003. © Crown copyright. All rights reserved. Licence number 100022765.

Phase 1 Evaluation

Subsequently, because of advanced proposals to extend and undertake development on the site, Bellway Homes Limited (West Lancashire) approached the University of Manchester Archaeological Unit to formulate a project design which would address the archaeological issues and meet planning requirements. The agreed initial strategy included a revised desk-based assessment of the whole development site, a geophysical survey, a metal-detecting survey and a programme of evaluation by trial trenching. All work was timetabled to accommodate a two-phase development programme. The Phase 1 evaluation initially comprised the excavation of 33 trial trenches (1-XXXIII) across the development site and two 1m square test pits (TPI and TP II) in a former walled garden on the site's southern periphery, and was carried out between October and December 2001 (Fig 2.1).

The geophysical survey was disappointing, with the majority of the features identified proving to be of geological origin, but the results of the trial trenching, supported by the metal-detecting survey, produced evidence for the survival of archaeological remains within the development site. The main focus of this activity was centred in the south-east of the site, where evidence for the survival of a Roman road and associated negative features in the form of pits and postholes was observed. To refine understanding of the extent of the remains within this part of the development site, and following consultation with the Archaeological Planning Officer for Cheshire County Council, UMAU carried out the excavation of additional trial trenches (XXXIV-XXXVIII) (Fig 2.1). A watching brief was also maintained during earth-moving operations within the Phase 1 corridor of a new access road into the site. It further refined the results of the earlier trenching by mapping the presence or otherwise of archaeological features and artefacts associated with Roman activity.

Phase 2 Excavation

Through discussions between Cheshire County Council, Bellway Homes Limited (West Lancashire) and UMAU, it was recognised that a further programme of archaeological work would be required. A strategy was agreed for an excavation targeted at areas which the evaluation had shown to have the highest potential for the survival of Roman remains. This Phase 2 work was carried out between January and September 2002.

The main Phase 2 excavation site comprised an area measuring *c* 85m east-west by *c* 110m north-south, located to the north of the former walled garden (Fig 2.1). This area was divided into five parts, identified from south to north as areas A-E, which were excavated in stages as the areas became available (Fig 2.2). In addition four separate trenches, numbered 44-47, were excavated on the west and north sides of the main excavation site, which was itself extended to join with Trenches 44 and 46; areas F and G were excavated to the north-east towards the west bank of the River Weaver; and four evaluation trenches (40-43) were opened along the Phase 2 access road corridor on the east bank of the river.

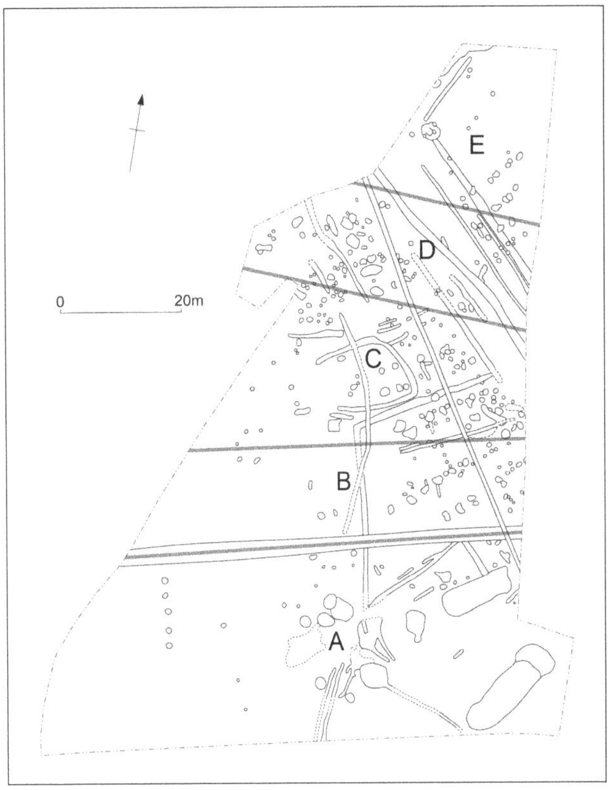

Figure 2.2: Location of excavation areas A-E.

All of the evaluation and excavation trenches were mechanically dug under archaeological supervision, followed by the manual excavation and recording of identified features.

Post-excavation Analysis and Reporting

The initial stage of the post-excavation analysis was an assessment of all classes of data generated by the excavation at Kingsley Fields. This enabled an interim excavation and finds report on the site to be produced in November 2004, along with a project design for a programme of more detailed analysis and publication (Power *et al* 2004). The present volume represents the completion of that process.

Site Topography and Geology

The Kingsley Fields development site encompassed an area of *c* 30ha. Although varied, the topography and land-use predominantly comprised gently undulating rough pasture characterised by streams and hedge and fence lines, with ground level at an average height of *c* 35m AOD. The superficial geology of the development site comprises glacial boulder clays, which are largely overlain by river terrace deposits and alluvium associated with the River Weaver. These overlie a solid geology of the Wilkesley Halite Member (formerly known as Upper Keuper Saliferous Beds). The main Phase 2 excavation area lay on sands of the river terrace deposits.

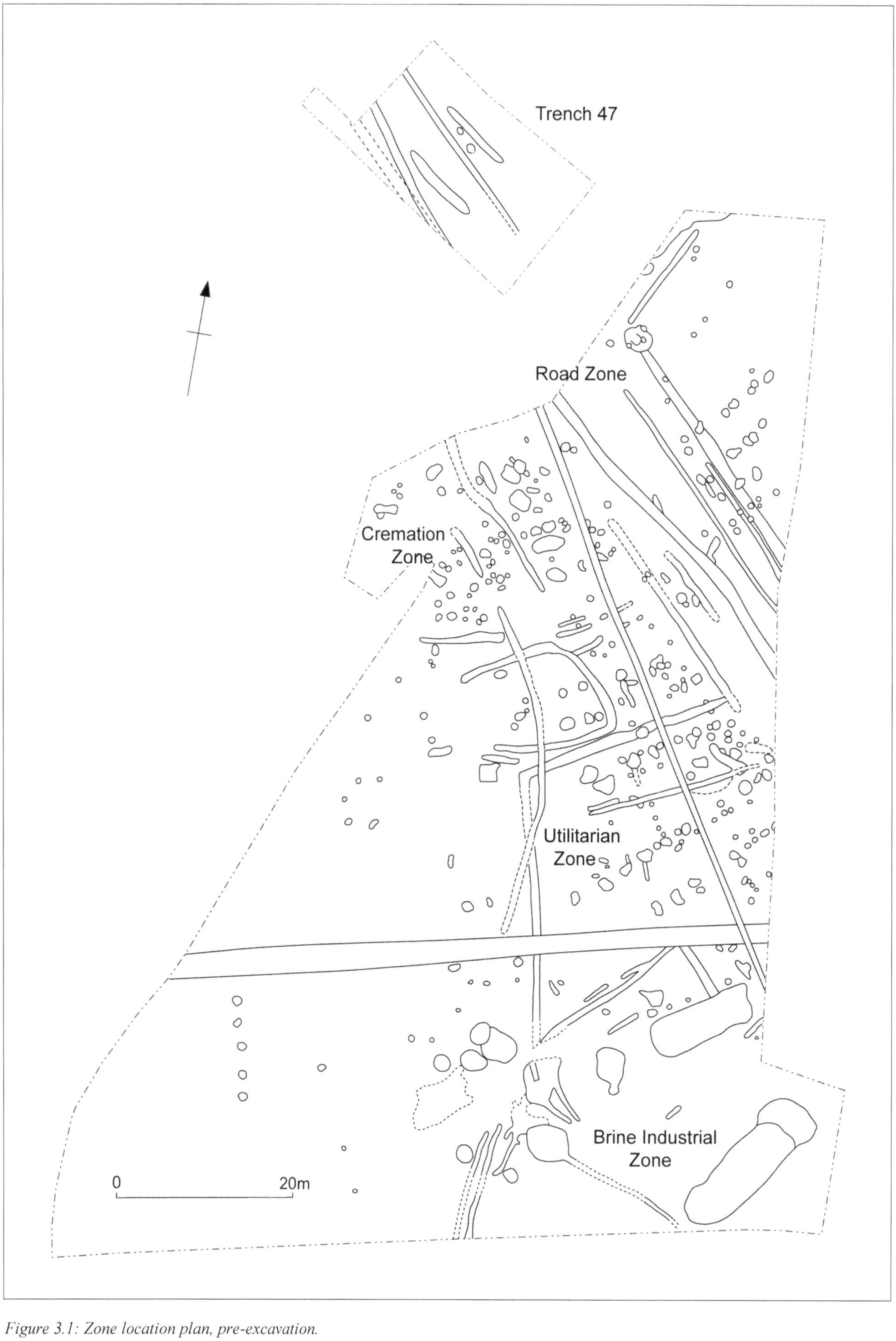

Figure 3.1: Zone location plan, pre-excavation.

3. The Excavation

Peter Arrowsmith, Simon Askew, Peter Connelly & David Power

The main area of excavation was examined in a series of five strips running from south to north, each of which was assigned its own code from A to E and a sequence of context numbers prefixed as follows: A - 1, B - 2, C - 3, D - 4 and E - 5. Two detached areas excavated to the north-east continued this sequence as F - 6 and G - 7 (Figs 2.1 & 2.2). Other trenches were assigned a context sequence number with the prefix 8. This system permitted the allocation of four-figure context numbers across the site and, allowing for some slight overlapping, enables the instant general location of a context to a particular excavation area.

Below the topsoil and subsoil, which were common across the site, Roman deposits were found to have been truncated by at least *c* 0.3m by later ploughing. This had largely removed the horizontal stratigraphy across the site and meant that surviving Roman deposits were mostly in the form of features cut either into the natural sands or into other features.

Roman Activity on the Site

The Roman occupation of the site represents the earliest and most significant period of activity. The archaeological remains of this period included evidence of groupings of buildings, cut features which included waterlogged structural remains and three cremation burials, and a road which provided a focal point of activity on the site. Most features lay alongside or close to the road, structural features were typically positioned at an approximate right angle to it and a number of linear features were set either at a similar angle or ran parallel to the road's alignment.

The spatial arrangement of the features shows a degree of organisation and planning to the settlement and the feature types also suggest that the site was arranged into discrete areas of separate activity. Following an earlier summary report on the site (Connelly & Power 2005), these have been grouped into four zones as follows (Fig 3.1):

1) the Road Zone, comprising the Roman road and features on its north side (largely corresponding with Trench 47, excavation area E and the eastern part of area D).

2) the Cremation Zone, located on the south side of the road in the north of the main excavation area and characterised by a small number of cremation burials and a larger number of pits. (This zone largely corresponds with the western part of excavation area D).

3) the Utilitarian Zone, on the south side of the road in the centre of the main excavation area, characterised by large timber-post buildings, hearths and enclosures (largely corresponding with excavation areas B and C).

4) the Brine Industrial Zone, on the south side of the road in the south of the main excavation area, characterised by two clay and timber-lined tanks, believed to have been used for the storage of brine, and a group of features including wicker-lined pits and pits containing wooden troughs. (This zone largely corresponds with excavation area A).

The Road Zone

The Roman Road

The remains of the Roman road and associated ditches were exposed across the northern part of the site, running on a north-west - south-east alignment (Fig 3.2; Plate 3.1). Within the main area of excavation the road itself had been largely removed by later activity, leaving the roadside ditches, but evidence of both was found to have survived to the north-west within the detached Trench 47. Including the part between the excavated areas, the road was recorded for a length of *c* 80m. There was evidence of three phases of road construction, in the second of which the road was narrowed and in the third its line shifted to the south.

Plate 3.1: The roadside ditches viewed from the north.

Trench 47 contained clear evidence of a road construction (8057) *c* 4m wide, laid in a shallow cut [8056] 0.27m deep (Fig 3.3). (8057) consisted of a layer of compact river-worn cobbles, mostly between 40mm and 60mm in diameter although there were occasional larger rounded

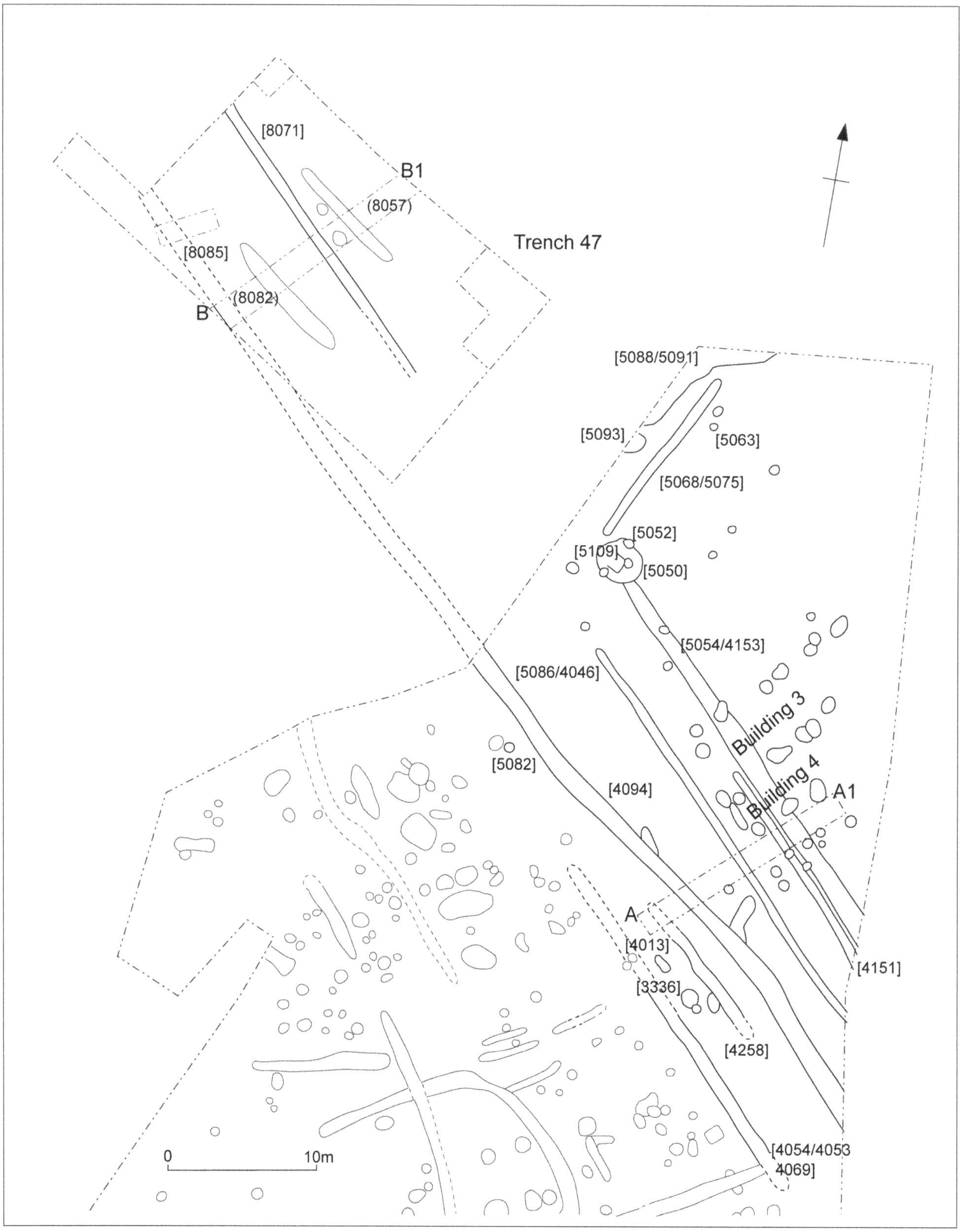

Figure 3.2: Road Zone plan.

cobbles measuring 0.1-0.15m. This road construction was bordered on the south-west by ditch [8071], 1.32m wide and 0.4m deep. At a distance of *c* 5.2m to the south-west of [8071] was a second ditch [8085], 1.6m wide and 0.44m deep. Between [8071] and [8085] were further remnants of a road construction. This had been damaged by later ploughing and a modern storm drain but was represented by a layer of compact cobbles (8082).

In the main area of excavation the road was marked by a series of parallel roadside ditches, aligned north-west - south-east. The most northerly of these [5054]/[4153]

was *c* 1m wide and 0.3m deep and extended from the eastern limit of the excavated area for a distance of *c* 28m, terminating on the north-west at the position of a well [5052]. For part of its length, [5054]/[4153] was bordered on the south-west by a smaller ditch [4151], which was 0.5m wide and 0.15m deep and extended from the eastern edge of the excavated area for *c* 15m. Its fill contained pottery of the early to mid 3rd century. At a distance of *c* 3.6m to the south-west of [5054]/[4153] ran a narrow ditch [5086]/[4046], measuring 0.37m wide and 0.24m deep. It extended from the eastern edge of the excavated area for *c* 29m, and terminated on the north-west at roughly the same point along the road as ditch [5054]/[4153].

Lying at a maximum of *c* 5.5m to the south-west of [5086]/[4046] was ditch [4094] which ran across the width of the excavated area, with a recorded length of *c* 38m. At a distance of *c* 10m from the north-west edge of the excavated area, the alignment of [4094] deviated to the east to a maximum of *c* 2.7m, before trending back towards its original alignment which may have resumed beyond the eastern edge of the excavated area. A section was cut across [4094] within this deviation and found that the feature measured *c* 2.5m wide and 0.8m deep and mostly contained a single fill (4137) above a basal layer with frequent cobbles (4145) (Fig 3.3).

To the south-west of the deviation of [4094] was ditch [4258] which measured *c* 11m long and 0.6m wide. To the south-west of [4258] in turn was ditch [4054]/[4053]/[4069], which measured 1-1.4m wide and 0.62m deep and extended for *c* 25m before terminating at both ends within the area of the excavation. To the east of the deviation of [4094] were the probable remains of a cobbled road surface in layers (4123) and (4127) (Fig 3.3). Layer (4127) sealed linear feature [4116], which was 0.9m wide and 0.5m deep, contained an accumulation of several fills and seems to have been a continuation of ditch [4258].

On the northern side of the road, ditches [5054]/[4153] and [4151] were cut by the postholes of possibly two later Roman buildings (Buildings 3 and 4, see below).

The Development of the Road

When originally built, the road was probably defined on its northern side by ditch [5054]/[4153] and on the south by ditch [4094] in the main excavation area and by ditch [8085] in Trench 47, which roughly followed the same alignment, giving a road width of *c* 9.25m (Fig 19.1). It is likely that the eastern deviation of [4094], which narrowed the road, represents a secondary phase and that the road here was originally bordered by ditch [4258] which roughly continued the straight alignment of [4094] and [8085]. On the north side of the road, ditch [4151] may have formed part of the same narrowing process, the effect of which would have been to reduce the road width at this point by roughly half, from *c* 9.25m to *c* 4.5m (Fig 19.2).

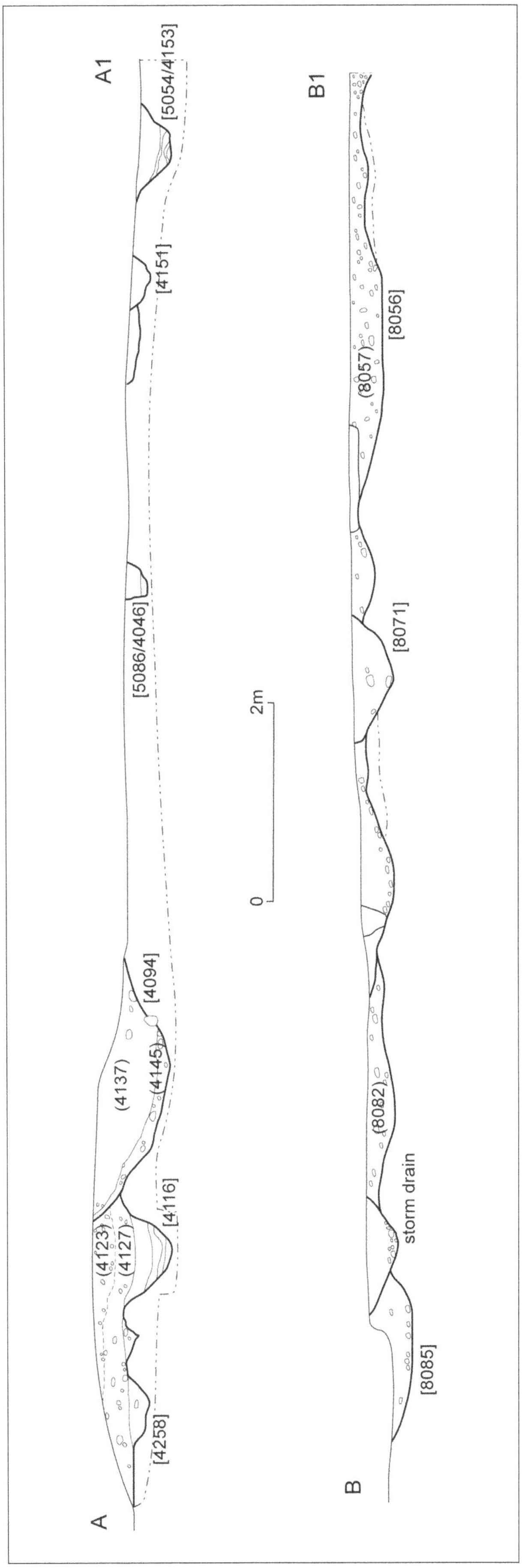

Figure 3.3: Sections through the road and associated ditches.

In the final phase the road appears to have been widened and shifted to the south. The northern roadside ditch [5054]/[4153] was infilled and Buildings 3 and 4 were constructed across its line and the adjoining road surface. The northern side of the road here was now probably defined by ditch [5086]/[4046], which appears to have continued as [8071] in Trench 47. On the south, the road was extended over the line of both [4094] and [4258]/[4116]. Ditch [4054]/[4053]/[4069] would appear to have demarcated the south edge of this phase of road, giving a final road width of *c* 8.2m (Fig 19.3).

The Well and Possible Trackway

Sited at the north-west end of the primary northern road ditch [5054] was the cut for a well [5052], measuring *c* 3.4m wide. The well was excavated by machine and was found to be *c* 3.5m deep. Its collapsed basal remains [5109] and primary fill (5108) were recovered in situ. The base was lined with timber planking, *c* 1m square (Plate 3.2), and contained a primary fill which included 2nd-century glass and a small amount of leatherwork (see Chapters 7 & 9). Above this deposit, the well had been infilled with sand similar to the surrounding superficial geology. Insect analysis of the primary fill implies a slow infilling of the well after it had gone out of use (see Chapter 18). The absence of any upper timber lining suggests that this may have been deliberately removed prior to backfilling.

Plate 3.2: Well [5109] viewed from the north.

Running north-eastwards from the well for a distance of *c* 12.5m was linear feature [5068]/[5075], 0.62m wide and 0.12m deep. Its northern end terminated close to curving feature [5088]/[5091] which extended under the edge of the excavated area but had recorded dimensions of *c* 10m long, 1.25m wide and 0.18m deep. Directly to the south of [5088]/[5091] was pit [5093]. This also extended under the edge of the excavated area and was 1.9m long, at least 0.9m wide and 0.35m deep. [5088]/[5091] contained 3rd-century pottery and [5093] pottery which may date to the late 3rd to early 4th century.

The termination of both the primary and the final phase northern roadside ditches, [5054]/[4153] and [5086]/[4046], to the south-west of this point suggests that [5068]/[5075] may have been associated with a trackway leading off the Roman road, with well [5052] being situated at the junction (Figs 19.1 & 19.3). This trackway could have led north-eastwards towards the river and an area in which Roman material was found in excavation areas F and G. This was mostly retrieved from the topsoil and subsoil and included brick, rooftile and flue tile, suggesting the presence of a heated building in the locality. Area F also contained an irregularly shaped pit [6017], 1.55m long by 0.9m wide and *c* 0.18m deep, which produced 3rd-century pottery as well as a fragment of flue tile.

Buildings 3 and 4

During the final phase of the road, building took place on its northern side, straddling the disused primary ditch. At least one building, Building 3, appears to have been erected here, identifiable by postholes which suggest a rectangular structure set at a right angle to the road (Fig 3.4). Building 3 seems to have been defined on the north-west side by postholes [5003], [5017], [5009], [5011], [5015] and [5013], on the south-east by postholes [4174], [5022]/[5025]/[5044], [5036], [5027]/[5034] and [5020], and on the south-west by posthole [5005] and possibly [4269], giving a structure *c* 11m long by *c* 4.6m wide. The postholes measured 0.65-1.43m in length, 0.36-1.08m in width and 0.16-0.49m in depth. Postholes [5022]/[5025]/[5044] and [5027]/[5034] showed recuts, suggesting the replacement or repair of timber posts.

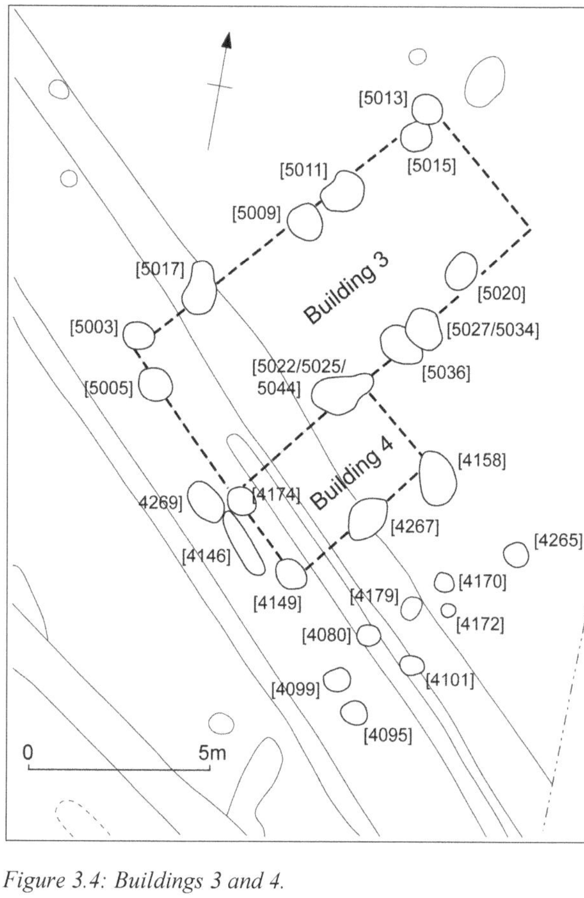

Figure 3.4: Buildings 3 and 4.

Running parallel to, and *c* 3m from, the western end of the south-east wall of Building 3 was an alignment *c* 5m long of three postholes, [4149], [4267] and [4158]. These were of a similar size to the postholes of Building 3, measuring 0.85-1.16m in width, with a maximum length of 1.48m and depth of 0.4m. Their position and size suggests that they formed an annex, Building 4, to Building 3. Running parallel to, and *c* 2m from, this last line of postholes was a rough alignment of smaller subcircular postholes, [4099], [4080], [4179], [4170] and [4265], measuring *c* 0.36-0.85m wide and 0.16-0.45m deep. They suggest that the annex was in turn abutted on the south-east by a less substantial structure, possibly a lean-to. On the south-east, postholes [4095] and [4101] may also have formed part of this last structure and the general building group. Running parallel to, and immediately to the west of, the south-west side of Building 4 was a linear cut [4146], *c* 2m long, 0.35m wide and 0.22m deep, which again may represent a structural part of the group.

Discussion

The road excavated on the site almost certainly represents a branch off the Whitchurch to Middlewich Roman road whose course is known to have run to the north and west of Nantwich, bypassing the town. When the alignment is projected to the south-east, this branch road would have reached the River Weaver close to the present bridge which crosses the river at the east end of Welsh Row. Archaeological work at Welsh Row subsequent to the Kingsley Fields excavation has confirmed the continuation of the alignment and has also shown that the Roman road remained in use into the medieval period (see Chapter 19).

The original construction of this road is likely to date to the very late 1st or the early 2nd century, when the coin evidence suggests that Nantwich first developed as a settlement (see Chapter 6).

The second phase, involving the narrowing of the road, lacks direct dating evidence. While the precise purpose and extent of this change are uncertain, the course of ditch [4094] within the excavated area suggests that the road narrowed over a relatively short length before widening again. The width and depth of [4094] at this point are the greatest for any of the recorded ditch sections and suggest that the overriding need, for whatever reason, was to control the movement of traffic.

The final phase of the road involved its rewidening and southward shift and appears from pottery in the fill of ditch [4151] to have occurred in the early to mid 3rd century. This date corresponds with early 3rd-century pottery found in pit [4013], which is likely to have been sealed by the widening of the south side of the road.

The infilling of [4151] also provides the earliest possible date for Buildings 3 and 4 but the pottery from Building 4 itself only gives a broad date range of the 2nd century or later. No direct evidence was found for the function of these buildings, whose posthole method of construction is also found at Kingsley Fields in 2nd-century Buildings 1 and 2 (see below).

The Cremation Zone

This zone lay on the west side of the Roman road, in the north-west of the main excavation area (Fig 3.5), and included three cremation burials (see Chapter 8) and numerous pits and possible postholes. The area is defined by roadside ditch [4094] on the north-east, while linear feature [3379] provides an approximate boundary on the south-west.

Crossing this area and separating a number of these features was linear feature [4131], which was aligned approximately south-east - north-west and measured 0.6-1.25m wide and 0.13m deep. It ran for *c* 20m from the north edge of the excavated area before terminating. The alignment of this feature appears to have been influenced by the orientation of the nearby road. The ditch produced a single sherd of samian ware of Hadrianic-Antonine date.

On the east side of [4131] was a pit containing an unurned cremation burial [4245]. This feature was subcircular in plan, *c* 0.75m wide and 0.17m deep (Fig 3.6). It held the remains of a 20-30 year-old male, together with pyre debris. This included a fragment of blue bead, believed to date from the late 2nd to 3rd century, and probable fragments of Wilderspool white-slipped fabric suggesting a 2nd-century date. The feature had been partially cut by pit [4242], which had an elongated oval shape, *c* 4m long, 0.93m wide and 0.33m deep. [4242] produced a single sherd of Hadrianic-Antonine samian ware.

Two other cremation burials lay to the west of [4131]. One was an urned burial within a circular pit [4003] *c* 0.2m in diameter and 0.13m deep. It contained the remains of a *c* 20-25 year-old female in a black-burnished ware vessel of mid to late 2nd-century date. The other was an unurned burial within a subcircular pit [4250], *c* 0.64m wide and 0.27m deep (Fig 3.6). It contained the remains of a >45 year-old female, together with pyre goods which give a likely 3rd-century date (see Chapter 7). [4250] had been partially cut by a rectangular pit [4231], *c* 2.6m long, 0.8m wide and 0.3m deep, which contained mid 3rd-century pottery. Also west of [4131] was a badly truncated circular pit [4239], 0.38m in diameter and 50mm deep, whose fill (4238) contained a small quantity of human bone. The feature included no evidence of pyre debris and the bone may represent the incidental inclusion of material from one of the other burials.

In the north-west of the zone, fill (4232) in pit [4237] contained a partly worked sandstone block (see Chapter 12), together with mid to late 3rd-century pottery.

There were also a large number of other pits and possible postholes within this zone to which no particular function

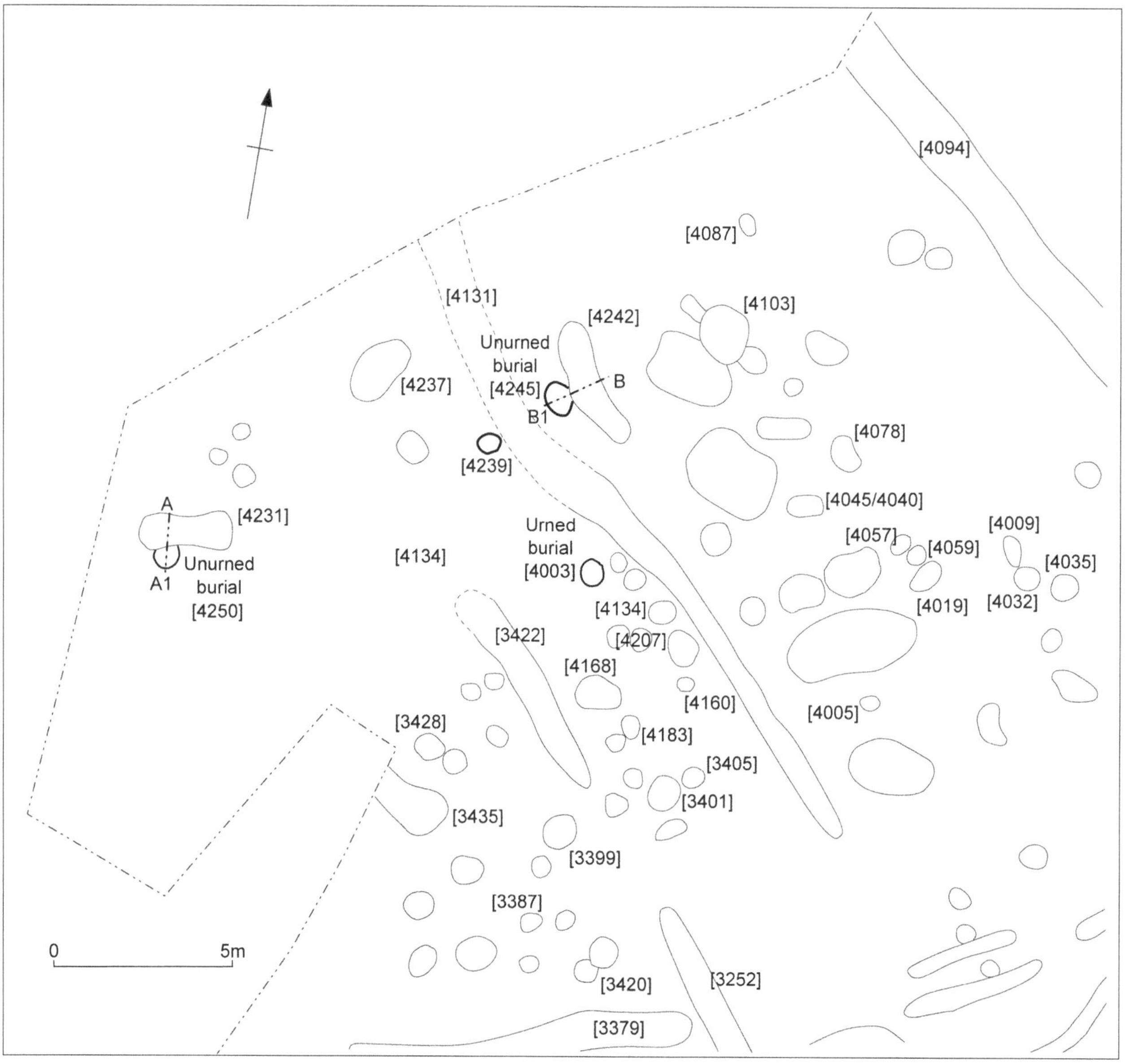

Figure 3.5: Cremation Zone plan.

could be ascribed, other than perhaps as pits dug for the disposal of refuse. From the pottery, these features variously date to the 2nd and 3rd centuries.

Discussion

The pottery evidence suggests that this zone was originally utilised during the mid to late 2nd century when it was partly used as a cemetery. Following a hiatus at the turn of the century, a second main phase of activity took place in the early to mid 3rd century which saw at least one other cremation burial.

The function of most of the pits within this zone is uncertain but the presence of the small group of cremation burials tallies with the wider archaeological evidence. Together they indicate that this cemetery area lay alongside the Roman road into Nantwich and on the north-west fringe of the settlement, with roadside features petering out beyond this point.

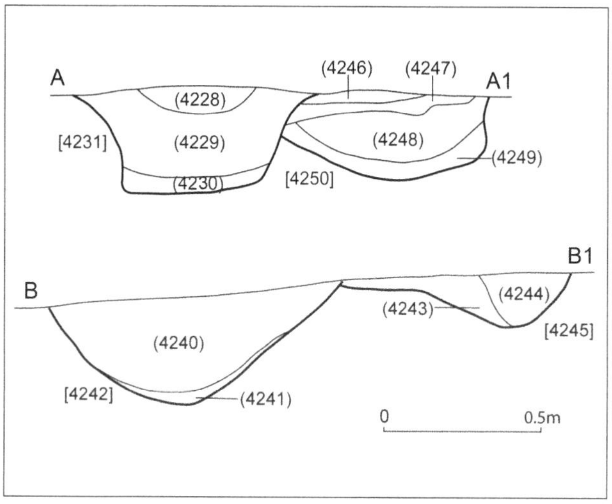

Figure 3.6: Sections through cremation burials.

The Utilitarian Zone

This area lay to the west of the road and occupied the centre of the main area of excavation (Fig 3.7).

Buildings 1 and 2, and the Large Enclosure

The southern half of the zone contained evidence of two substantial buildings, aligned at a right angle to the road.

Building 1

The more southerly and larger of these was Building 1 (Fig 3.8). This was indicated by two distinct rows of postholes which defined the extent of a post-built rectangular structure aligned roughly north-east - south-west. It is possible that the north-eastern extent of the building lay beyond the limits of the excavated area. A post-medieval ditch had probably removed the western end of the south-eastern row of postholes.

The north-western row of postholes consisted of [2156]/[2158]/[2177], [2172]/[2136]/[2170], [2091], [2100], [2047], [2039], [2023], [2021], [2071], [3076] and possibly [3066]. These postholes measured 0.37-1.45m long, 0.3-1.35m wide and 0.19-0.5m deep. The south-eastern range of postholes consisted of [2198], [2054], [2015], [2112], [2083], [2077], [2079] and [2197]. These measured 0.3-1.06m long, 0.24-0.82m wide and 0.05-0.31m deep. The full extent of [2197] extended beyond the edge of the excavation, while [2198] was not excavated. Within the eastern half of the building there was evidence of an inner row of smaller postholes represented by [2005], [2089], [2056], [2062], [2049] and possibly [2009] and [2086]. These were 0.33-0.95 in length, 0.2-0.6m in width and 0.07-0.34m in depth.

The outer postholes give a building measuring *c* 24m in length and *c* 5.5m in width. Differences in the layout of the postholes suggest that this substantial structure was formed by two adjoining buildings of roughly equal

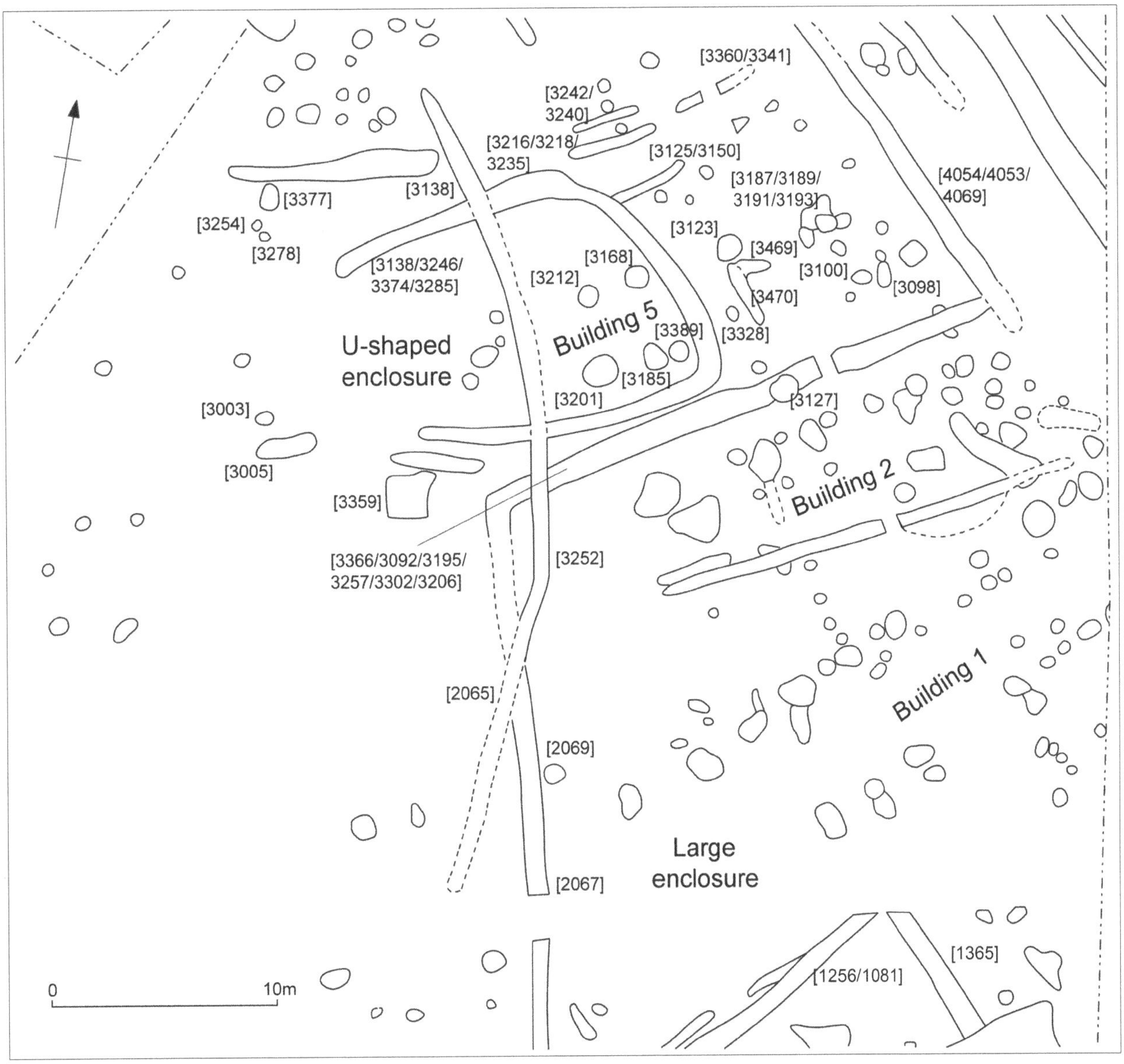

Figure 3.7: Utilitarian Zone plan.

length. The western part appears to have been a four-bay structure, *c* 12m long, defined by opposing principal posts set in [2156]/[2158]/[2177], [2172]/[2136]/[2170], [2091], [2100] and [2039] on the north-west and surviving postholes [2198], [2054] and [2015] on the south-east. The surviving evidence of postholes in the eastern half of the range seems less complete, with a number appearing to have been removed through later truncation, but it does suggest that here the posts were more tightly spaced. This eastern part of the range also includes the inner row of postholes noted above.

Pottery from Building 1 implies a mid 2nd-century date for its construction. Posthole [2039], however, contained a sherd of early 3rd-century pottery alongside residual Antonine sherds. Plant microfossil analysis found that charred cereal remains and charcoal were abundant in (2157), (2159) and (2160), which formed the fills of posthole [2156]/ [2518]/[2177] at the north-west corner of Building 1. The remains suggest that Building 1 had suffered damage by fire and that at this time it was possibly being used as a granary (see Chapter 18). From pottery in (2159) that episode may have occurred in the late 2nd to early 3rd century.

Building 2

Building 2 was located to the north of Building 1 (Fig 3.8). This building had a subrectangular plan, narrowing towards the west, which was defined by a line of

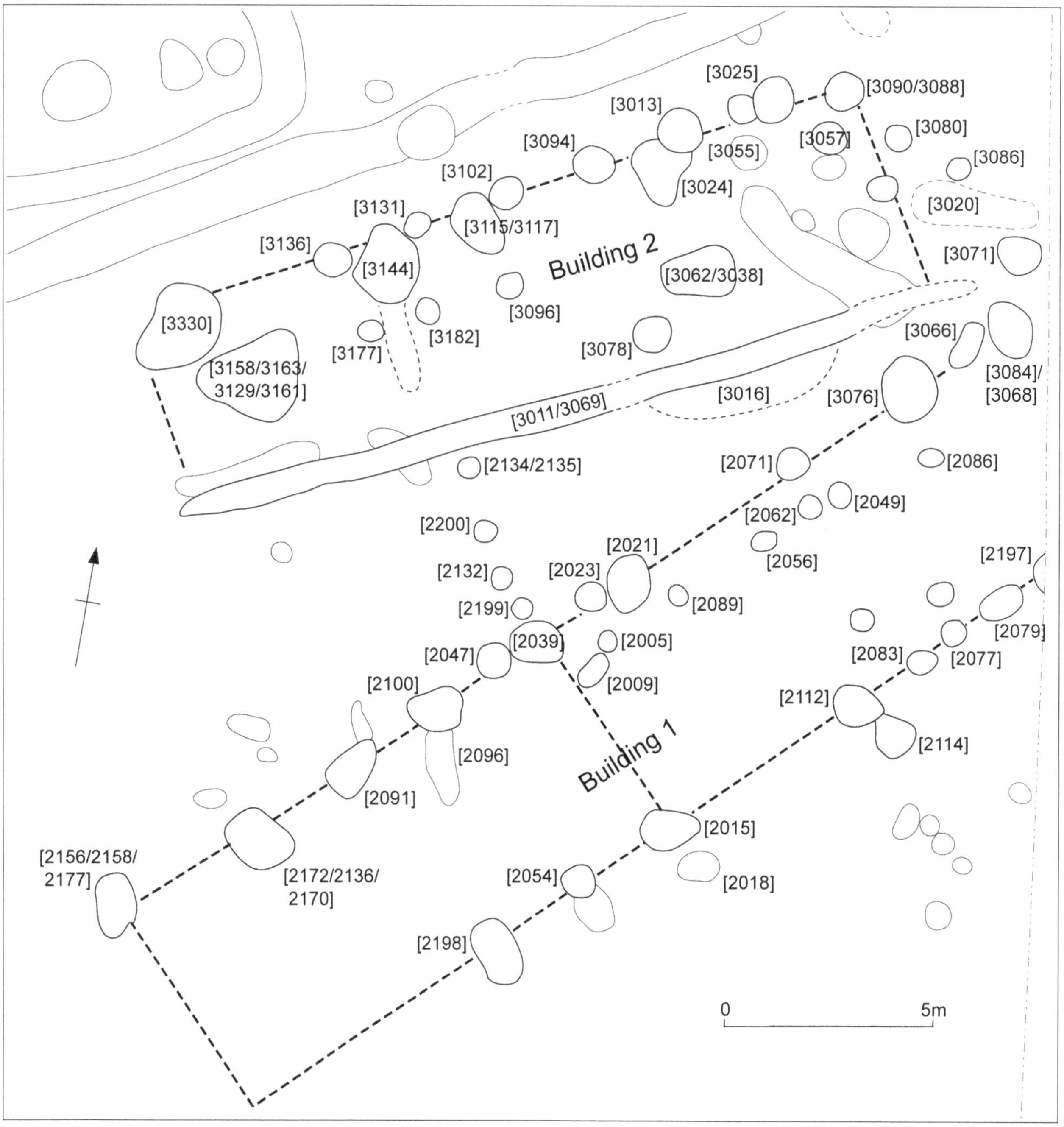

Figure 3.8: Buildings 1 and 2.

postholes on the north-west and a linear construction trench on the south-east. The building was aligned approximately east-north-east - west-south-west and measured *c* 18m long and *c* 4.5-5.2m wide. It was less well-defined than Building 1, suggesting the possibility that it was more temporary in nature. As with Building 1, the pottery from Building 2 was mainly Antonine.

The north-west side of Building 2 may be principally represented by a fairly evenly spaced row of postholes, comprising [3136], [3131], [3102], [3094], [3013], [3025] and [3090]/[3088], measuring 0.64-1.2m long, 0.59-1m wide and 0.24-0.46m deep. Adjacent to these, postholes/pits [3144], [3115]/[3117] and [3024], which measured 1.1-1.39m long, 0.75-1.3m wide and 0.35-0.66m deep, may also represent structural elements of the building. The south-eastern side of the building was defined by a straight construction trench [3011]/[3069], measuring *c* 20m long, 0.4-0.6m wide and 0.23-0.45m deep. Within the body of the building a line of postholes, [3177], [3182] and [3096], 0.36-0.7m long, 0.28-0.57m wide and 0.2-0.5m deep, may be evidence of a central row of posts. Possibly also associated with Building 2 were pit [3030], located at the building's north-west corner and measuring 1.67m long, 1.25m wide and 0.36m deep, and pits [3158]/[3163]/[3129]/[3161], [3062]/[3038] and [3078] which were situated within the interior of the building and were 0.68-2.61m long, 0.55-1.47m wide and 0.1-0.47m deep.

Running between Buildings 1 and 2 was a row of subcircular postholes, [2134]/[2135], [2200], [2132] and [2199], aligned approximately south-east - north-west and measuring *c* 0.5m wide. The pottery evidence suggests a 2nd-century date for this feature which seems to represent a fence line, enclosing a yard between the two buildings. At its southern end the alignment was continued by posthole [2039] in Building 1, providing a further possible indication that this last posthole formed part of a division between a western and eastern structure within that range.

The Large Enclosure

The area containing Buildings 1 and 2 was surrounded by ditches which appear to have formed the northern, western and southern sides of a large enclosure (Fig 3.7).

A series of excavated sections generated several cut numbers for these ditches: [1081], [1256], [2067], [3092], [3195], [3206], [3257]/[3267], [3302]/[3306]/[3308] and [3366]. Sections within the central part of the northern ditch of the enclosure revealed evidence of recutting. Here [3302] was a recut of [3306], of which there was only partial evidence, and was in turn recut by [3308]. In addition [3267] was recut as [3257]. Other sections showed only a single cut.

The sections through the ditches also revealed pits [3127], [3199], [3197], 0.5-1.1m long, 0.44-1m wide and 0.15-0.2m deep, which were cut into the top fill of the northern ditch.

The northern side of the large enclosure measured *c* 23.5m long and was aligned approximately east-north-east - west-south-west. The west side was aligned roughly north-south and was *c* 30m long, while the south side was aligned roughly north-east - south-west and was *c* 20m long. The excavated sections revealed the width of the ditches as between 0.71m and 3.2m, although widths of over 1.5m were possibly the result of heavy erosion which had naturally widened the feature. The depth varied between 0.17m and 0.8m.

Probably associated with the large enclosure was ditch [1365], which ran between the north-eastern end of the southern ditch of the enclosure and the northern brine tank [1182]. This ditch was not excavated but measured 7m long and *c* 0.75-1.3m wide.

The northern ditch of the large enclosure ran parallel to the northern side of Building 2, with the two being separated by a distance of *c* 1.5-2m. The southern ditch of the enclosure likewise appeared to follow the long alignment of Building 1, while the alignment of the western ditch seems to have been related to the respective western ends of the two buildings. This spatial relationship, the presence of 2nd- and 3rd-century pottery in the ditch fills and the evidence of recutting all suggest that the enclosure ditch was constructed during the same period as Buildings 1 and 2, rather than postdating their demise as was previously suggested (Connelly & Power 2005, 34-5).

Two pits situated at the eastern end of Buildings 1 and 2 produced pottery of the 3rd century or later. One was an oval pit [3071], 1.62m long, 1.07m wide and 0.24m deep. The other was a circular pit [3084], 1m in diameter and 0.2m deep, which had in turn cut through [3068], an oval pit 1.76m long, 0.8m wide and 0.2m deep. Pit [3084] contained three quernstone fragments within its fill (3085). These pits lay close to the Roman road and may represent roadside activity postdating Buildings 1 and 2.

The U-shaped Enclosure, Building 5, the Planked Pit, Hearths, Pits and Linears

The U-shaped Enclosure

Immediately to the north of the large enclosure was a second, U-shaped, enclosure (Fig 3.7). It was defined by a curvilinear ditch [3138]/[3246]/[3374]/[3285], *c* 36m long, 0.5-1.15m wide and 0.14-0.25m deep. This enclosed a space which was *c* 12.5m long and *c* 7-9.5m wide and was open-sided on the west. The eastern arm of this ditch ran roughly north-west - south-east and the other two arms east-north-east - west-south-west, with the southern arm turning inwards. An excavated section [3138] revealed evidence of at least one post setting, with the post observed in situ within the cut. On the pottery evidence the U-shaped enclosure may date from the early to mid 3rd century. An excavated section showed that the southern ditch of the U-shaped enclosure was contiguous with the northern ditch of the large enclosure, with no clear indication of one ditch cutting the other.

Contained within the south-east corner of the U-shaped enclosure was evidence of a possible four-post structure, Building 5, in the form of pits [3212], [3201], [3168] and [3389], measuring 1.12-1.8m long, 0.75-1.32m wide and 0.3-0.5m deep. Features [3168] and [3201] showed signs of being recut by pits [3170] and [3203] respectively. Subcircular pit [3185], 0.7m long, 0.45m wide and 0.21m deep, may also be associated with this structure and have been a replacement for [3389]. On the pottery evidence, Building 5 would seem to have a late 2nd- to early 3rd-century date. Their spatial relationship suggests that this building and the U-shaped enclosure were contemporary.

Directly to the south of the open end of the U-shaped enclosure was a rectangular pit [3359], *c* 2m wide and 0.45m deep. It had a clay lining (3368) *c* 0.2m thick, upon the base of which were laid five oak planks (Fig 3.9). Around this planked base was evidence of an upstanding wicker lining. From the pottery evidence, a 3rd-century date seems likely for this feature.

To the east of the U-shaped enclosure was a subrectangular area defined on the south-east by the ditch of the large enclosure and on the north-east by roadside ditch [4054]/[4053]/[4069]. The southern part of this area contained a group of pits/postholes with no discernible pattern and with no obvious function, other than as possible refuse pits. Among these, pit [3191], 0.85 long, 0.48m wide and 0.64m deep, produced a substantial quantity of fired clay (8182g), comprising hearth or oven structure and furniture and representing the single largest deposit of such material found during the excavation (see Chapter 13). [3191] cut pit [3193], 1.1m long, 0.81m wide and 0.37m deep. It also cut pit [3189], 0.41m long, 0.26m wide and 0.13m deep, which in turn cut pit [3187], 0.86m long, 0.57m wide and 0.2m deep.

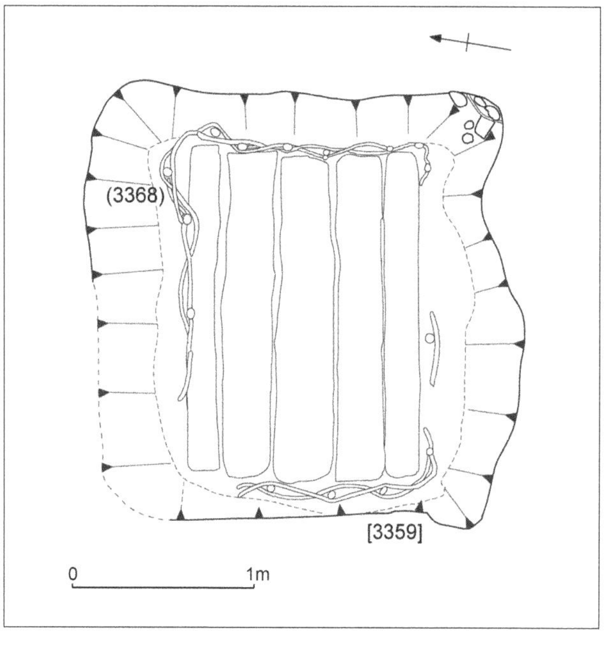

Figure 3.9: Planked pit [3359].

Two short linear features were also contained within this area. Linear feature [3470] was aligned roughly north-west - south-east and measured 3.6m long, 0.96m wide and 0.32m deep. Linear feature [3469] was aligned roughly east-west and was 2m long, 1.12m wide and 0.42m deep. [3470] produced 2nd-century pottery.

Hearths

To the north was an area which contained a number of hearths (Figs 3.10 & 3.11). Two hearths, [3242]/[3240] and [3216]/[3218]/[3235], lay adjacent to each other in the north-west of this area and were both aligned approximately north-east - south-west. Hearth [3242]/[3240] was *c* 2.8m long, 0.6m wide and 0.24m deep. It contained evidence of firing, with upper fills (3244) and (3245) being characterised by an abundance of charcoal. Hearth [3242]/[3240] cut a pit which extended to the north of the feature but was not excavated.

Hearth [3216]/[3218]/[3235] was *c* 3.8m long, 0.6m wide and 0.25m deep, and contained evidence of firing and multiple reuse, with fills (3217), (3219), (3221) and (3222) all rich in charcoal. Hearth [3216]/[3218]/[3235] cut pit [3237], *c* 0.5m wide and 0.16m deep, which extended to the north of the hearth in a similar arrangement to hearth [3242]/[3240]. It contained a single fill (3238), lacking charcoal, suggesting that it was infilled prior to the hearth coming into use. These hearths contained pottery of a mainly 2nd-century date.

Situated to the east and positioned on the same alignment was hearth [3360]/[3341]. This was cut by a modern storm drain but had a total length of *c* 4m and was 0.6m wide and 0.2m deep. Its secondary fill (3343) was rich in charcoal. This feature contained mid 2nd-century pottery.

A fourth hearth [3125]/[3150] lay to the south of [3360]/[3341]. This measured *c* 3.5m long, 0.5m wide and 0.07m deep. Its secondary layer again contained abundant charcoal.

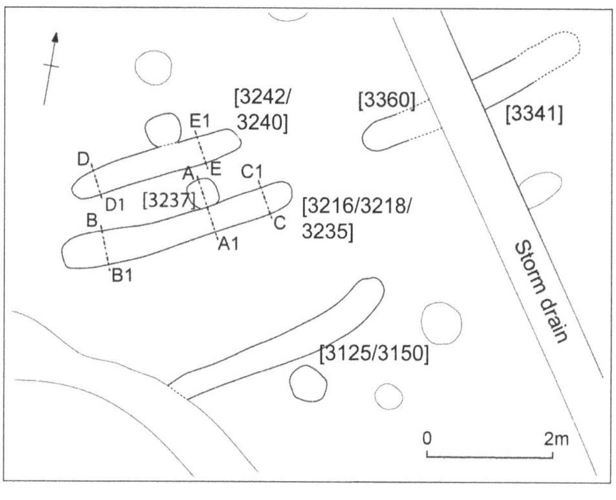

Figure 3.10: Hearths plan.

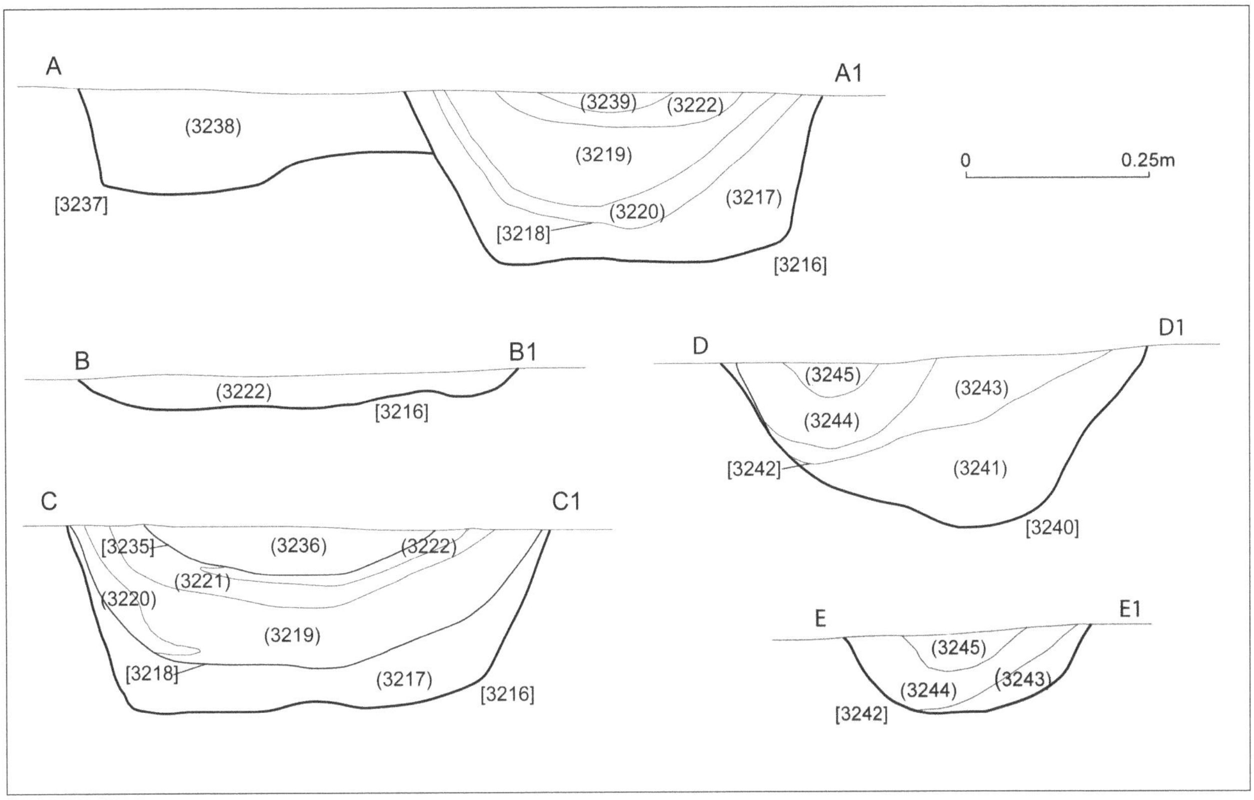

Figure 3.11: Sections through hearths.

The Western Features

To the west of the main Utilitarian Zone, away from the Roman road, features were sporadic (Fig 3.7). They included subrectangular pit [3005], 3m long, 0.9m wide and 0.5m deep, and circular pit [3003], 0.68m wide and 0.2m deep, containing pottery of the 2nd to 3rd century.

Ditch [3252]

Running roughly north-south across the centre of the site for a distance of *c* 36m was ditch [3252], measuring *c* 0.75m wide and *c* 0.1m deep. In the north of the site its alignment was continued by ditch [3422] in the Cremation Zone (Fig 3.5). Both ditches suggest some form of enclosure or plot division, with a possible entrance defined by the space, *c* 3.6m wide, between the two. [3252] crossed the northern and southern ditches of the U-shaped enclosure and, on the evidence of an excavated section, cut the northern ditch. To the south, [3252] also crossed the northern and western ditches of the large enclosure. Ditch [3252]/[3422] appears to represent a late and possibly post-Roman feature.

Discussion

The Utilitarian Zone was dominated by the two large 2nd-century Buildings 1 and 2, aligned at a right angle to the Roman road and measuring respectively *c* 24m by *c* 5.5m and *c* 18m by *c* 4.5-5.2m. Building 1 is likely to have been a storehouse. While the charred grain found within posthole [2156]/[2158]/[2177] at its north-west corner suggests a possible use as a granary, the building may well have housed more than one product. The process of salt production which is evident on the site would have required warehousing for the salt following evaporation of the brine and Building 1 could have been used in that role (Connelly & Power 2005, 38-9). Building 2 differs from Building 1 in size and method of construction, with one long side using postholes and the other a foundation trench, which could have contained either posts or sleeper-beams (cf Hanson 1982, 170-1). Its function is less clear, although a link with salt production again seems likely.

To the north of Buildings 1 and 2, hearths [3242]/[3240], [3216]/[3218]/[3235], [3360]/[3341] and [3125]/[3150] were almost certainly used for the evaporation of brine to produce salt and were built with a long narrow flue comparable to examples found at Middlewich (Bestwick 1975; Dodd 2005, 27-9; Williams & Reid 2008, 14-15) and Whitchurch (Jones & Webster 1968, 210). In this same area of the site, pit [3191] produced a significant deposit of fired clay, comprising hearth or oven structure and furniture, which may be derived from these hearths. Other fired clay deposits were found in the brine tanks, [1207] and [1182], at the southern end of the site and it is possible that the main group of hearths lay between the tanks and the Roman road, outside the area of excavation.

It is likely that Buildings 1 and 2 were bounded on the north, south and west by the large enclosure ditch. The extension of this boundary to brine tank [1182], by means of ditch [1365], and the apparent ease of access between the large enclosure and the brine tank support Buildings 1 and 2 being part of the industrial operation.

From the pottery evidence, activity within the zone effectively began in the mid 2nd century, to which period probably belong the construction of Buildings 1 and 2, the large enclosure surrounding those buildings, and the hearths. The level of activity within this zone appears to have rapidly declined towards the close of the century and to have not subsequently recovered. This decline suggests the end of the use of the zone for salt production. Building 2 was probably dismantled or destroyed before the end of the century. There is some slight evidence to suggest that Building 1 may have remained standing into the early 3rd, and it appears that the large enclosure ditch continued in use into that period. On its north side perhaps in the early 3rd century the U-shaped enclosure was constructed, containing in one corner the possible four-post Building 5. A change in the function of the zone is strongly implied by the orientation of this enclosure, which turned its back on the road, and by its form, which is suggestive of an animal pen. It is possible from the small number of 3rd-century features within the large enclosure that this was also used during that period as a paddock.

The Brine Industrial Zone

This zone lay in the south-east of the main excavation area (Fig 3.12). It was characterised by two large rectangular clay and timber-lined tanks situated close to the Roman road, in the east of the area, and by several wicker-lined pits and two large pits containing wooden troughs, in the west. The timber-lined tanks are believed to have been used for the storage of brine and represent the most intact examples excavated from Roman Britain.

The two brine tanks, [1207] and [1182], were both aligned roughly at a right angle to the road. Although they differed in size and in some structural details, they shared three main constructional elements, these being a possible construction cut, the clay lining and the timber post and plank structure itself.

The recording of the stratigraphy within the tanks was subject to the constraints of time and conditions on the site, in that the clay-lined tanks underwent continual flooding from the surrounding area. Two baulks were left across the width of each tank and provided the most secure stratigraphical profiles. Between these baulks the upper fills were removed by a combination of machine

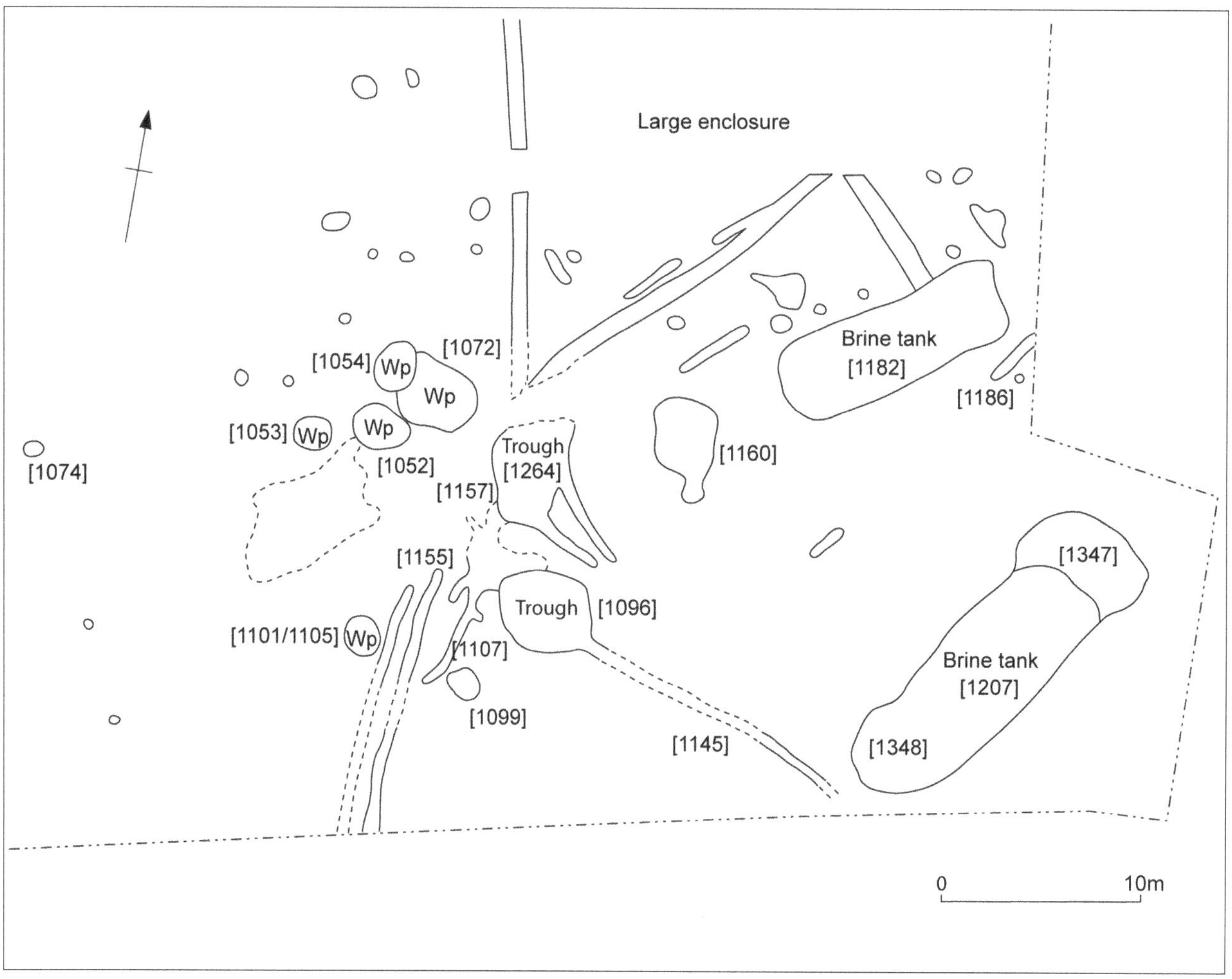

Figure 3.12: Brine Industrial Zone plan.

Plate 3.3: Brine tank [1207] viewed from the north, with chamber [1347] in the foreground.

and hand excavation, and the lower fills by hand. The infill was removed in spits between the baulks but it was not possible to record these spits in plan and excavated fills were referenced to the baulk sections, where this was believed to be applicable.

Brine Tank [1207] and Associated Features

The Construction of [1207]

The larger brine tank [1207] and associated features [1347] and [1348] (Fig 3.13; Plate 3.3) were situated in the south-east corner of the main excavation area and were orientated roughly north-east - south-west. The various elements of the tank had a common method of construction, comprising an outer clay lining and an inner timber-built structure. The outer lining (1241) was made from a homogeneous light brownish red plastic clay up to 0.6m thick. Excavation found the base of the tank to be formed by the natural boulder clay, and the sides were made up of similar material probably extracted from the same source.

Machine trenching across the southern half of tank [1207] revealed a possible outer cut [1376] within which the tank may have been constructed. This possible cut extended at least 3.5m beyond the clay lining on the north-west side of the tank and contained a mid- to light brown silty sand (1377). Within [1376] it was noted that the level of the natural clay dropped away from the base of the tank, and this change in level could be indicative of where the clay used to form the tank's upper lining was extracted.

The machine trenching across the tank also revealed cattle remains, including a head, embedded within the clay lining (1241). The remains of a second animal, again including the head, were found within the upper clay lining at the tank's southern end (context (1290)). These remains were excavated from the clay by hand but because of time constraints not all may have been recovered.

The inner timber-built structure of tank [1207] measured *c* 11.2m long by *c* 3.6m wide at the base. At the north-eastern end of this tank was a separate timber-lined chamber [1347], measuring *c* 1.75m by *c* 3.6m. At the south-western end was a failed timber-built wall [1348], *c* 3.5m wide. Both [1347] and [1348] shared the outer clay lining (1241) of the main tank, from which they were separated by continuations of that same feature.

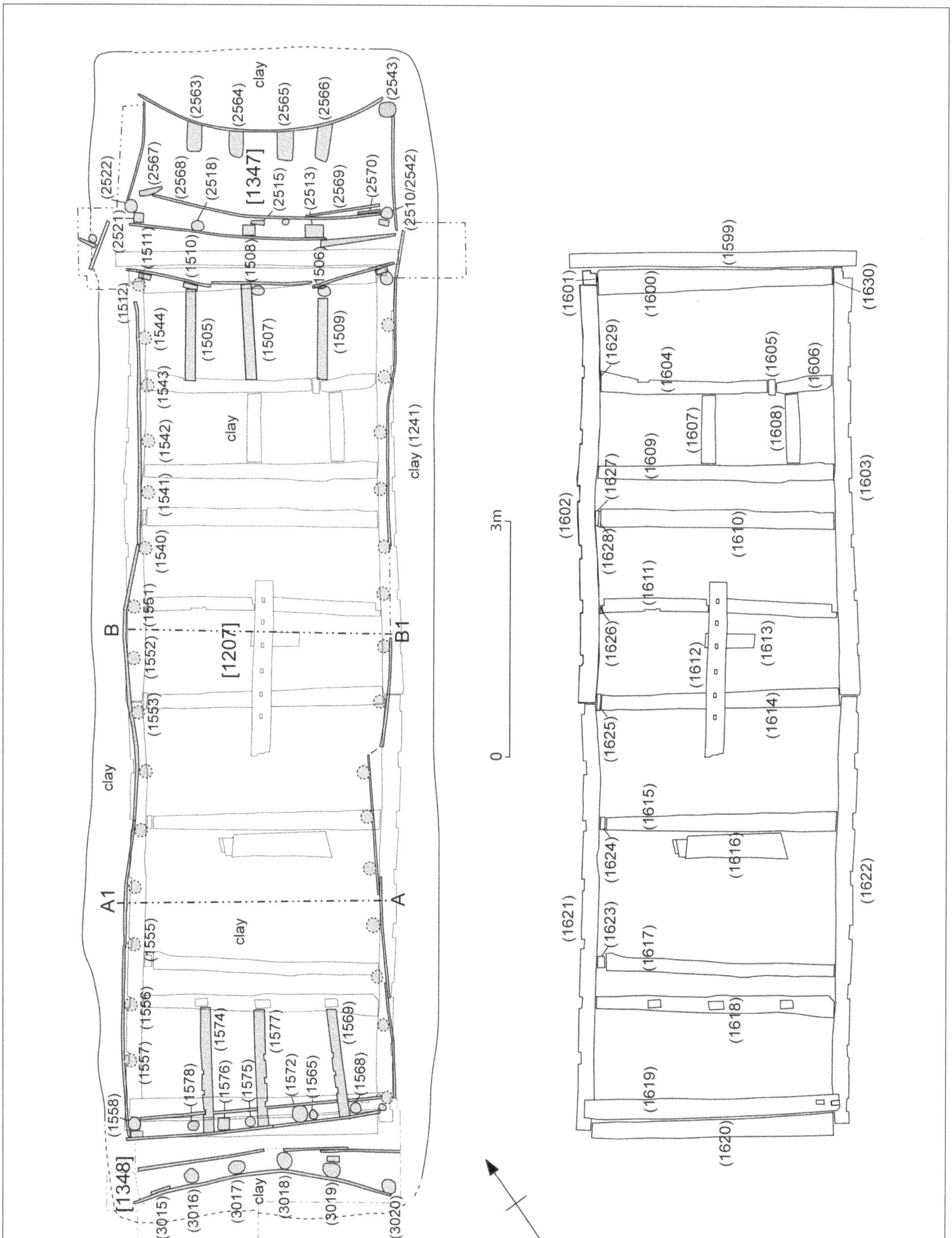

Figure 3.13: Left, plan of brine tank [1207], failed wall [1348] and chamber [1347]. Right, plan of base timbers of [1207].

Plate 3.4: The base of brine tank [1207] viewed from the south-west.

The largest structural timber elements of tank [1207] were baseplates (1621), (1602), (1603) and (1622) which ran along the north-western and south-eastern elevations, two on either side butting end to end (Fig 3.13; 16.1, 1622). The longest of these timbers (1602) was 5.48m in length, while the remaining three measured 5.4m. All were hewn from oak and were sub-square in section with a width ranging between 0.24m and 0.3m. Mortices were cut along the outer edge of these timbers, positioned at regular intervals *c* 0.5-0.54m apart, into which tenoned upright timber posts were set. At the north-eastern end of the tank, timbers (1602) and (1603) both had a redundant lap/scarf joint.

The baseplates at the north-eastern and south-western ends of the tank, (1600), (1599), (1619) and (1620), were set in a double arrangement, comprising an inner and outer baseplate, separated by small timber wedges. Three of these end baseplates were oak, while the outer baseplate at the north-eastern end was Scots pine (1599). This last timber extended slightly beyond the width of the tank. The adjacent inner baseplate (1600) ran between the side baseplates, with a small wedge, (1601) and (1630), inserted at either end. At the tank's south-western end, the inner baseplate (1619) (Fig 16.2) was a reused timber with lap-jointed ends which were set into mortices in the side baseplates. The outer baseplate (1620) was positioned between those side timbers. Of the four end baseplates, (1599) and (1619) were squared, while the others were whole timbers, unconverted apart from (1620) being partly chamfered at one end.

Internally the width of the tank was spanned in eight locations by oak timbers running between the side baseplates. Five of these, (1609), (1610), (1614), (1615) (Fig 16.2) and (1617), were unconverted timbers. Timber (1611) (Fig 16.2) showed clear evidence of reuse. The most southerly of the cross-timbers (1618) contained three mortices which held angled bracing (see below) but which may have originally contained vertical posts and be indicative of reuse. The most northerly of the cross-timbers was a composite of three parts, (1604), (1606) and wedge (1605), and was strengthened by two short timbers, (1607) and (1608), which ran between it and the neighbouring cross-timber to the south. In six locations along the north-west side of the tank, wedges ranging in thickness from 10mm to 0.12m had been inserted between the cross-timbers and the baseplates, ie (1623), (1624), (1625), (1626), (1627)/(1628) and (1629).

The cross-timbers at the north-eastern and south-western ends each supported three oak braces which rose at an angle of approximately 40 degrees from the vertical to meet the end timbers. Two of the braces at the north-eastern end of the tank were unconverted timbers, (1507) and (1509), and the third (1505) (Fig 16.6) was reused. Those at the south-western end were also reused timbers, (1569) (Fig 16.6), (1577) and (1574).

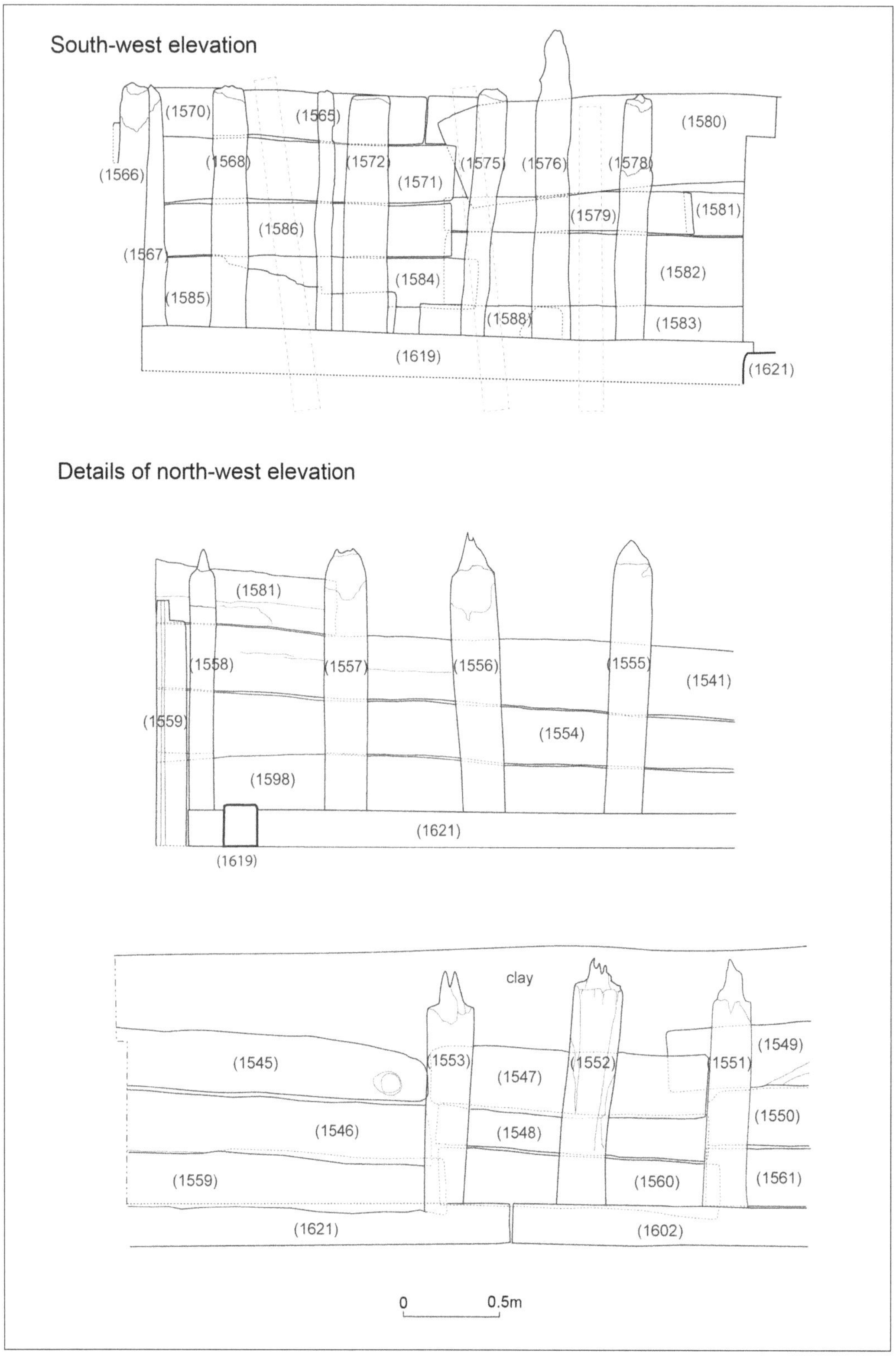

Figure 3.14: Brine tank [1207] elevations.

The baseplates and cross-timbers in [1207] sat proud of the surface of clay lining [1241]. Within the centre of the tank, reused timber (1612) (Fig 16.3) rested across two neighbouring cross-timbers and was further supported by a small timber (1613). A tenoned beam (1616) lay within the base of the tank adjacent to cross-timber (1615).

A total of 45 timber uprights were used in the construction of [1207], all of oak. The uprights survived to a height of *c* 1-1.5m above the baseplates and were mostly between 0.2m and 0.25m in width (Fig 3.14). Sixteen posts supported each of the long, south-eastern and north-western, elevations. All were whole timbers, unconverted apart from having a bare-faced tenon which slotted into the morticing on the outer edge of the baseplate timbers (eg Fig 16.5, 1556).

Eight uprights supported the south-western end of the

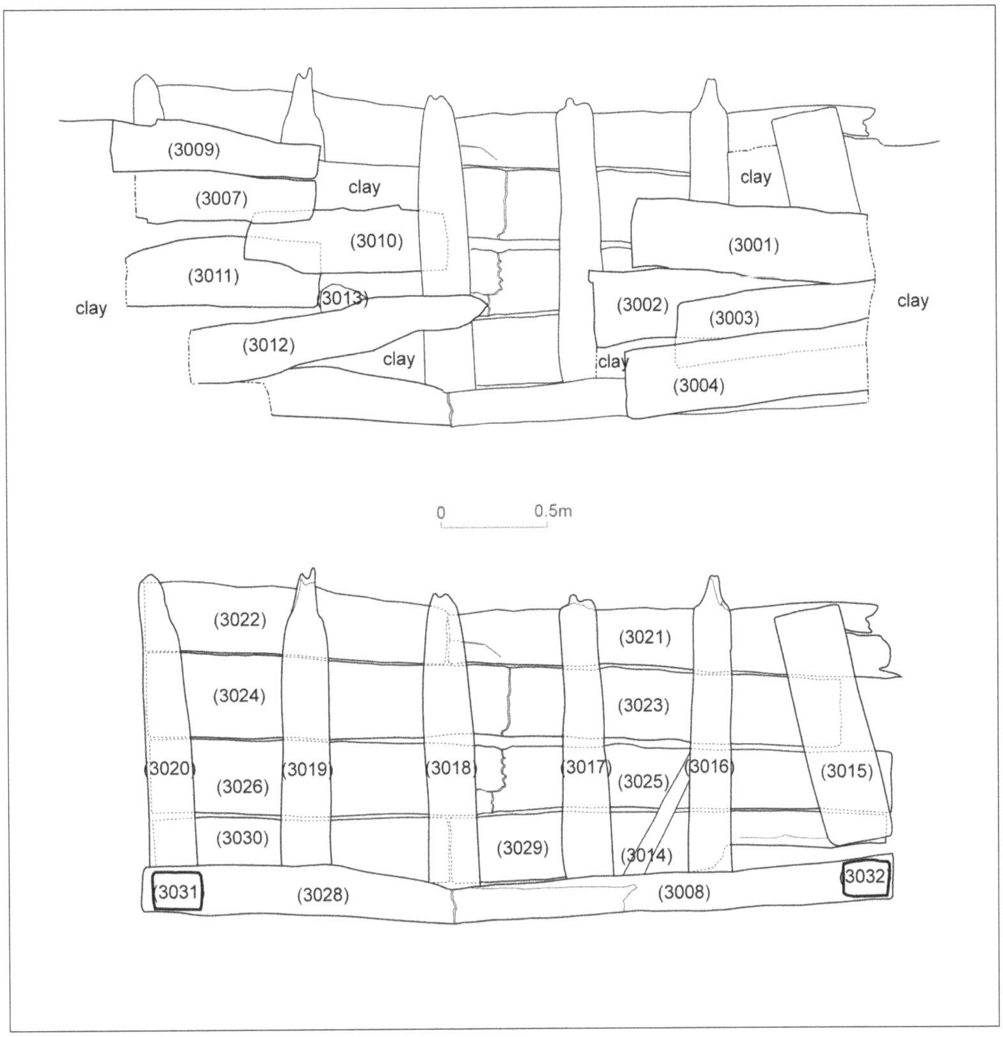

Figure 3.15: Above, elevation of failed wall [1348]. Below, the elevation following the removal of the outer planking.

tank (Fig 3.14). They included four whole timbers again unworked apart from a tenon at the base, (1568), (1572), (1575) and (1578). This end also contained squared timbers (1559) and (1576), roundwood post (1567) and a reused timber (1565) (Fig 16.7). The north-eastern end of the tank contained four main uprights, comprising unworked timbers (1506) and (1508), squared timber (1511) and converted timber (1510) (Fig 16.5). At both ends, the bases of the uprights seem to have been set between the double arrangement of baseplates.

The sides of both the long and short axes of the structure were made up of horizontally aligned oak planks (Fig 3.14), firmly positioned between the timber upright posts and the clay lining. These planks were tangentially faced and up to 4m in length, with the majority having a width of *c* 0.3-0.4m and a depth of 30-40mm. They were laid one on top of another, either edge to edge or with some overlapping, and survived to a height of four courses. From the height of the posts, at least one other course had been lost.

The Failed Wall [1348]

Immediately to the south-west of [1207] and separated by clay lining (1241) was failed post and plank construction [1348] (Figs 3.13 & 3.15). Its remains consisted of two shortened side baseplates and, making up the south-western end, a baseplate, six uprights and a planked wall. All these components were surrounded by clay lining (1241). The timbers were all of oak. The side baseplates (3031) and (3032) projected from the end baseplate by *c* 0.47-0.54m and had been shortened by means of axing. The end baseplate (3008)/(3028) was a rounded timber with a width of 0.2m and had snapped in antiquity at a weak point where it had been morticed to accept a timber upright (3018). The side baseplates butted against the end baseplate but (3032) also had a redundant lap joint. Of the six timber uprights, the north-westernmost (3015) was a squared post. The others, (3016), (3017), (3018), (3019) and (3020), were whole timbers, unworked apart from a bare-faced tenon at the base which slotted into mortices on the outer edge of the baseplate (Fig 16.5, 3017). The posts survived to a maximum height of 1.35m above the baseplate. The wall was made up of tangentially faced planks with a maximum length of 3.1m and width of 0.4m, and an average depth of 40mm. These planks had fractured in line with the break in the baseplate.

On the north-east side of [1348] was an irregular arrangement of short interlocking planks (3001), (3002), (3303), (3004), (3007), (3009), (3010), (3011), (3012),

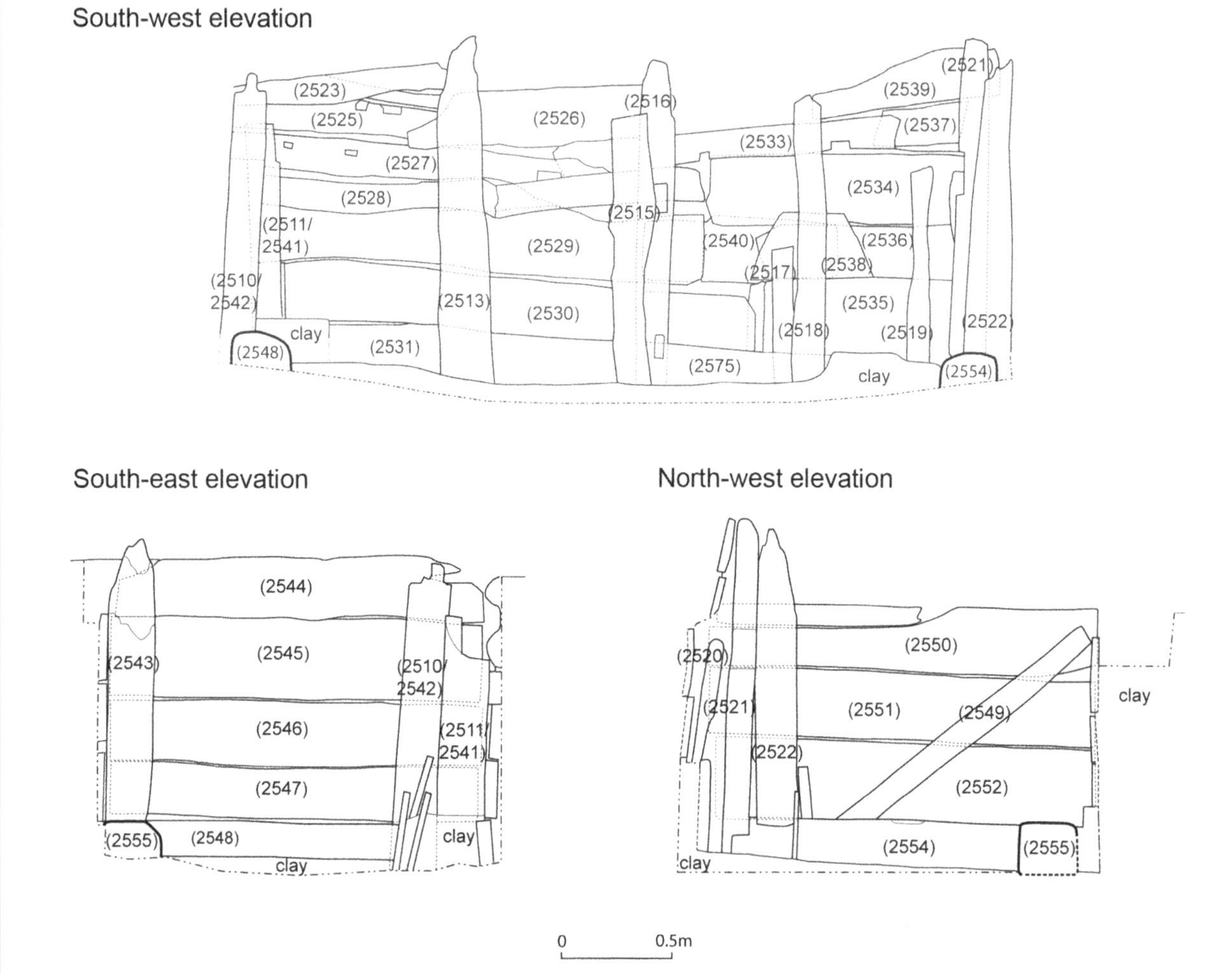

Figure 3.16: Elevations of chamber [1347].

(3013) and possibly (3006), which were partially supported by the posts and lay against the clay lining separating [1348] and tank [1207] (Fig 3.15). Beneath that lining, a centrally placed wedge (1305) ran between the base of the failed wall and the outer baseplate (1620) of [1207].

Chamber [1347]

At the north-east end of brine tank [1207] was feature [1347], a clay and timber-lined chamber (Figs 3.13 & 3.16; Plate 3.5). This was of a generally similar construction to [1207] but differed in some details. The timbers were all of oak, apart from a post and a plank which were of birch. The north-eastern wall sat on a baseplate (2555), *c* 3.6m long, with mortices on its outer edge. Linked to this by lap joints were side baseplates (2548) and (2554) (Fig 16.1), 1.6m in length, each with two mortices set 0.47-0.5m apart. Both of the side baseplates appear to have been deliberately shortened. The four walls of [1347] were of a post and plank construction. With the exception of the south-western wall, the planking was laid in a similar arrangement to [1207] and again survived to a height of four courses.

The uprights of the north-eastern wall comprised four whole timbers, (2563), (2564), (2565) and (2566), unconverted apart from each having a bare-faced tenon set into the mortices of the baseplate. When excavated, both the uprights and the planks making up this wall were found to lean forwards at an angle of *c* 45 degrees. The shift from the vertical was evidently due to pressure caused by the weight of the clay lining. The tops of all the uprights in this wall had been sawn off, presumably after the collapse had rendered the wall redundant.

The south-eastern side wall contained two uprights, (2510/2542) and (2543), and the north-western side wall a single upright (2522). Of these (2510/2542) was a squared timber and the others whole timbers, and all had a bare-faced tenon jointed into the baseplate. Redundant mortices in the baseplates implied that three other uprights had been removed.

The south-western wall was of a more irregular construction. Unlike [1207] and the other walls of [1347], it did not include a baseplate and its supporting uprights were set directly into the clay. A total of seven uprights were used in its construction and were of a variety of forms. Two were whole unconverted timbers, (2513) and (2518). Two were radially cleft stakes, (2515) (Fig 16.6) and (2519). Uprights (2516) and (2521) were converted timbers containing mortices from a previous use. The final upright, (2511/2541) (Fig 16.2), was a converted timber which was roughly hewn at one end, with a bare-faced tenon at the other and with a mortice on one side. Its form suggests that it was a reused part of a baseplate.

Positioned to the front of these uprights was a single horizontal plank (2568), set on its edge and reinforced by shorter planking at each end, (2567), (2569) and (2570) (Fig 3.13). These planks were secured by stakes driven into the clay. Of these, stake (2517) (Fig 16.6) contained nails and nail holes suggestive of reuse. The whole arrangement was packed on either side with clay and provided additional support for the south-western wall.

The planking of the south-western wall was also more irregular than the other walls of the chamber, comprising a poorly constructed arrangement of variously shaped and sized timbers. A high proportion were reused, including timbers (2525), (2527), (2530), (2533), (2535) (2538), (2573), (2575) and possibly (2528) (Figs 16.3 & 16.4).

Plate 3.5: Chamber [1347] viewed from the south-east.

The Relationship of [1207], [1347] and [1348]

The position of the failed wall [1348] immediately to the rear of the clay lining at the south-west end of brine tank [1207], its failure in antiquity and the truncated remains of side baseplates (3031) and (3032) all strongly indicate that [1348] was originally intended to be the end wall of the brine tank. With the failure of the end baseplate and consequent fracture of the plank wall, this build was abandoned and the end wall of the tank brought forwards.

A further and greater shortening of the tank appears to have taken place at the opposite end, again as a result of the failure of the end wall. Here the lack of a baseplate in the south-west wall of chamber [1347] and the irregular nature of the wall's construction strongly suggest that this wall was a secondary feature. This in turn implies that the other walls of the chamber represent the original north-eastern extent of brine tank [1207], a conclusion which is supported by the apparent shortening of the side baseplates of the chamber, (2548) and (2554), and the redundant lap/scarf joints on side baseplates (1602) and (1603) in [1207].

The base-frame of the tank as originally planned would thus have included four long baseplates of roughly equally length with two smaller baseplates jointed to these at the north-east end, an arrangement also found in brine tank [1182] (see below). In the case of the larger tank, other short side baseplates were found at the tank's south-west end. These survived only as truncated timbers (3031) and (3032) in failed wall (1348) but originally would have butted against the ends of long side baseplates (1622) and (1621) respectively.

It is possible that end baseplates (1620) and (1600) in [1207] originally served as cross-timbers within this larger tank and that, when each of the end walls failed, the decision was made to establish a new end wall by utilising those cross-timbers, and adding a second timber to create a setting for the uprights. At the south-west end, mortices were cut into existing baseplates (1621) and (1622) to take this timber (1619). At the north-east end, timber (1599) was laid across the ends of baseplates (1602) and (1603). Other cross-timbers were inserted to support the angled braces which now served to strengthen the new post and plank end walls.

The construction of the new walls shortened the tank by *c* 1m on the south-west and by *c* 2m on the north-east. Both walls included an outer clay lining, which at the south-west end was built up against a facing of planks laid across failed wall [1348]. At the opposite end, the lining was laid against a roughly built post and plank wall, enclosing chamber [1347]. Probably either during or immediately following the construction of this wall the chamber was backfilled. At both ends of the tank, the construction of the new post and plank walls and their clay lining was preceded by the removal of sections of the side baseplates, one of which seems to have been reused as a post in the south-west wall of [1347].

The failure of the original end walls of the tank, due to thc latcral pressure of the clay lining, very probably occurred while the tank was still under construction and before sufficient components were in place to give the structure the required rigidity. The insertion of the new end baseplates would have been difficult, although not impossible, once the side walls of the tank were in place at these points.

Dendrochonology provides a possible date of AD 114-42 for failed wall [1348], derived from timber (3012) (see

Chapter 17), which formed part of the wall's plank facing (Fig 3.15). The plank is chamfered and it is uncertain whether the chamfering dates from its incorporation within the wall or is evidence of reuse. However, the dendrochonological analysis also shows that timber (3001), from the same plank facing of [1348], and timber (2550), part of the north-west elevation of [1347], were derived from the same tree. These timbers may confirm that the failure of [1348] occurred during construction.

It is likely that the upper part of the tank contained a superstructure matching the base timbers, that is the upright posts were tenoned into horizontal top plates running along the edge of the tank, with these top plates kept rigid by a series of cross-timbers. Confirmation of the need for such a superstructure was provided during the excavation, when removal of the fills of [1207] caused the sides of the tank to begin to collapse inward, requiring supports to be put in place. No remains of a timber superstructure could be identified with certainty but the tops of many of the tank's posts had suffered damage, probably through later agricultural activity which could also have removed other upper parts of the structure. The function of the reused beam (1612), which was located roughly within the centre of the tank and was aligned along its axis, is uncertain.

The Fill

As noted above, the difficult conditions of excavation means that the principal record of the stratigraphy of the infill of [1207] comprises the sections provided by the two baulks left across the body of the tank (Figs 3.13 & 3.17). Other contexts were noted as the infill of the tanks was removed in spits.

The primary fill comprised two thin bands, (1336) and (1335), which may have accumulated during the use of the tank for the managed storage of brine. These were overlain by (1337) which was in turn sealed by (1239), a dark grey silty sand. Fill (1239) was of note for the quantity and condition of finds contained within it, including intact pottery vessels, metalwork and unusually well-preserved organic material. This and other finds-rich

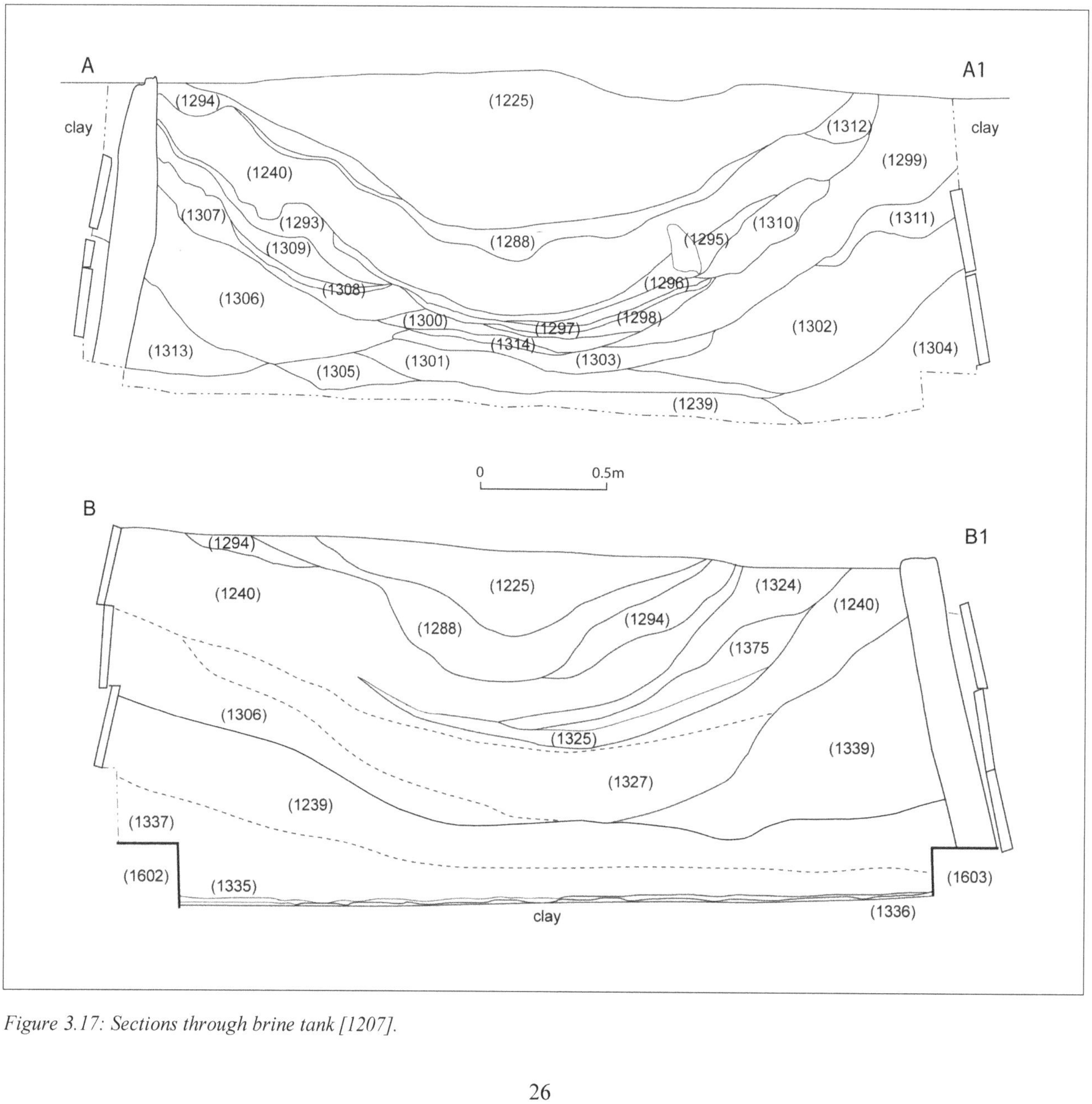

Figure 3.17: Sections through brine tank [1207].

contexts in brine tanks [1207] and [1182] contained animal bones, most frequently of cattle but also of sheep/goat, pig, horse, dog, red deer and domestic fowl, which included both articulated remains and certain parts of the skeleton, such as cattle heads (see Chapter 18). The examples of the latter from context (1239) included one skull perforated by projectile holes. Also among the varied assemblage from (1239) were small finds which included iron tools and a complete lead alloy casserole, wooden objects which included two brine paddles, a turned bowl and a large perforated object which may be a lid or seat, and an upper and lower quernstone which are probably a matching pair (see Chapters 7, 10 and 11). The unusual nature of the overall assemblage implies a strong element of ritual deposition as well as the more casual disposal of waste. From the pottery evidence it is likely that (1239) was deposited *c* AD 180-90. The dating is complicated by the assignment of pottery of a 3rd-century date to the same context. However, this later material seems to represent a relatively minor 3rd-century intrusion into earlier deposits, not identified on site due to the conditions under which the tank was excavated.

In section B-B1, (1239) was overlain by sand (1339), sandy silt (1327) and clay (1306), which were sealed by (1240). Section A-A1 shows a more complex sequence between (1239 and (1240). Here (1239) was overlain by silty sand (1313), peat-like deposit (1305), and silty sands (1301) and (1304). Within the centre of the section these were overlain by thin bands of sandy silts, (1303), (1314), (1300), (1298) and (1297), sand (1296) and silt (1295), while to each side were broader bands of sands and clays, (1306), (1309), (1293), (1302), (1311), (1299) and (1310), as well as sandy silts (1307) and (1308). Of the various contexts between (1239) and (1240), only uppermost fill (1295) contained pottery, including a Wilderspool mortarium of Antonine date.

Fill (1240) was a mixed deposit of pink clay and brown sand which from the pottery seems to have been laid down in the mid 3rd century. Like (1239), (1240) contained a significant collection of animal bone (see Chapter 18). The quantity, however, was smaller than in (1239), and the variety and number of other finds within (1240) were also less.

In section A-A1, (1240) was sealed by clay (1288) and, to either side, sands (1294) and (1312). These were in turn overlain by peat deposit (1225). In section B-B1, (1240) was partly overlain by and partly overlay sandy silts (1324) and (1325), and sand (1375), which would appear to be part of the same episode of deposition. This group of contexts was overlain by sand (1294), clay (1288) and peat (1225).

Above (1225) and recorded during the initials stages of excavation was a further sequence of thin bands of accumulated sands and silts, comprising, from the lowest to the uppermost, (1236)/(1237), (1224), (1233), (1209), (1232), (1231), (1230), (1229) and (1208).

North-east chamber [1347] contained a horizontally banded sequence of fills. Lower fills (1349) and (1351) contained pottery consistent with the chamber being infilled in the mid 2nd century.

The Phasing of the Tank

Brine tank [1207], associated features [1347] and [1348], and their fills show several phases of development. Phase 1 was the constructional phase of the tank, involving the building of the clay outer lining and the post and plank inner structure. This phase also saw the two foundation deposits of cattle being included within the clay lining. It is likely that the structural failures evident in [1347] and [1348] occurred during construction and resulted in modifications which shortened the length of the tank. Dendrochronology gives a possible date for the building of the tank between AD 114 and 142 (see above). A fragment of samian ware recovered from the tank's clay lining (1241) gives a somewhat later *terminus post quem* of *c* AD 150 (see Chapter 5, D7).

Phase 2 represents the period of use of the tank for the managed storage of brine. This phase probably gave rise to thin basal deposits (1336) and (1335). The lack of a significant deposition in this phase suggests that the base of the tank was periodically cleaned. Analysis of the diatom assemblage from (1335) found that it derived mainly from slightly brackish water with relatively high concentrations of dissolved salts. The phase may end with (1337), which insect analysis found to be dominated by taxa characteristic of rotting plant matter.

Phase 3 relates to the closure of the tank and is principally represented by deposit (1239). During this phase some domestic waste was thrown into the tank but much material was also placed here in a ritual deposition. Insect analysis of (1239) suggests the presence of stagnant water. This closure took place *c* AD 180-90.

Phase 4 is represented by the deposits above (1239) and below (1240). During this period, sands and clays were washed or eroded into the tank creating a central watery basin in which silts developed. Samples from fills (1303) and (1295) which belong to this phase had very few charred plant remains, which may indicate that the tank was now used much less for waste disposal. The Phase 4 deposits have likewise produced very little pottery. Analysis of (1295), the uppermost deposit in this phase, found it to be dominated by slow-water taxa.

Phase 5 appears to have been a second phase of ritual deposition, in the mid 3rd century, which is principally represented by (1240). This deposition was on a lesser scale to that in the late 2nd century but again involved animal remains and intact pottery vessels. It was probably during this phase that some intrusion took place into Phase 3 deposit (1239).

Phase 6 saw the shallow central basin infilled by peat and the tank finally covered by an accumulation of sands and silts. This process seems to have continued from the mid 3rd century into the 4th and possibly later.

Brine Tank [1182]

The Construction of the Tank

Brine tank [1182], like [1207], was of a post and plank construction, with an outer clay lining (1314) (Fig 3.18; Plate 3.6). The lining was again made from a homogeneous light brownish red clay. It was found to be of an average thickness of 0.7m and, unlike in [1207], did not appear to incorporate the underlying natural deposit within this area of the site. There was again evidence of a possible construction cut. This was much less distinct than that of [1207] but measured *c* 15m south-west - north-east and *c* 7.60m south-east - north-west and had a maximum recorded depth of 2.15m. This possible construction cut contained deposit (1355), a mid- to light brown silty sand.

The post and plank tank itself [1182] was 10.5m long and 3m wide (Fig 3.18). All the timbers were of oak. The long south-eastern and north-western sides each contained three baseplate timbers, four of which, (2059), (2061), (2062) and (2063), were 4.2-4.44m in length, and the others, (2058) and (2064), 1.8-1.9m. They were all sub-square in section with a width ranging between 0.2m and 0.3m (Fig 16.1, 2061). They were jointed with lap/scarf joints. All had mortices along the outside edge, regularly spaced at intervals of 0.5-0.6m, into which the tenoned upright timber posts were set.

The south-western and north-eastern ends of the tank were each crossed by a single baseplate, (2056) and (2066), again with mortices to accept uprights. Baseplate (2056) was set between lap joints in side baseplates (2061) and (2062). Baseplate (2066) butted against the ends of side baseplates (2058) and (2064).

The tank was spanned transversely from south-east to north-west in five locations by timbers (2051), (2052), (2053), (2054) and (2055). Of these, (2051) was a whole unconverted timber, while the others were sub-rectangular in section. Cross-timber (2053) was 0.22m shorter than the others and was chocked at its western end by a timber block (2060), into which it had been jointed with some precision using a bird's-mouth recess. It also contained two mortices and would appear to have been reused. The baseplates and cross-timbers in [1182] were all firmly embedded into the clay lining (1314), unlike those in [1207] which sat proud of the surface.

A total of 29 upright timber posts survived in situ (Fig 3.19). Thirteen of these were positioned along the baseplates on the north-western side and fourteen on the south-eastern side. Two posts, (2001) and (2066), survived at the north-eastern end and the baseplate (2057) here also contained mortices for two other posts. At the south-western end, baseplate (2056) contained the possible stub of a post and the mortices for two others. Eight of the uprights, (2004), (2005), (2012), (2013), (2022), (2026), (2028) and (2030), were whole timbers unworked apart from each having a bare-faced tenon at the base which slotted into the morticing of the baseplates (Fig 16.5, 2012). The remainder were roughly worked and subrectangular in section and again had bare-faced tenons at their base (Fig 16.5, 2027 & 2029). The uprights generally survived to a height of up to 0.75m above the baseplates and were mostly between 0.15m and 0.23m in width.

The sides of the tank were made up of planks firmly set between the upright posts and the clay lining. The planks were tangentially faced with a maximum length of *c* 3.8m and width of *c* 0.4m, and a depth of 30-40mm. They mostly survived to a height of two courses but there were also slight remains of a third.

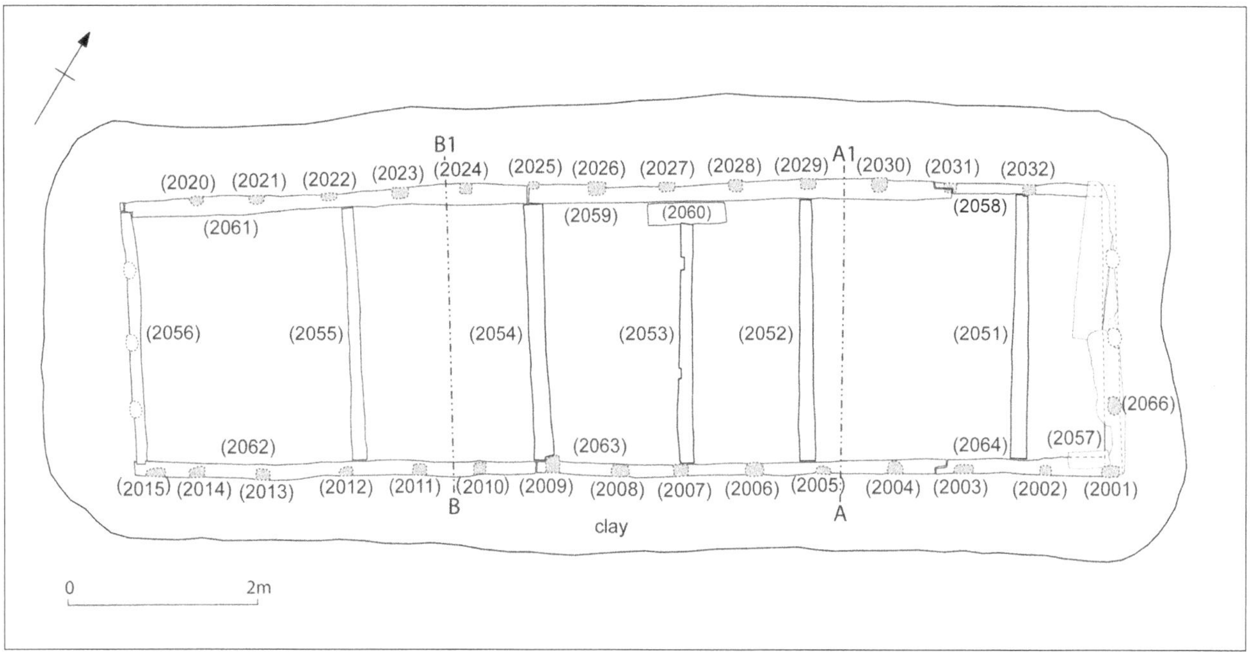

Figure 3.18: Plan of brine tank [1182].

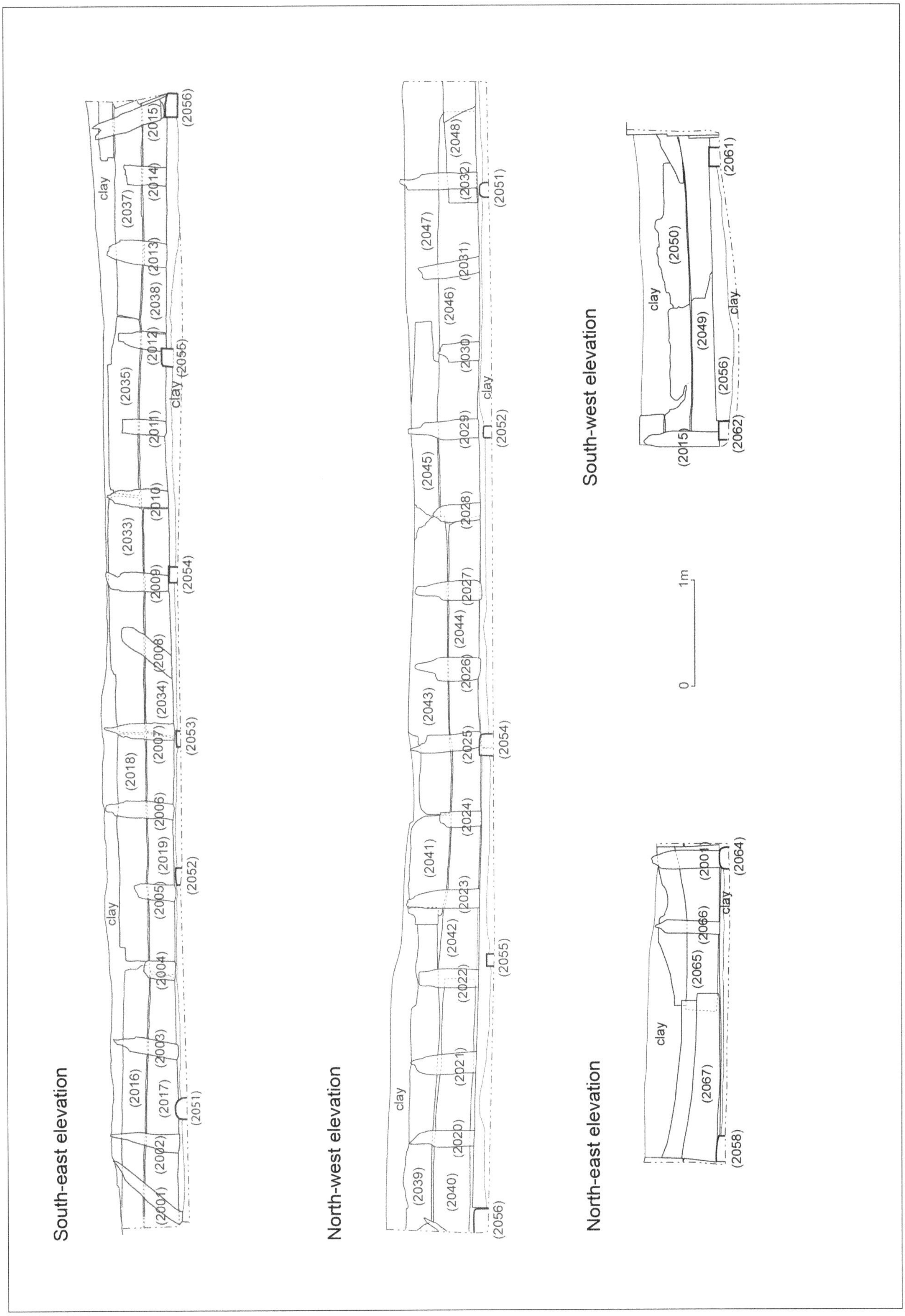

Figure 3.19: Elevations of brine tank [1182].

Plate 3.6: Brine tank [1182] viewed from the south-west.

The Fill

As with brine tank [1207], two baulks were left across the body of the tank and provided the principal record of the stratigraphy of the infill (Fig 3.20).

Common to both sections was a thin basal clay deposit (1286). In section A-A1, this was sealed across the tank by (1285), a dark brown silty clay with an abundance of organic material comprising leaves and twigs. In section B-B1, organic deposit (1333) and sand and clay deposits (1353) and (1292) partly underlay (1285). Fill (1285) contained a large and varied assemblage, comparable to that associated with (1239) in brine tank [1207] and again including intact pottery vessels and unusually well-preserved organic material. Among the finds were wooden objects which included a single-piece shovel, the blade of a composite shovel and a perforated stave, all of which were possibly used in brine production, a staved vessel and a possible musical instrument, and other small finds which included an iron drill bit and a creaser (see Chapters 7 and 11). Again a large quantity of animal bone was present in (1285) and the overlying deposit (1284) (see Chapter 18). Fill (1285) also contained several shoes and waste leather (see Chapter 9).

As with (1239), from the pottery evidence it is likely that (1285) was deposited in the Antonine period. Again some 3rd-century pottery has been assigned to this context and seems to represent a later intrusion which was not identified during the excavation.

Above (1285) was organic deposit (1283) which formed the interface with a series of overlying bands mostly of sand and clay deposits, (1280), (1281), (1282), (1330) and (1332), but also peat deposit (1331). In section B-B1, the deposits were sealed by a sequence comprising clay and peat deposit (1279), sand (1287), clay and sand (1278), peaty deposit (1277) and clay (1220). In section A-A1, deposits (1282), (1281), (1280) and possibly (1283) appear to have been cut, with (1279), (1278), (1277) and (1220) being fills of that cut.

Fill (1283), the overlying bands (1280), (1281) and (1282), and clay and peat deposit (1279) all contained animal bones, although in each case fewer individual specimens were identified than in the lower deposits (1285) and (1284) (see Chapter 18). Contexts (1283) and (1279) contained shoes and waste leather, although again in lesser quantities than in (1285), and (1279) also

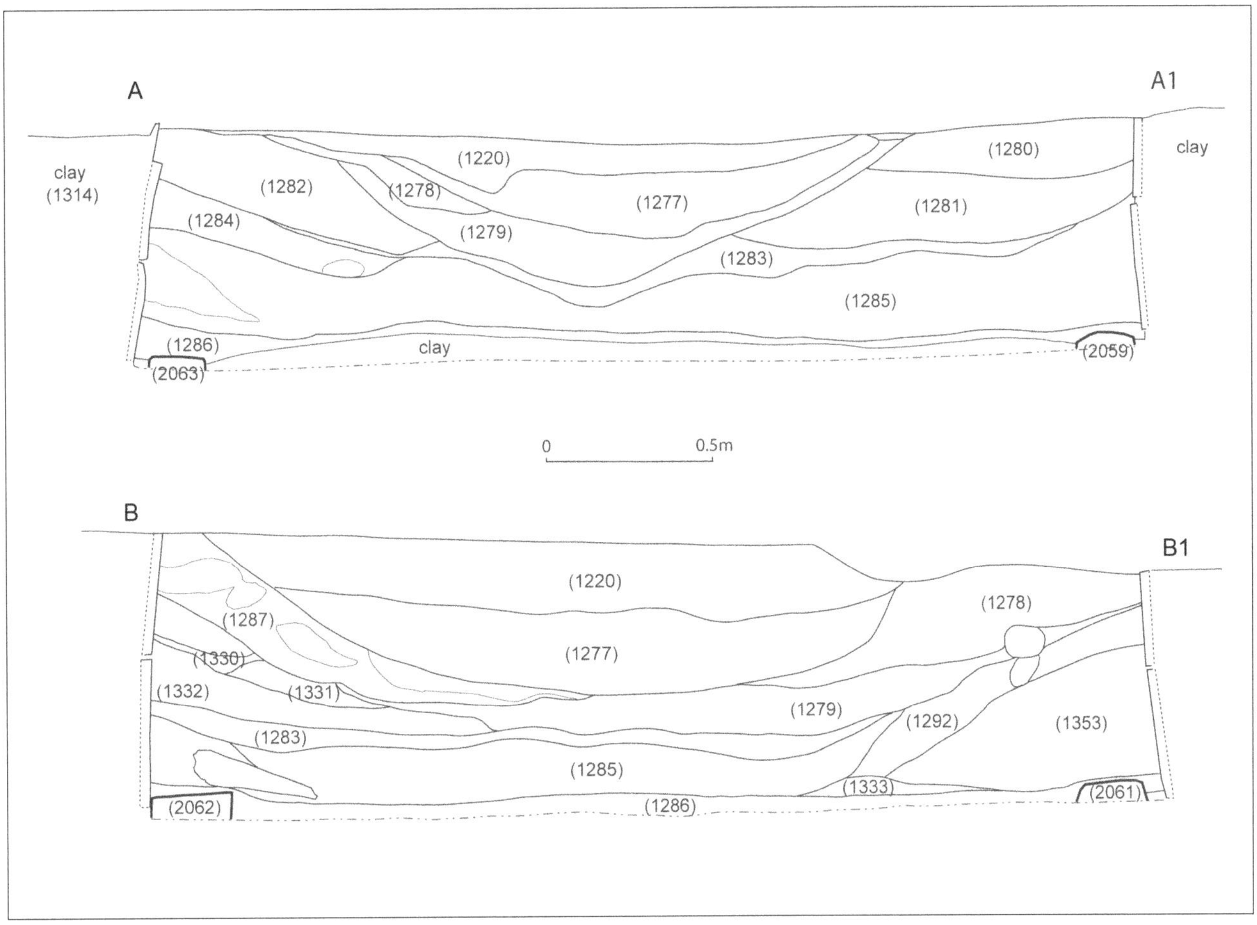

Figure 3.20: Sections through brine tank [1182].

included several hobnails. The upper deposits produced an upper and a lower quernstone which appear to be a matching pair and which at the time of excavation were respectively assigned to contexts (1277) and (1280) (see Chapter 11). The pottery from (1280) and (1281) implies a 2nd-century date of deposition, that from (1279) and (1277) a 3rd-century date.

Above (1220) but not recorded in the sections were two further deposits, (1188) and uppermost fill (1183).

The Phasing of the Tank

Like [1207], brine tank [1182] saw several phases of development. While the sequences show a general similarity to [1207], there are also differences and the equivalent phases need not be exactly contemporary.

Phase 1 was the construction of the tank for which dendrochronology gives a date range of AD 130-7 in the late Hadrianic (see Chapter 17). This dating derives from timber (2029), a post in the north-western elevation, and (2049), a plank at the south-west end. Both timbers were found in situ and neither had evidence of reuse.

Phase 2 is represented by the thin basal deposit (1286) which probably relates to the tank's period of use for the managed storage of brine. As with [1207], the lack of significant deposits during this phase suggests periodic cleaning of the bottom of the tank.

Phase 3 is largely represented by (1285), which contained a ritual deposit parallel to (1239) in [1207]. The pottery indicates a mid to late Antonine date.

As in [1207], Phase 4 saw a deposition of sands and clays, chiefly represented by (1281), (1282), (1283) and (1284). However, the Phase 4 deposits contained a much greater quantity of pottery than the equivalent contexts in [1207] and this is mainly of an Antonine date. Again in contrast to [1207], the deposition of animal remains continued within the Phase 4 contexts. Insect analysis of a sample from (1284) shows a slow-water environment with rotting plant material (see Chapter 18).

Phase 5 involved deposits, including (1279), which appear to have partly filled a cut into the Phase 4 layers. The deposition of animal remains is again found. From the pottery evidence Phase 5 seems to date from the early to mid 3rd century. It was probably during this phase that some intrusion took place into Phase 3 deposit (1285).

Phase 6, as in tank [1207], saw an accumulation of peat (1277) within a central basin, in turn sealed by clay (1220).

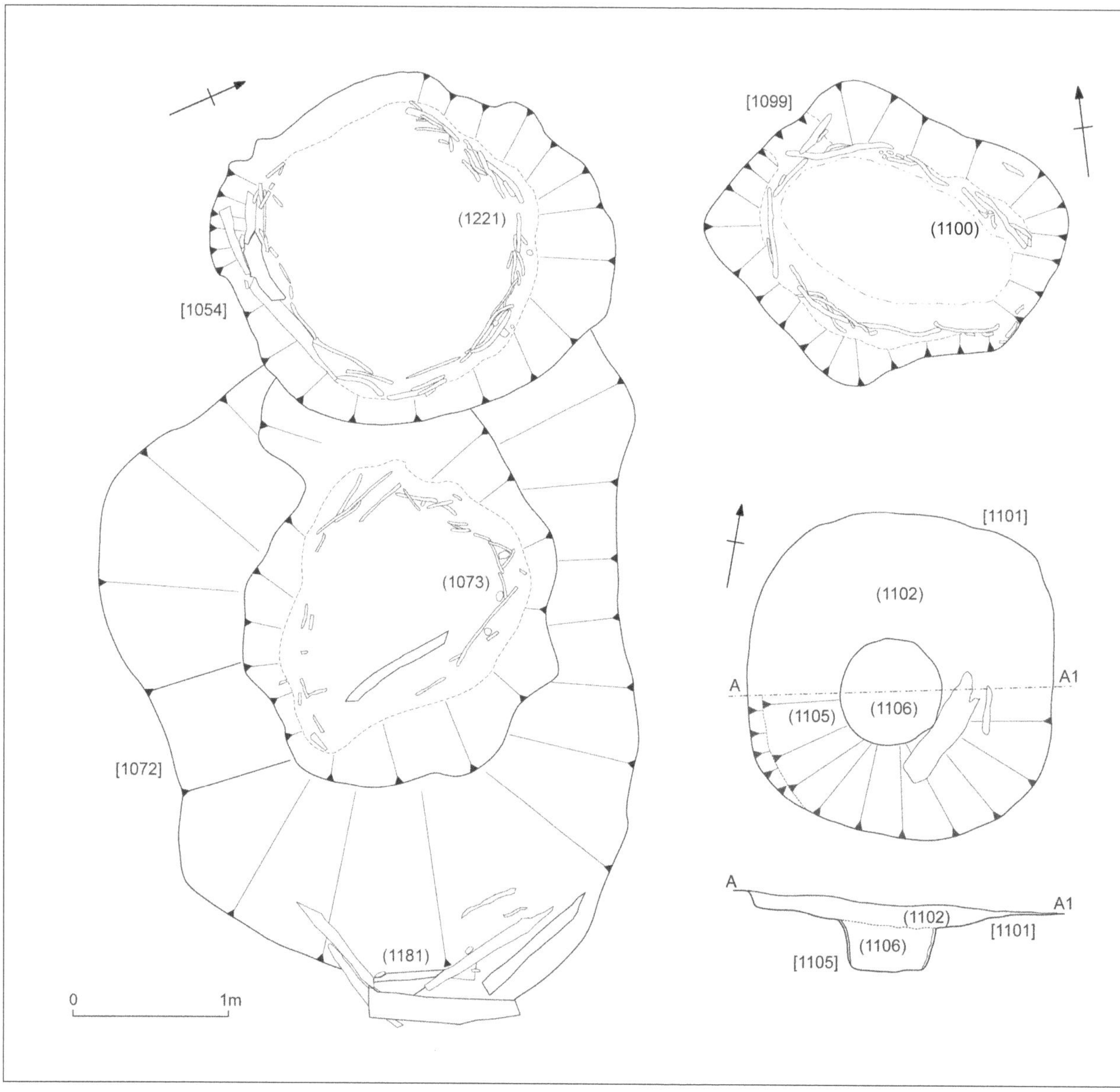

Figure 3.21: Pits [1054], [1072], [1099] and [1101].

Wicker-lined Pits, Pits, Troughs and Ditches

Located to the west of the two brine tanks, [1182] and [1207], was a collection of features which included pits and linear features (Fig 3.12). Several pits were wicker-lined or contained the remains of wooden troughs. The wicker-lined pits principally comprised features [1052], [1053], [1054], [1072] and [1099]. Their wicker lining was constructed from roundwood vertical staves driven through the base of the cut of each pit and interwoven with roundwood withies. In some examples the diameter of the wicker lining was substantially smaller than the cut of the pit and the space between the wicker lining and the cut had been purposely backfilled.

Feature [1052] was a subcircular pit 2.8m long, 2.3m wide and 0.25m deep. It contained the partial remains of a wicker lining from which no overall dimensions could be taken, although the lining appeared to be substantially smaller than the pit. Feature [1053] was a subcircular pit 2.28m long, 2.21m wide and 0.42m deep. It had a wide gradual U-shaped profile which contained a wicker lining (1222) measuring 0.96m long, 0.92m wide and 0.38m deep. Feature [1054] was a subcircular pit 2.72m long, 2.24m wide and 0.48m deep with a wide gradual U-shaped profile. It contained wicker lining (1221), measuring 1.88m long, 1.68m wide and 0.42m deep (Fig 3.21). Feature [1054] appeared to cut the edge of [1072] which was an oval pit 4.2m long, 3.35m wide and 0.58m deep (Fig 3.21). Feature [1072] contained two structures, (1181) and (1073). (1181) appeared to be the remains of one side of a wooden polygonal retaining structure, consisting of small horizontal planks held in position by vertical stakes. (1073) was the remains of a subcircular wicker lining which measured 1.83m long, 1.67m wide and 0.35m deep (Plate 3.6). Feature [1099] was a pit 2.4m long, 2m wide and *c* 0.35m deep, which contained an oval wicker lining (1100) (Fig 3.21).

Within the same area, pit [1101] was a subcircular shallow feature measuring 2.11m long, 2m wide and 0.16m deep (Fig 3.21). Positioned centrally within [1101] was a smaller circular wicker-lined pit [1105], which measured 0.68m wide and was a further 0.28m deep.

Wooden troughs were found within features [1096] and [1157]. Pit [1157] was positioned immediately to the east of pit [1072] and contained the timber-built trough [1264]. [1157] was a subcircular/oval pit which measured 5.1m long, 4.36m wide and *c* 0.95m deep and had a widely flaring funnel-like profile.

Trough [1264] was located 0.35m below the top of [1157], was subrectangular and measured *c* 2.5m long, *c* 0.8m wide and 0.6m deep (Fig 3.22; Plate 3.8). The remains of the trough, which had partially collapsed after falling out of use, consisted of eight upright and two fallen staves. These were a combination of variously sized roundwood and worked timbers, behind which had

Plate 3.7: Wicker lining (1073) viewed from the west.

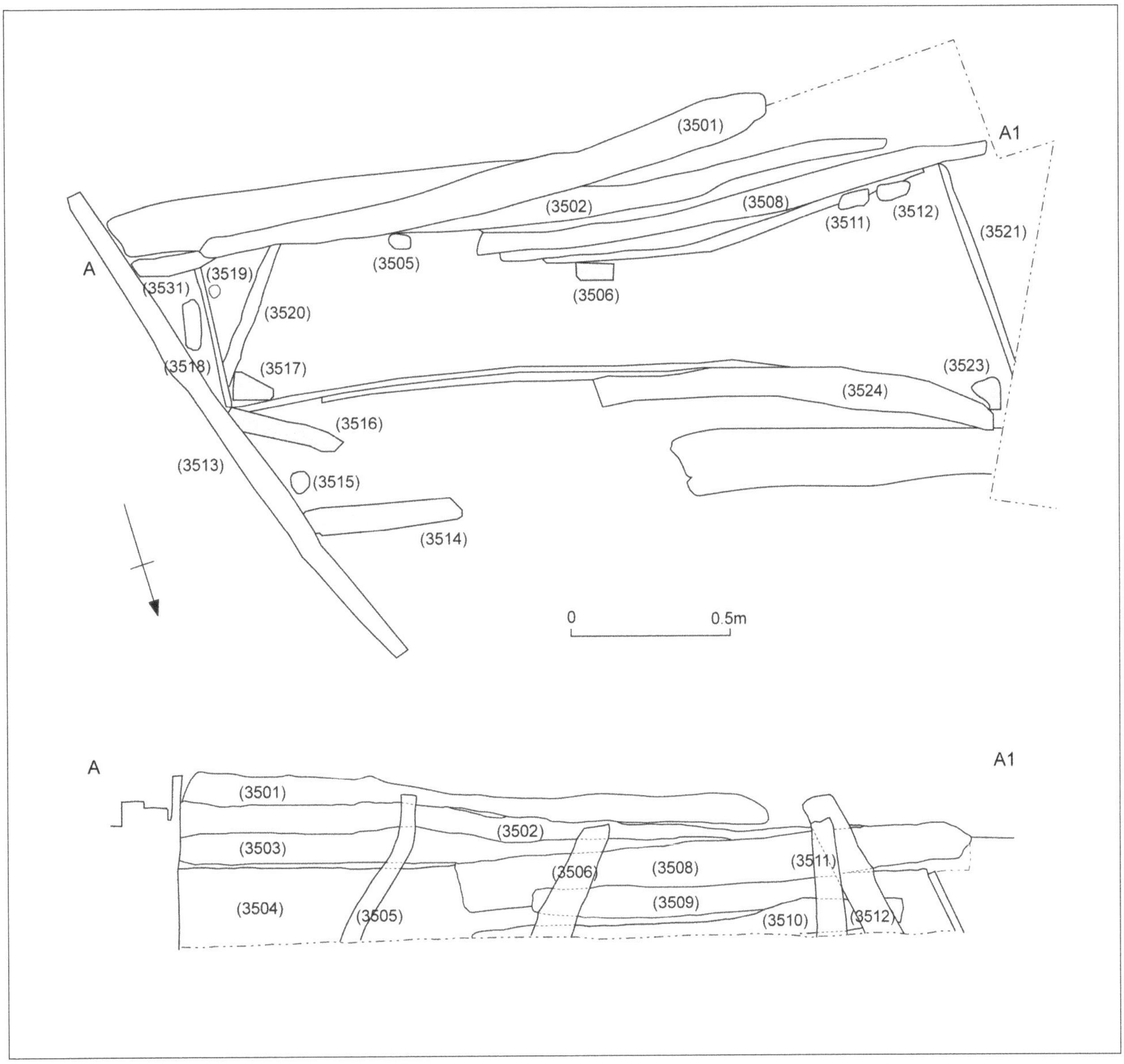

Figure 3.22: Trough [1264], plan and elevation.

Plate 3.8: Trough [1264] viewed from the west.

been placed roughly worked overlapping planks of no consistent size. There were traces of possible channels leading from pit [1157] towards brine tank [1207].

Pit [1096] measured 7m long, 4.1m wide and *c* 1.2m deep. Because of the waterlogged nature of the feature, the final *c* 0.4m was excavated by machine removing 50mm spits at a time. The basal fill contained the remains of a wooden trough of post and plank construction. A row of four roundwood posts aligned east-west, with a diameter of 0.12m and height of *c* 0.45m, survived in situ. On the south side of this row were the collapsed remains of planking, comprising fourteen pieces 0.3-1.84m in length and 0.05-0.3m in width. The fills of [1096] included 250 pieces of waste leather, representing the largest single group of waste recovered from the excavation, along with at least one leather shoe (see Chapter 9).

An arrangement of small linear gullies appeared to extend to the south-east from pits [1157] and [1096]. Linear [1145] which ran from [1096] was 0.27-0.56m wide and 0.08m deep. It had been partly truncated but its line could also be traced immediately to the west of brine tank [1207].

Positioned to the south-west of pit [1157] were parallel gullies [1155]/[1131] and [1119], running on a roughly north-south alignment which continued to the edge of the excavated area. Both were recorded for a length of *c* 13m and were 0.8m wide and *c* 0.1m deep. Gully [1155]/[1131] contained a possible fragment of wooden planking 1.22m long and 0.1m wide.

Feature [1107] was a north-east - south-west aligned gully which measured at least 5.2m long, and was 0.5-1.05m wide and 0.15m deep. The south-west terminus of [1107] may have been truncated by feature [1155].

Other features within this area were of a more irregular form. Cut feature [1055]/[1064]/[1058] measured 8.7m long, 4.5m wide and 0.14-0.56m deep. This feature produced late 3rd- to mid 4th-century pottery but had been heavily disturbed by root action. [1160] was a large irregularly shaped pit which measured 5.5m long, 4.1m wide and 0.83m deep. The pottery suggests an Antonine date.

Phasing

The pottery evidence implies that these features were associated with three main phases of activity. The first involved deposition in the 2nd century and included pit [1096] containing a wooden trough, pit [1160] and possibly wicker-lined pit [1099]. This phase was contemporary with the use of tanks [1182] and [1207] for the managed storage of brine. The second phase saw deposition in the early to mid 3rd century; trough [1264] in pit [1157] can be dated to this phase. The third phase dates to the late 3rd to 4th century and included wicker-lined pits [1052], [1053], [1054], [1072] and [1105]. In broad terms, these phases suggest a shift of activity from east to west.

Discussion

It is clear that this zone as a whole was dedicated to the collection and storage of brine. The wicker-lined pits and wooden troughs were evidently located on a natural spring line. During excavation water still welled into these features from the natural sand and had to be frequently pumped out (Connelly & Power 2005, 38). Similar wicker-lined pits have been excavated at King Street, Middlewich, and followed a fault in the bedrock where brine rose to the surface (Williams & Reid 2008, 19-23, 30-1). The wicker lining would have given some stability to the sides of such pits and may have also helped to filtrate the brine.

Brine tanks [1207] and [1182] would have been used to store the brine prior to the process of heating to evaporate the water, leaving salt. Such storage would have allowed heavier particles to settle out of the brine, producing a cleaner solution. It would also have allowed some natural evaporation during the summer months, increasing the salinity (Connelly & Power 2005, 38).

The operation of brine tanks [1207] and [1182] was probably linked with the earliest of the three phases of activity represented by the various pits and other features located to the west. The principal collection point within the area of excavation in this phase seems to have been the trough in pit [1096]. Linear [1145] running south-eastwards from that feature suggests the course of a channel through which brine could have been transferred to [1207].

A later phase, between the early and mid 3rd century, involved the use of trough [1264] in pit [1157]. Although this phase postdates the use of the tanks for the managed storage of brine, there is some evidence for a channel system linking this collection point with brine tank [1207], possibly in connection with the tank's ritual use. The final phase is characterised by the use of wicker-lined pits which also represent some of the latest features excavated on the site.

Linears [1155]/[1107]/[1131] and [1119], one including a fragment of a possible timber lining, ran southwards from the second and final phase collection points and imply the existence of a brine storage and processing area outside the area of excavation. Their alignment perhaps most closely corresponds with the position of trough [1264], suggesting that, in addition to any ritual function, this feature served an industrial role.

The presence of leather within the fills of the excavated brine collection points, in the greatest quantity in pit [1096], led to an initial suggestion that they may also have been used for leather processing or finishing (Connelly & Power 2005, 39). However, analysis of the leatherwork found no evidence for the use of these or any other features in the tanning process (see Chapter 9). Analysis of insect remains from pits [1054], [1096] and [1157] and trough [1264] also found little in the way of direct evidence of leather production (see Chapter 18).

Post-Roman

No evidence was uncovered to indicate occupation of the site following the abandonment of the Roman settlement.

A north-south alignment of five postholes on the western edge of excavation area A (Fig 3.1) contained a single sherd of 13th/14th-century pottery. These postholes may represent a fence line associated with the agricultural use of the site in the medieval period.

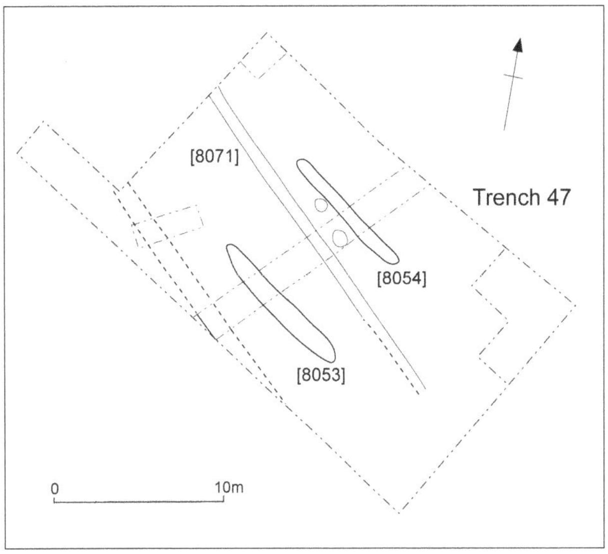

Figure 3.23: Linear features [8053] and [8054].

Trench 47 contained two shallow parallel linear features, [8053] and [8054], running roughly north-west - south-east (Fig 3.23). The more southerly of these produced two sherds of medieval pottery. These features lay *c* 5.6m apart and were situated to either side of Roman road ditch [8071]. They also roughly followed the alignment of that road and may relate to its continuing use in the post-Roman period.

4. Coarse Pottery

Philip Mills with Jeremy Evans

Some 3919 sherds of pottery, including 588 of samian ware, were presented for examination. The stratified assemblage totals 3301 sherds, including samian ware, weighing 102.2kg. The pottery was recorded following the Oxford Archaeology/Warwickshire Museum system (Booth 2000), onto an Access 2000 database. The material was examined by context, with sherds sorted by fabric, determined by x20 microscopic examination.

These sherd families were recorded by number of sherds (NoSh), weight (Wt), minimum number of rims per context (MnR), rim equivalent (RE) and base equivalent (BE). Surface deposits, such as sooting, were coded separately, as were stamps, decoration and other observations of interest. For full analysis the samian ware has been incorporated into the coarseware database. The full assemblage is used for dating evidence but only the stratified material has been otherwise used in the analysis. The full pottery archive is given in Appendix 4.1.

This report initially presents the dating evidence from the pottery, first for the whole site, then by the individual zones, and features with good evidence within them. The following section deals with the taphonomic profile of the site and the zones. Then information is presented about the supply of pottery to the site, divided into basic ware types and the individual forms within each ware and fabric type. The next section deals with the functional analysis of the assemblage, again first by the whole site and then by zones, providing insight into the site's status. Then other forms of information relating to sooting, graffiti, cross-joins and complete vessels are presented. The conclusions from the analysis of the assemblage are presented in the closing discussion.

Following standard procedure, the 'early Antonine' is taken as *c* AD 140-60, 'mid Antonine' as *c* AD 160-80 and 'late Antonine' as *c* AD 180-200. SG, CG and EG are used for Southern, Central and Eastern Gaulish samian ware respectively. In the tables the following abbreviations are used for functional types: A - amphorae; B - bowls; Bk - beakers; CJ - constricted-necked jars; Cu - cups; D - dishes; F - flagons; J - jars; Ju - jugs; L - lids; M - mortaria; Tk - tankards; SJ - storage jars; WMJ - wide-mouthed jars; O - other.

Chronology by Site

The date distribution of all the pottery, by RE, is shown in Figure 4.1. This clearly indicates that the main focus of activity relates to the 2nd century, with a peak in the mid century. There is a marked decline in the early 3rd century, followed by a slight increase in the mid 3rd, with relatively little activity in the 4th century. Slightly more precise date distribution can be achieved by examination of specific ware types.

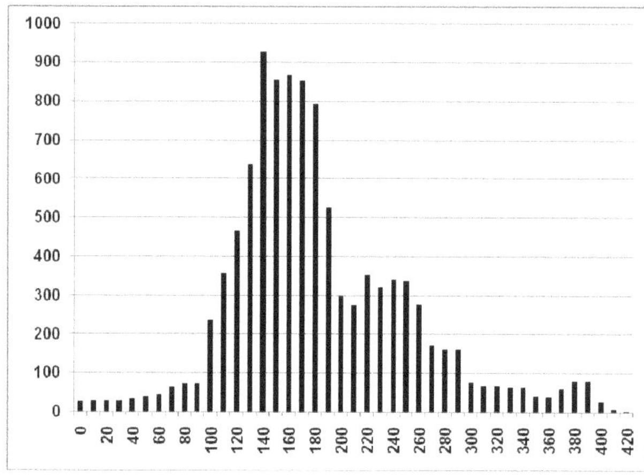

Figure 4.1: Date distribution of entire assemblage by RE.

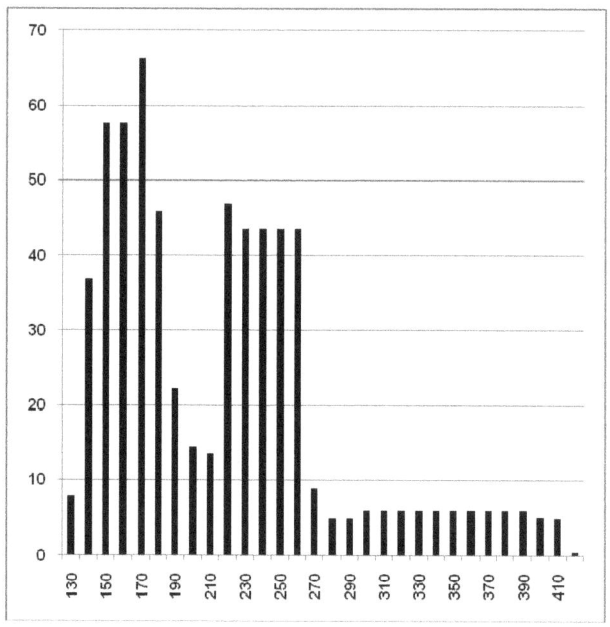

Figure 4.2: Date distribution of mortaria by RE.

The date distribution, by RE, for mortaria shows a peak in the mid 2nd century, followed by a trough in the very late 2nd and early 3rd, and then a sharp rise to a plateau in the mid 3rd century, in turn followed by a minimal presence from the mid 3rd onwards (Fig 4.2). Unusually the date range plotted by MnR (Fig 4.19) shows a different picture, with a much higher peak in the mid 3rd century than in the mid 2nd, indicative perhaps of some very broken vessels in the later period.

The date distribution for samian ware, by NoSh, not surprisingly shows a much tighter date range, beginning at *c* AD 60 and with samian ware virtually absent after *c* AD 190 (Fig 4.3). The main period of samian ware dates is Hadrianic-Antonine and there is a peak in the mid 2nd century, reflecting the stamp and decoration types from this period, before a slightly lower peak which ends *c* AD 180-90, reflecting the most likely closure date of the brine tanks.

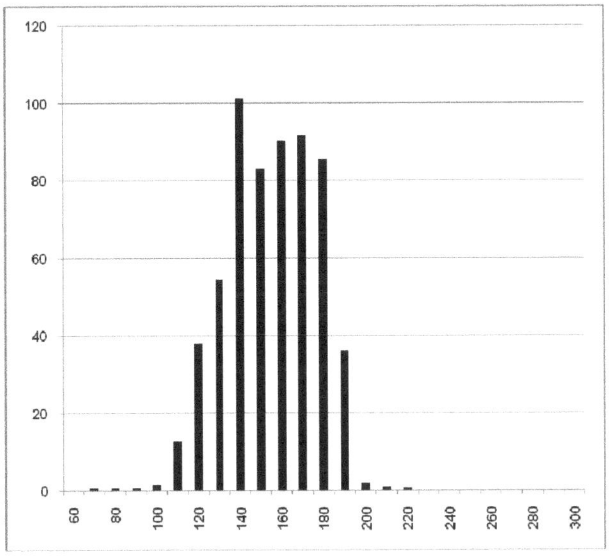

Figure 4.3: Date distribution of samian ware by NoSh.

The commonest types are Dr 37, Dr 31, Dr 33, Dr 18/31 or 31, Dr 18/31R, Dr 18/31R or 31R, Dr 31 or 31R, Dr 31R, Dr 38 and Dr 27, in that order for fully identifiable types. Dr 31s generally date to after *c* AD 150 and Dr 31Rs to after *c* AD 160. Samian mortaria are generally reckoned to date to after *c* AD 170 but there is a noticeable lack of these and Dr 31 bowls outnumber Dr 31Rs by 3.9:1. The ratios of samian form types fit quite well with Willis's data for an assemblage of *c* AD 160 (Willis 1998, 97 ill 1).

The evidence of both the samian ware and the mortaria suggest the possibility of a gap in occupation in the very late Antonine or the Severan period.

Chronology by Zone

By far the majority of material derives from the Brine Industrial Zone, amounting to some 72% of the overall stratified assemblage (Table 4.1). A further fifth of the assemblage comes from the Utilitarian Zone, with about 5% from the Cremation Zone and 3% from the Road Zone.

The other excavation areas and trenches account for a small amount of pottery evidence, but not enough to include in the major analysis in this report.

	NoSh%	Wt%	MnR%	RE%	BE%
Road Zone	2.8%	1.7%	3.1%	1.4%	2.5%
Cremation Zone	5.3%	3.3%	5.0%	4.0%	1.9%
Utilitarian Zone	21.4%	11.3%	19.1%	11.3%	14.8%
Brine Industrial Zone	69.8%	82.9%	72.3%	82.7%	79.2%
Area F	0.2%	0.6%	0.4%	0.5%	1.5%
Trench 44	0.1%	0.0%	-	-	-
Trench 45	0.1%	0.1%	0.2%	0.2%	-
Trench 46	0.2%	0.1%	-	-	-
Trench 47	0.1%	0.0%	-	-	0.2%
Total	**3301**	**102161g**	**523**	**8876**	**8581**

Table 4.1: Pottery by zones, excavation areas and trenches.

The Road Zone

The Road Zone shows a pattern which contrasts with the main assemblage (Fig 4.4). The small quantity of material from this zone means that it is largely swamped by the larger data set from the Brine Industrial Zone in the overall date distribution. There is deposition in the 2nd century which peaks in the mid century and a further period of deposition in the 3rd century with a higher peak in the early to mid century, but very little activity after the mid 4th century.

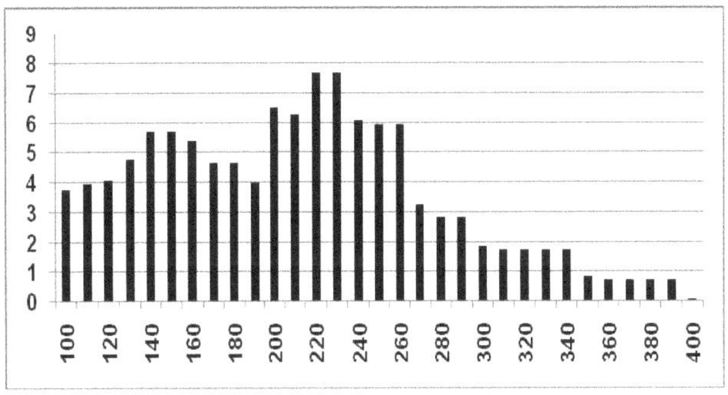

Figure 4.4: Date distribution for the Road Zone by RE.

By Feature

The earliest features, with a late 2nd-century date, include pit [3336], with a Mancetter-Hartshill whiteware mortarium, M13.1, dated AD 170-200, and a CG Dr 33 of Antonine date. Pit [4028] has a CG Dr 31R bowl, with a date range of AD 160-200. Also dating to this period is material from posthole [5017], Building 3, which includes a 2nd-century red-slipped bowl, F05.1, and a Severn Valley ware wide-mouthed jar, O04.16.

The next sequence of deposition dates from the early to mid 3rd century. This includes pit [4013], with an early 3rd-century black-burnished ware jar, B01.10, feature [5088] which contains an early to mid 3rd-century black-burnished ware bowl, B01.22, a 3rd-century black-burnished ware dish, B01.27D, an early 3rd-century Mancetter-Hartshill whiteware unreeded hammerhead mortarium, M13.9, Severn Valley ware flagon O03.1 and constricted-necked jar O04.5, and a sherd of Dressel 20 amphora. Also in this period is roadside ditch [4151], with a black-burnished ware bowl, B01.22, with a date range of early to mid 3rd-century. There is a 2nd- to 3rd-century wide-mouthed jar, O04.13, in pit [5098].

The latest datable pottery fragment is a late 3rd-century to mid 4th-century black-burnished ware bowl, B01.23, from pit [8103]. Pit [5098] has an undiagnostic Severn Valley ware wide-mouthed jar, O04.13.

The dating of Building 4 is suggested by the presence of a sherd of Dressel 20 amphora (1st-3rd century) from posthole [4101] and a Mancetter-Hartshill mortarium fragment from posthole [4099], a sherd of Wilderspool oxidised ware from posthole [4149] and Severn Valley wares from postholes [4099], [4149] and [4158], from which nothing closer than a broad 2nd-century or later range can be given.

Other features with pottery include roadside ditch [4094], with a body sherd of Severn Valley ware O03, and pit [4030], with a sherd of probable Wilderspool oxidised ware O02 (2nd to early 3rd century). Well [5052] has a CG samian bowl rim fragment of Hadrianic-Antonine date, as well as one sherd each of Severn Valley ware fabrics O03 and O04, a single sherd of probable Wilderspool fabric O02, and one sherd of white-slipped fabric Q01. Pit [5093] has a medieval shell-tempered fabric, some residual fragments of Nene Valley ware F01, and Wilderspool mica-dusted ware F031.

The Cremation Zone

The date distribution for the pottery from the Cremation Zone is shown in Figure 4.5. Unlike the main assemblage, there are two very clear and similar peaks of deposition, one in the mid 2nd century, contemporary with the main period of deposition for the whole site, and a slightly higher one in the mid 3rd century.

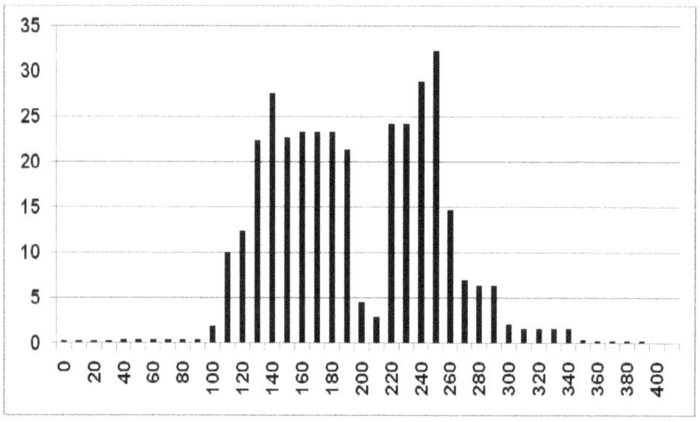

Figure 4.5: Date distribution for the Cremation Zone by RE.

By Feature

A complete urn with cremation fill was recovered from pit [4003]. This urn is a black-burnished ware B01.2, with a date range of the mid to late 2nd century.

There were only a few other sherds presented from grave fills, including undiagnostic sherds of R01 and O03 from (4238) (a possibly redeposited cremation burial) and probable fragments of Wilderspool white-slipped fabric Q01 in (4243) and (4244) in pit [4245], giving a likely 2nd-century date.

Pits with earlier pottery include [4078] with Hadrianic-Antonine CG Dr 27 cup. [3428] and [4087] both have Hadrianic-Antonine BB1 jar rims. [4057] has a rim fragment of a black-burnished ware beaker, with a Hadrianic-Antonine date. [3440] has amphora rim A01.1, dated AD 110-50, and white-slipped ware dish Q01.5. [4242] has a single fragment of Hadrianic-Antonine CG samian. [4207] has a sherd of CG samian, Hadrianic-Antonine+. Linear cut [4131] has a single fragment of CG samian of Hadrianic-Antonine date.

There are a number of pits with mid to late Antonine pottery. These include [3405], with jar B01.2; [4330], with a Dressel 20 amphora, A01.1, along with a Wilderspool white-slipped dish, Q01.5; [4160] with a jar B01.4; [4168] with two examples of jar B01.2, and with four sherds of CG 18/31R of Trajanic to early Antonine date; [4185] with jar B01.2; and [4103] with dishes B01.25A and B01.27A, the latter decorated with intersecting arc burnished lines, with a date of AD 160-80+. Pit [3399] has a sherd of CG Dr 37 decorated bowl (D2), dated *c* AD 135-70, with a sherd of white-slipped ware Q01.

Two features have evidence of late 2nd- to 3rd-century deposition. [4139] has a Wilderspool jar O02.4 and bowl O01.4. [3401] has a Severn Valley ware jar, O03.6. [4237] has a residual example of B01.27A and an early to mid 3rd-century mortarium, M13.8, a mid 3rd-century jar, B01.11, and a mid to late 3rd-century Nene Valley colour-coated beaker, F01.2. Pit [4231] has a residual stamped (S3) CG Dr 31R, dated *c* AD 150-80, and a mid 3rd-century black-burnished ware jar, B01.12.

Other features with pottery include pit [3387] which contains a fragment of Wilderspool fabric O01, of the late 1st to early 3rd century. Pit [3420] has a sherd of black-burnished ware B01. Pit [3435] has four sherds of Dressel 20 amphora, A01, as well as a sherd of B01 and O02 suggesting a Hadrianic *tpq*. Posthole/pit [4005] has a sherd of O01. Pit [4009] has three sherds of O04. Pit [4032] has a sherd of O02. Pit [4035] has a sherd of O04. Pit [4040] has a sherd of O03. Pit [4059] has a micaceous greyware jar, R31.1, as well as one sherd each of F12 (2nd century?) and B01. Pit [4081] has a sherd of O02. Pit [4134] has one sherd each of B01 and R32. Pit [4183] has two sherds of O01. Pit/tree bole [3397] has a mid to late 2nd-century flagon, Q01, and a body sherd of a possible Dr 37 bowl of Hadrianic-Antonine date.

The Utilitarian Zone

The date distribution for this zone is shown in Figure 4.6. It indicates minimal activity before the 2nd century, with increased values until the mid 2nd, followed by a rapid decline in the very late 2nd and little activity in the 3rd century and onwards, considerably less than for the site assemblage as a whole.

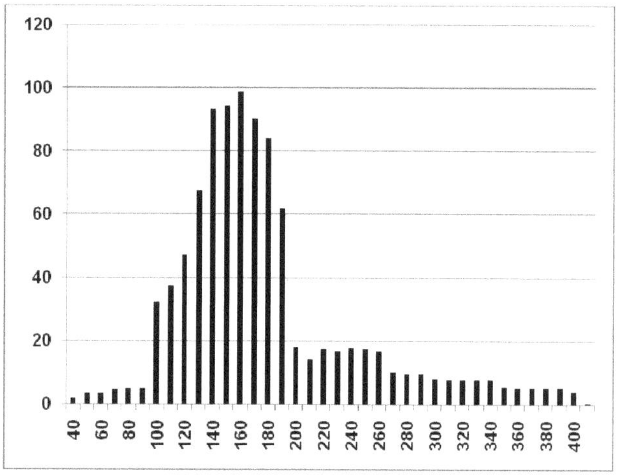

Figure 4.6: Date distribution for the Utilitarian Zone by RE.

Figure 4.7 shows a stacked bar chart for the date distribution of the sub-zones in the Utilitarian Zone. Building 1 has its peak of deposition starting in the 2nd century and declining in the last couple of decades of the 2nd century, with a small quantity of deposition occurring afterwards. Building 2 follows a similar pattern, but with much less pottery present and a decline which starts in the mid 2nd century. The large enclosure also peaks in the 2nd century and pottery remains deposited there in the 3rd after it has ceased in Building 2. Building 5 has a later peak than the other two buildings but deposition dwindles sharply in the early 3rd century. The U-shaped enclosure has a slight 2nd-century peak, but has a minimal presence in both the 2nd century and the 3rd.

By Feature

The majority of the pottery evidence from features from Building 1 is dated to the mid to late Antonine period. From the postholes are a black-burnished ware beaker, B01.17, from [2005]; an Antonine CG Dr 31 from [2009]; a jar, B01.2, and a fragment of a late 2nd-century jar rim from [2021]; a Severn Valley ware bowl, O03.26, and a CG samian Dr 37 from [2089]; an Antonine Dr 31 from [2100]; a Dr 31R (Antonine) from [2112]; an Antonine Dr 31 CG samian bowl from [2156]; a Severn Valley ware wide-mouthed jar, O03.13, a Hadrianic-Antonine CG Dr 30 dish and Antonine Dr 31 bowl from [2177]; and Gauloise amphora A02.1 and black-burnished ware jar B01.2 from [3076]. Of less certain date there is micaceous greyware jar R36.1 from [2018], and polished greyware flagon R43.1 from linear cut [2114]. Posthole [2023] has two sherds of B01, and posthole [2062] has a sherd of O01. Posthole [2096] has two sherds of B01 and two of O04. Of a possible later date is posthole [2158], with a late 2nd- to early 3rd-century black-burnished ware bowl, B01.6, and a 2nd- to 3rd-century Severn Valley ware bowl, O04.26, although these vessels would fit with the date range from the above postholes.

Two features which lie on the line of the north-west wall of Building 1 have later pottery. Pit/posthole [2039] has an early 3rd-century black-burnished ware jar rim, B01.10, alongside residual Antonine examples of B01.21A, B01.26b, B01.2 and a CG Dr 37 bowl. Pit/posthole [2136] has a mid 3rd-century black-burnished ware bowl, B01.12, as well as residual B01.17, B01.25A, B01.25C, and CG Dr 31 and Dr 37 bowls.

The dating evidence from Building 2 is mainly Antonine, with a small mid Antonine component. This includes a Severn Valley ware jar, O04.18, from posthole [3057]; a black-burnished ware dish with acute lattice decoration, B01.27C, from posthole [3094]; a black-burnished ware dish with pointed arc decoration, B01.25C, and a Wilderspool roughcast beaker, F11.1, from posthole [3117]; an Antonine CG Dr 31 from [3088]; and an Antonine CG Dr 37 bowl from [3017]. Linear cut [3011]/[3069] has two sherds of A11, one of B01, one of M03, and two of O02 suggesting a 2nd-century date. Posthole [3013] has four sherds of O03; posthole [3024] has a body sherd of Antonine CG Dr 37. Posthole [3080] has three sherds of B01 and one of O04; posthole [3086] has a sherd of O03. Posthole [3094] has Hadrianic-mid Antonine dish B01.27C. Posthole [3102] has a 2nd-century sherd of roughcast Wilderspool fabric F12.

Of the features associated with Building 2, Antonine dates can be given to pit [3062], with a black-burnished ware acute lattice decorated bowl, B01.21B; pit [3074], with a CG Dr 33 cup; pit [3129], with a CG Dr 30 bowl; posthole [3144], with a 2nd-century Severn Valley ware jar, O04.19, and a Hadrianic-Antonine black-burnished ware dish, B01.25A; and pit [3331] with a Wilderspool oxidised ware jar, O02.1, and a bowl, O02.6. Slightly later dates of AD 160-80+ are suggested by the rim sherd of black-burnished ware intersecting arc decorated simple rimmed dish, B01.27A, from pit [3078]. There is also a late 2nd- to early 3rd-century black-burnished ware jar, B01.6, from pit [3020]. Linear cut [3016] has thirteen sherds of A01, three of B01, one of O01 and five of O04, suggesting a Hadrianic date. Pit [3158] has a sherd of a CG Dr 31 or 18/31 bowl, of Hadrianic-Antonine date. Pit/posthole [3202] has two sherds of O01. Stake hole [3055] has a fragment of an Antonine CG Dr 31 samian bowl.

The large enclosure has Antonine material coming from [3092] in the form of Wilderspool red-painted beaker F61.1; a Severn Valley ware 2nd-century bowl, O04.27, from [3206]; and Antonine CG Dr 31 bowl and a Dr 33 cup, with illegible stamp (S17), from [3366]. [3195] includes an early to mid 3rd-century whiteware Mancetter-Hartshill hammerhead mortarium, M13.8, as well as residual examples of M01.1, B01.4, B01.2 and CG Dr 30, all Antonine, a Hadrianic black-burnished ware bowl, B01.29b, and a Flavian-Trajanic mica-dusted jar, F41.2. [1081] has a Severn Valley ware constricted-necked jar, O03.4, with a date range of the mid 1st to 4th century. [3111] has a sherd of fabric O02. [3257] has a sherd of O02 and a sherd of O04. The latest dated pottery from the large enclosure includes an Oxford type P24 parchment ware bowl (Young 1997), dated AD 240-400+, W03.1, from ditch [2067].

Hearth [3216]/[3218]/[3235] produced an Antonine black-burnished ware jar, B01.3, as well as a white-slipped beaker, Q01.4, and a Severn Valley ware bowl, O04.11. Hearth [3240]/[3242] has evidence of a similar date with a rim fragment of a black-burnished ware jar, B01.1. Hearth [3360]/[3341] has a single sherd of B01.

Building 5 seems to have a late 2nd- to early 3rd-century date. Pit/posthole [3168] has an early to mid 3rd-century black-burnished ware bowl, B01.21, and a late 2nd- to early 3rd-century jar, B01.5, as well as a 1st- to 3rd-century Severn Valley ware jar, O03.7, and residual examples of B01.25C and F11.1. Pit/posthole [3212] has a mid to late 2nd-century wide-mouthed jar, O03.12, a Severn Valley ware wide-mouthed jar, O03.14, and a black-burnished ware dish with intersecting arc decoration, B01.27A, with a date range of mid 2nd century+, as well as F031.1 and two examples of Antonine CG Dr 31.

The U-shaped enclosure is possibly of early to mid 3rd-century date. This is suggested by the Severn Valley ware bowl, O03.23, from ditch [3374] and an early to mid 3rd-century black-burnished ware jar rim, and a Severn Valley ware bowl, O04.23, from ditch [3138]. [3246] has five sherds of B01 and one sherd each of O02, Q01 and Q02.

The earliest period of deposition among other features would appear to be Antonine, with evidence including a black-burnished ware dish, B01.26b, Wilderspool beaker, F031.17, and a Wilderspool mortarium, M01.1, from linear cut [3470]. Posthole [2015] has a sherd of a CG Dr 37 bowl of probable Antonine date and a sherd of O01. Posthole [2021] has a mid to late 2nd-century jar, B01.2, as well as a late 2nd-century black-burnished ware jar rim and a fragment of CG samian. Pit [2028] has sherds of B01 and two of O02, giving a Hadrianic *tpq*. Pit/posthole [3098] has a late 2nd-century black-burnished ware interlocking arc decorated dish, B01.25D. Pit/posthole [3100] has a sherd of B01 and F12, giving a 2nd-century date. Pit/posthole [3181] has sherds of A01, B01, O01 and CG samian suggesting an Antonine date. Pit/posthole [3187] has a rim fragment of a mid to late 2nd-century flagon, W01.1. Pit [3254] has a sherd of a CG Dr 37 bowl, dated AD 165-200. Pit [3377] has sherds of A01, three of B01, two of O02 and O04 and one of R02, suggesting a Hadrianic *tpq*.

Also possibly of this date were ditch [2003], with an oxidised Wilderspool bowl, O02.4, and pit [3066], with an oxidised bead and flanged mortarium, M04.2, of perhaps early to mid 2nd-century date. Pit [3003] contained a Severn Valley ware bowl, O03.24, with a date range of the 2nd to 3rd century, which may belong to this period of deposition or the next.

Cut feature [3150] has a sherd of Nene Valley ware F01 indented scale beaker in brown colour coat, of late 2nd- to late 3rd-century date, with sherds of B01, O02, R01, R52 and W01. Ditch [3252]/[2065] has sherds of O01, O04 and Q01, suggesting a 2nd- to 3rd-century date.

Early to mid 3rd-century deposition is evidenced in pit [3084] by the presence of a black-burnished ware bowl, B01.22, and a presumably residual dish, B01.29A. Pit/posthole [3193] has a rim fragment of a mid 3rd-century black-burnished ware jar, B01.11, and pit [3317] has a fragment of a 3rd-century black-burnished ware jar rim. Pit [3071] has a fragment of Oxfordshire mortarium fabric M06, suggesting a date of AD 240+. Oxfordshire colour-coated ware bowl F21.1 of Young's (1977) type C45, dated AD 240-400+, is present, along with examples of Severn Valley ware jar O05.11 and Mancetter-Hartshill whiteware hammerhead mortarium M13.8, in pit [3359].

Posthole [2134]/[2135] has one sherd of B01, six of O02, two of O01 and one of Q01. Pit [3005] has a sherd of B01, O04 and Q02. Pit/posthole [3123] has a sherd of Q01. Pit [3278] has a sherd of O02. Pit [3328] has sherds of B01, O04 and R32. Posthole [3127] has a sherd of B01. Pit [3273] has a sherd of Q01. Pit [3290] has four sherds of B01 and one of O02. Pit [3293] has a sherd of B01. Linear cut [3300] has a sherd of O03 and R52. Cut? [2032] has a sherd of O02. Tree bole [3142] has a sherd of O02 and tree bole [3175] a sherd of a CG Dr 30 bowl of Antonine date.

The Brine Industrial Zone

Figure 4.8 shows the date distribution for the pottery from the Brine Industrial Zone. It is not surprising, given the disproportionate amount of material from this zone, that its profile broadly follows that of the site as a whole. It shows a major deposition in the 2nd century, peaking in the mid century and declining from the later 2nd century, but with a slight recovery in the mid 3rd century.

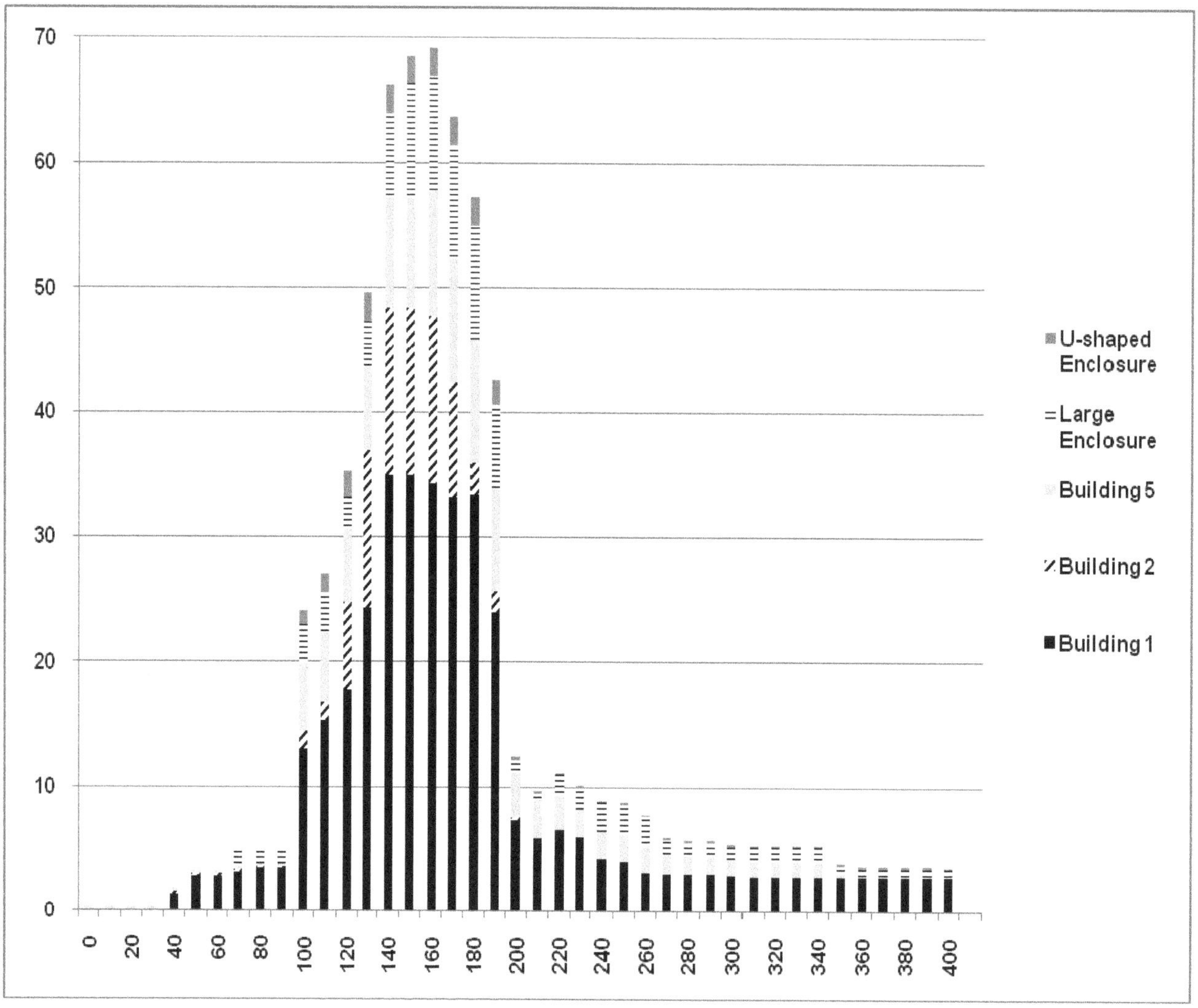

Figure 4.7: Date distribution by sub-zone in the Utilitarian Zone by RE.

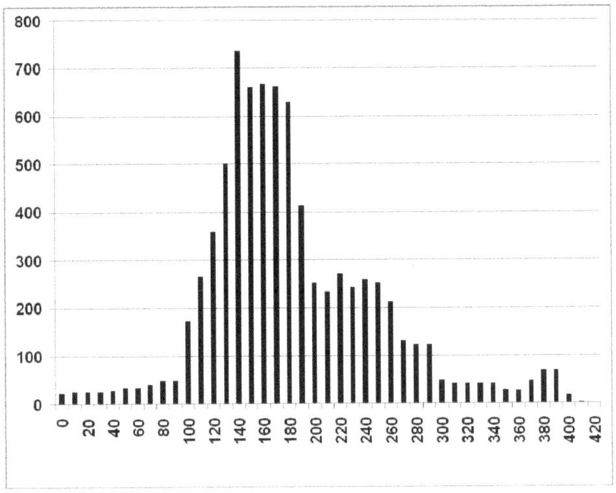

Figure 4.8: Date distribution for the Brine Industrial Zone by RE.

The Brine Tanks

The majority of the material from the zone, and in fact the site, derived from brine tanks [1182] and [1207]. The date distribution, by RE, for all the material from the two tanks clearly shows that they have a similar depositional pattern over time (Fig 4.9). The majority of material dates to the 2nd century, with a peak in the mid 2nd, after which the material declines to a plateau in the early to mid 3rd century, which is in turn followed by a rapid decline in the later 3rd and a small 4th-century tail-off. From the pottery record it is clear that the 3rd-century deposition of material intruded into earlier deposits.

Brine Tank [1182] by Context

The main sequence in [1182] begins with basal deposit (1286) which contained Wilderspool oxidised bowl type O02.4 of the 2nd century. Above this, (1285) has an example of a mid 3rd-century+ Mancetter-Hartshill mortarium, M13.5, and a 3rd-century storage jar in oxidised Wilderspool fabric, O02.6. It also has examples of a late Antonine Dressel 20 A01.1 with a stamp (St21), B01.2, B01.3, B01.4, B01.19, B01.20, B01.26C, F03.1, F12.1, F12.2, F32.1, M01.1, O01.1, O03.2, O03.23, O04.11, O04.21 and O11.1. The samian comprises CG Dr 30, Dr 31, two Dr 33 (one stamped (S16)), Dr 37 and a Curle 15. Above (1285), context (1282) has a mid 2nd-century black-burnished ware pointed arcade decorated dish, B01.27B, and a stamped (S12) CG Dr 33, dated *c* AD 160-90. Context (1281) has an intrusive example of a medieval flanged rim buff jar with patches of thin glaze, in a mainly Antonine assemblage comprising Dressel 20 amphora lid A21.1, black-burnished ware forms including two of B01.24, B01.25A, B01.25B and B01.26C, Wilderspool white-slipped flagon Q01.3, with CG samian forms including two Dr 37 bowls (including decoration D18) and a Dr 31 bowl with an illegible stamp.

Above these, context (1279) contained jar B01.10, dated to AD 200-40, wide-mouthed jar O06.3, dated to the 2nd to early 3rd century, and CG samian forms including two Dr 31, one stamped (S5), and a stamped Dr 33 (S14). Above this, context (1278) has a Severn Valley ware bowl, O03.19, and context (1287) a 2nd-century Wilderspool beaker, F03.1.

Context (1277) has an early 3rd-century black-burnished ware jar, B01.10, and a probably 3rd-century black-burnished ware simple rimmed dish, B01.27D. Also present are a CG Dr 27, Dr 31, Dr 33 and a Dr 37 bowl (with decoration D19). This context was equivalent to (1206) which has two examples of mid 3rd-century black-burnished ware jar B01.11, with a B01.18, and CG Dr 31 and 31R.

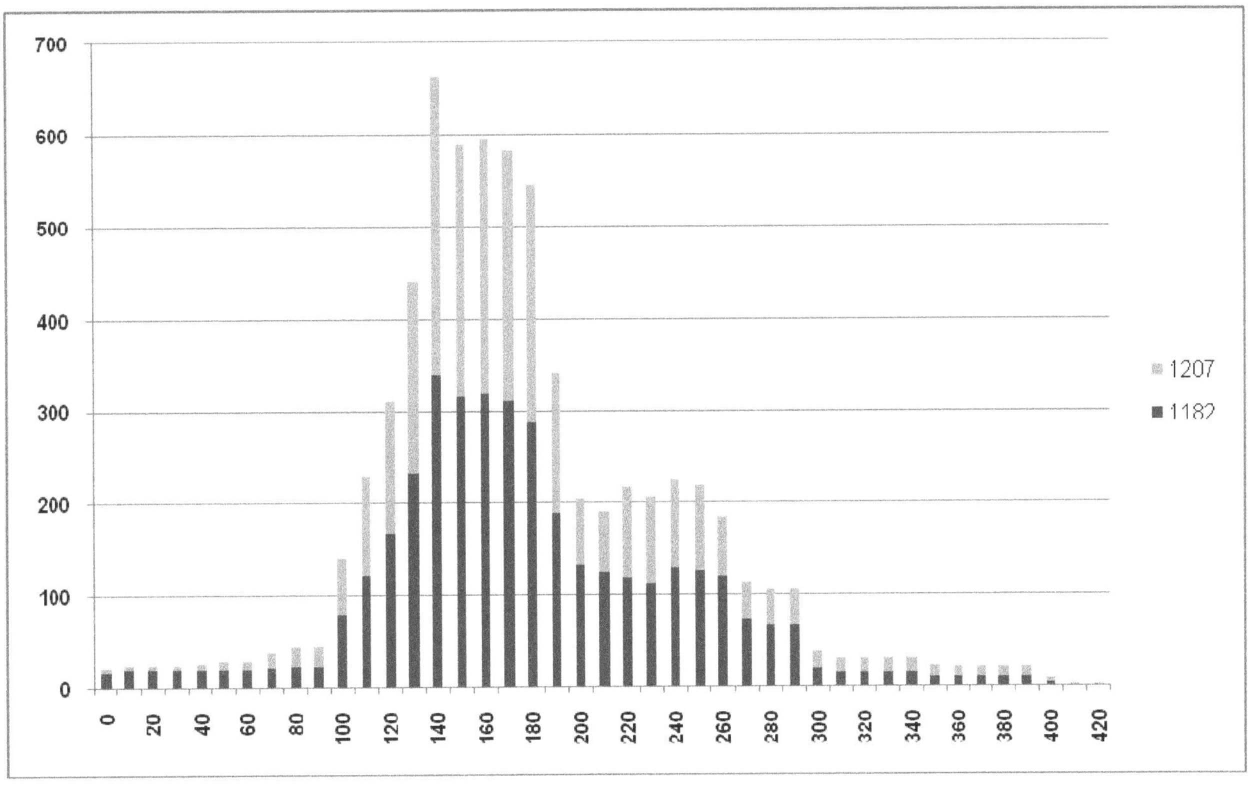

Figure 4.9: Date distribution from brine tanks [1182] and [1207] by RE.

Above this, context (1205) has a mid 3rd-century black-burnished ware jar, B01.11, red-slipped 'Castor box' F05.2, of late 2nd-century or later date, and Catterick vicinity mortarium M21.1. There are also two CG examples of Dr 31, including an example with a stamp (S7) dating to *c* AD 160-200.

A separate sequence includes contexts (1276), (1275), (1274) and (1273), of which only the upper context (1273) has pottery in the form of a stamped (S13) CG Dr 38, dated *c* AD 160-200, with an Antonine CG, Dr 33.

Both sequences are sealed by (1220) which has two mid 3rd-century black-burnished ware jars, B01.11, and one B01.12, early to mid 3rd-century jar B01.10 and early to late 3rd-century bowl B01.22, and an example of a Young (1977) type C45 Oxford colour-coated bowl, F21.1, dated AD 240-400+. Residually there are two B01.27A dishes and a triple vase fragment, O03.29, with a cross-join to context (1188). Also present are a mica-dusted bowl, F44.1, Severn Valley ware forms O03.7, O03.12, O03.22, O06.1, O06.4, O06.6, a Wilderspool white-slipped flagon Q03.1 and a greyware crucible R03.6. There are examples of CG samian Dr 31R, two examples of Dr 33, and five examples of Dr 31, one stamped (S15). There is also an EG samian Dr 18/31.

Above this, (1188) contains a black-burnished ware with intersecting arc decorated dish B01.27A, with a date of AD 160-80+, a black-burnished ware jar, B01.2, with a mid to late Antonine date range, a Severn Valley ware beaker, O03.17, as well as forms O06.1, O04.1, O08.1, O08.3, and relatively undiagnostic greyware R02.2 and crucible R01.1. The samian includes, all CG, Dr 31R and Dr 33 with an Antonine date and two Dr 31 bowls, one stamped (S7), dated to *c* AD 160-200.

The uppermost fill (1183) has one example of mid 3rd-century black-burnished ware jar, B01.11, a Nene Valley 'Castor box', F01.4, of late 2nd- to 4th-century date, 3rd-century Severn Valley ware constricted-necked jar O03.5, and 2nd- to 4th-century constricted-necked jar form O04.1. There are also B01.29A, O03.24 and O04.20, along with CG samian forms Dr 31, 31R and two Dr33s.

A small number of contexts cannot be located within the tank stratigraphy. Those that have mid to late Antonine material include (1213), which has an Antonine CG Dr 1; context (1215), which has a mid to late Antonine black-burnished ware jar, B01.2, an Antonine jar, B01.3, and a CG Dr 33 vessel; and context (1291), which has a Wilderspool constricted-necked jar, O01.1, a late 2nd-century Severn Valley ware tankard, O04.12, and a CG Dr 31R, with a date range of AD 160-200.

Brine Tank [1207] by Context

Clay lining (1241) has a CG Dr 37 (decoration D7), with a date of *c* AD 150-80.

Basal deposit (1336) has a brown Wilderspool? colour-coated beaker, F08.1 and a polished greyware Dr 33 bowl copy, R21.1, of 2nd-century date. Then context (1335) has a Hadrianic-Antonine black-burnished ware jar, B01.7, and a 2nd-century Wilderspool white-slipped flagon, Q03.1. This is followed by context (1337), which has a Hadrianic-Antonine example of black-burnished ware jar B01.7.

Above this context (1239) contains an early 3rd-century jar, B01.10, two examples of mid 3rd-century black-burnished ware bowl B01.11, dish B01.26D, a late 2nd- to 3rd-century Severn Valley ware jar, O03.6, a constricted-necked jar, O04.17, with a 2nd- to 4th-century date, and O04.2A beaker. Also present are forms A01.1, B01.2, B01.3, B01.18, B01.21B, B01.21C, B01.21D, B01.25B, B01.25D, B01.30, F12.1, O02.5, O03.1, O03.3, O03.11, O04.8, O04.22, Q01.1, R02.5 and R21.1. There are CG samian forms comprising three Dr 31 bowls, four Dr 33, one Dr 35, four Dr 37 and one Dr 31R. There is also a SG samian Dr 37 bowl with decoration D9 and stamp S10. Context (1295) has a Wilderspool mortarium, M01.2, with an Antonine date.

Context (1240) has two mid 3rd-century black-burnished ware jars, B01.12, and a 3rd-century Severn Valley ware jug, O04.9, and a bowl, O03.23, of mid 2nd- to 4th-century date. It also includes forms B04.3, B01.3, B01.21C, B01.20, B01.1 and B01.2, as well as CG samian forms including two Dr 33 cups, and one example each of Dr 31 and 38, and a fragment of a SG Dr 37 (decoration D4) bowl.

Context (1225) has a mid 3rd-century black-burnished ware jar, B01.11, an early 3rd-century jar, B01.10, a late 2nd- to 3rd-century Severn Valley ware jar, O03.6, a 2nd- to early 3rd-century wide-mouthed jar, O03.14, a 2nd- to 3rd-century bowl, O06.6, a 1st- to 3rd-century constricted-necked jar, O04.4, and a 3rd-century Wilderspool whiteware flagon, W02.1. Also present is a late Antonine Dressel 20 amphora, A01.3, with stamp (St20). Residually present are forms O03.27, O03.25, B01.18, B01.27B and CG samian comprising five Dr 31, Dr 33 (S8), Dr 31R (S1), three Dr 37 (decoration D9 and D8) and two Dr 38.

Context (1236) has 1st- to 3rd-century jar O03.7 and mid to late 2nd-century Severn Valley ware wide-mouthed jar O04.12. samian forms are CG bowls of Dr31R and Dr 31, both of Antonine date.

3rd-century material is found in context (1224) in the form of Severn Valley ware jar O04.10, with residual examples of B01.21A, O03.12 and a CG Dr 37 bowl, with decoration D10. Context (1233) has a 1st- to 3rd-century example of a Severn Valley ware constricted-necked jar, O03.2.

Context (1209) has a 4th-century Oxfordshire colour-coated ware bowl, F21.2, an early 3rd-century whiteware mortarium, M11.2, and residual mid 2nd-century forms which include black-burnished ware jar B01.3 and bowls B01.20 and B01.21C, an oxidised Wilderspool jar, O02.1, and a Severn Valley ware jug, O03.8. Context (1229) has a probable 3rd-century Severn Valley ware jar, O06.2.

The final context in the sequence, (1208) has an Oxfordshire colour-coated ware bowl, F21.1, dated AD 240-400+, along with a 3rd-century Severn Valley ware constricted-necked jar, O03.4, a bowl, O03.24, a wide-mouthed jar, O04.15, and early to mid 3rd-century black-burnished ware vessels B01.10, B01.15 and B01.22. Also present are residual forms A01.3, O04.14, O03.10 and CG samian forms including Dr 31 sherds (S6), two Dr 31R, Dr 37 (decoration D11) and Dr 36.

The north-east chamber of [1207], [1347] contains a mid 2nd-century black-burnished ware bowl, B01.20, and a 2nd-century flagon, Q03.1, along with a CG Dr 18/31R Hadrianic-early Antonine dish.

Other Features

The earliest deposition elsewhere within the zone appears to be Hadrianic-early Antonine. Pit [1048] has an example of a CG Dr 31 Bowl with an Antonine date. Pit [1160] has a black-burnished ware acute lattice decorated dish, B01.25B, and a mid 2nd-century acute lattice decorated dish, B01.26A. Pit [1227] has a Severn Valley ware bowl, O07.3, a mid 2nd-century black-burnished ware beaker, B01.18, and a stamped (S4) CG Dr 31 dated *c* AD 130-55. Linear cut [1003] has a Hadrianic-Antonine jar, B01.3.

A number of features show deposition in the late 2nd or 3rd centuries. Pit [1096] has 2nd- to 3rd-century Severn Valley ware jar O03.6, and mid to late 2nd-century examples of B01.2 and B01.13, mid 2nd-century dish with pointed arcade burnished lines B01.25C, B01.29A, M01.4, O03.12, and decorated (D1) CG Dr 37 bowl dated *c* AD 125-50. There are also Antonine dated forms B01.17, O01.3, O02.30 and CG Dr 31 and Dr38, and Hadrianic-Antonine B01.21A, O04.3 and O04.5. Wicker-lined pit [1099] has Severn Valley ware bowl O03.21, with a late 2nd- to 3rd-century date, and a residual CG Dr 27 cup.

Ditch [1131] has a 3rd-century Severn Valley jar, O03.9, and a Wilderspool bowl, O02.3. Posthole [1041] has a mid 3rd-century black-burnished ware jar, B01.12. Posthole [1040] has a mid 3rd-century jar, B01.12, and a BB1 sherd with obtuse lattice decoration, giving a 3rd-century date. Pit [1157] has a mid 3rd-century date suggested by black-burnished ware jars B01.8 and B01.12, and dish B01.28. There are also examples of F41.1, O03.12, O07.2, CG Dr 31 and Dr 30. Trough [1264] within this feature contains a mica-dusted complete face constricted-necked jar, F41.3, with an early 3rd-century date, jar O03.6, and 3rd-century Severn Valley ware wide-mouthed jar O03.15.

There are also features with evidence of deposition between the late 3rd and late 4th centuries. Linear cut [1068] has a greyware developed beaded and flanged bowl, R02.4, with a date of AD 270+, with residual examples of B01.1, O04.29 and W01.3. Ditch [1175] has sherds of late 3rd-century or later southern shell-tempered ware, C01, along with two of M13, two of O03, a sherd of R41, and an Antonine? EG samian bowl. Wicker-lined pit [1053] has a late 3rd- to mid 4th-century black-burnished ware bowl, B01.23, and Severn Valley forms O03.18 and O14.1. Pit [1055]/[1064]/[1058] has bowl B01.23, with a late 3rd- to mid 4th-century date, and residual examples of F031.4, O04.2 and R42.1. Wicker-lined pit [1072] has a late 3rd- to mid 4th-century black-burnished ware bowl, B01.23, a 3rd-century black-burnished ware dish, B01.27D, an early to mid 3rd-century Mancetter-Hartshill mortarium, M13.8, and a late 2nd- to 3rd-century Severn Valley ware jar, O03.6. Also present, residually, are A02.1, B01.13, B01.27A, F031.1, F031.5, O02.3, O03.16, R41.1, and CG Dr 31 and 33. Pit [1079] has a late 3rd- to 4th-century shell-tempered jar, C01.2, an early 3rd-century jar, B01.10, a sherd of late 3rd- to 4th-century pink grog tempered ware, G11, three examples of M13.8, early 3rd-century, examples of CG Dr 27 and Dr 33, and sherds of a Dr 37 bowl. Gully [1116]/[1125] has a southern shell-tempered ware jar, C01, of late 3rd- to 4th-century date.

Wicker-lined pit [1052] has a 4th-century date suggested by an Oxford colour-coated bowl, F21.2. It also includes residual beaker F031.1, F031.2, F031.6, M13.8 and O02.6. Pit [1105] has a Nene Valley colour-coated lid, F01.3, with a late 4th-century date. Gully [1113] has late 4th-century deposition with a Nene Valley colour-coated lid, F01.3, and a late 3rd-century Dales ware jar, C04.1, with residual material including a black-burnished ware jar, B01.11, and a CG Dr 31.

Wicker-lined pit [1054] has a 3rd- to mid 4th-century example of black-burnished ware dish B01.27D and a Mancetter-Hartshill plain hammerhead mortarium, M13.8, with a date range of AD 220-50. Linear cut [1004] has a possible 3rd- to 4th-century date evidenced by a southern shell-tempered ware jar, C01.1. Residually present are B01.3, B01.27A, F031.2, R31.2 and W01.2.

Ditch [1162] has four sherds of white-slipped fabric Q03, 1st to early 3rd century?. Ditch [1107] has sherds of M13, O01 and four of O03. Ditch [1120] has a sherd of O03 and a sherd of O04. Pit [1074] has a sherd of O02. Tree bole [1109] has lid O01.5. Posthole [1042] has a sherd of O02 and a sherd of Antonine CG samian. Pit [1038] has an undiagnostic Severn Valley ware constricted-necked jar rim, O04.6.

Pit [1101] has a sherd of medieval shell-tempered ware, Z21, and residual Antonine CG samian.

Other Areas and Trenches

Area F

The small quantity of material recovered from area F shows a main deposition date of the early 3rd century. Pit [6017] has an example of M13.4, dated AD 200-50 and a 2nd- to 3rd-century wide-mouthed jar, R02.4

Trench 44

The small amount of datable pottery from this area is of the first half of the 3rd century.

Trench 45

The small amount of material from this area is 3rd-century in date.

Trench 46

[8031] contains a sherd of A01 and two sherds of M13, giving it a 2nd- to 4th-century date.

Taphonomy

By Site

The breakdown of pottery recovered by stratified context is shown in Table 4.2. The largest proportion, amounting to 50% of all the pottery, derives from the brine tanks. The site had suffered from later truncation and as a consequence pottery from horizontal deposits, such as make-up layers, is absent. Comparison with the taphonomic profiles of other sites suggests that large quantities of material from horizontal stratigraphy are the norm on urban and higher status sites (Mills forthcoming; Martin 2003a & b; Evans & Mills forthcoming). However, for basic level rural sites it is normally ditches and gullies which contribute the most material. Even if the contribution of the brine tanks is excluded, the amount of material from linear features is low. It has been noted that levels from pits are more variable for rural sites. For example, at Haddon (Evans 2003, table 4.3), Ship Lane, Aveley (Martin 2003a), and Great Holts Farm and Bulls Lodge Dairy, Boreham (Martin 2003b) levels are 8.3%, 10%, 3%, and 22% respectively. The level here is somewhat higher at 32% (even higher if the material from the brine tanks is excluded). This could relate to the industrial nature of the site, echoing the pattern at King Street, Middlewich (Leary 2008).

Context type	NoSh%	Wt%	MnR%	RE%	BE%
Pit	31.57%	17.35%	25.62%	18.57%	22.85%
Feature - general	1.09%	0.89%	1.72%	0.75%	2.33%
Ditch	7.30%	4.94%	6.69%	4.28%	4.58%
Gully	0.33%	0.11%	0.76%	0.19%	-
Posthole	7.27%	3.20%	9.56%	4.60%	5.19%
Hearth	0.58%	0.17%	0.76%	0.37%	0.48%
Well	0.09%	0.01%	-	-	-
Tank	50.65%	72.61%	54.68%	70.82%	64.13%
Burial	0.91%	0.68%	0.19%	0.42%	0.24%
Total	**3301**	**102161g**	**523**	**8876**	**8581**

Table 4.2: Pottery deposition by context type.

By Zone

The breakdown of deposition type by the different zones shows variations in the pottery discard, notwithstanding the huge bias caused by the deposition of so much pottery in the brine tanks (Appendices 4.6-4.9). These variations to some degree reflect the relative frequency of feature types within the zones. Thus while pits are the second most important category on the site, in the Utilitarian and Road Zones there is also a strong showing from postholes associated with the several buildings. In the Cremation Zone pits are of particularly high importance for deposition and may reflect activity here other than that associated with the area's use as cemetery.

Overall the average sherd weight for the site is 30.9g, a high figure at the top of the range for urban and military sites in the north (Evans 1985). Similarly the average percentage of rim is high at 17.0%. However, these figures vary markedly across the site (Appendix 4.10). In the Brine Industrial Zone the average sherd weight is a very high 36.7g and the average percentage of rim 20.0%, but levels on the other zones are below 20g and 13% and fall comfortably within the usual northern military/urban pattern. The figures for the brine tanks reflect the considerable votive deposition of a relatively large number of complete and largely complete vessels (Table 4.19).

Supply

Wares

When the stratified assemblage is broken down by ware type, the most common are oxidised fabrics, black-burnished ware and samian ware (Table 4.3). The large number of heavy fragments from amphorae compared to other ceramic types is clearly demonstrated by the ratio of weight to NoSh to MnR. The data can be compared with that collected by Leary (2008) for sites including King Street, Middlewich, Barton Street, Manchester (Leary 2005a), Castlefield, Manchester (Leary 2004), Mellor, Stockport (Leary 2005b), Warrington (Wilderspool) (Lucas 2007), Tarbock (Jones 2000), Tarporley (Fairbairn 2002) and Rocester (Leary 1996) (Table 4.4).

Amphora levels, at 8%, are very similar to the 9% by count from Walton-le-Dale (Evans & Rátkai forthcoming a) and 8% from Middlewich (Leary 2008; Evans 2002). Similarly the high level of samian ware, over 13%, is closely comparable to the 11.1% from Walton-le-Dale and the 13.2% from Middlewich. These high levels of samian and amphorae both suggest a military related rather than civilian supply (Willis 1998; Evans 2001, fig 11). The same can be said of the high amphora levels from the two Manchester sites, and again these have high incidences of samian ware.

Code/Ware type		NoSh%	Wt%	MnR%	RE%	BE%
A	Amphora	8.2%	29.4%	1.9%	4.6%	-
B	Black-burnished	21.8%	14.4%	31.5%	27.9%	21%
C	Shell-tempered	0.4%	0.2%	0.8%	0.3%	-
F	Fine	6.9%	3.8%	7.5%	8.9%	12%
G	Gritted	0.4%	0.5%	0.2%	0.2%	-
M	Mortaria	2.4%	3.9%	4.0%	2.9%	3%
O	Oxidised	38.8%	32.5%	23.5%	28.0%	27%
Q	White-slipped	3.0%	1.8%	1.5%	5.0%	4%
R	Reduced	3.8%	2.7%	2.7%	3.9%	8%
S	Samian	13.1%	10.2%	25.4%	16.7%	24%
W	Whiteware	1.3%	0.6%	1.0%	1.7%	1%
Total		**3301**	**102161g**	**523**	**8876**	**8581**

Table 4.3: Pottery by ware type.

The putative mining administrative centre at Plas Coch, Wrexham, also produced high amphora levels (Table 4.4), consistent with those from sites with a military supply. While its level of samian ware, although within the urban range, was not as high as many of these north-western sites with a 'military supply', it did fall within the range of the probably marginal site at Warrington. In comparison, the Flavian fort at Brithdir has 10.5% amphorae by count and a very high 21.5% samian ware (Evans 1997), whilst the Antonine fort at Thornborough Farm, Catterick, produced 9.0% amphorae by count and 15.0% samian ware (Evans 2002). The higher level of amphora supply to military sites has been demonstrated by Carreras (MoL RFG conference 1992) in the density distribution of amphora stamps, first noted by Collingwood (Collingwood & Myres 1937). It is clear that the amphora supplies at Walton-le-Dale, Middlewich and Warrington, as well as those at Manchester, have a military basis, in contrast to those on the civilian sites at Tarbock, Mellor and Tarporley. This does not mean that all these sites were military (see below; Evans forthcoming a) but rather that they received the same supplies as the military.

The level of black-burnished ware, Dorset BB1 (Tomber & Dore 1998), in the assemblage is at the higher end of the regional range at 21.8%. This figure is reasonably comparable to the levels at Walton-le-Dale (20%) and Plas Coch, Wrexham (19%) but somewhat higher than that for Middlewich (10%), Warrington (10.5%), Manchester (8-14.5%) and Rocester (14%). Nantwich is also close to the range for Middlewich's early to mid Antonine phase 2, at *c* 19% (Leary 2008). This suggests that the relatively high level at Nantwich may in some part be due to the lack of any volume of pottery supply before the Antonine period. Chronologically the black-burnished ware assemblage falls into two parts, with what appears to be a gap between them. The Antonine wave has a mix of jars and tablewares but the 3rd-century supply is much more jar-dominated, with only a few tablewares supplementing it.

Class C shell-tempered wares, probably from Harrold, Bedfordshire, are present in very minor quantities on most of the sites in Table 4.4. At Manchester Castlefield the high level probably relates to the stronger levels of 4th-century activity here, while at Rocester the site ends too early to receive this fabric group. The perseverance of the fabric, albeit in very small quantities, is of note and might relate to it travelling as a container for some particular commodity.

Colour-coated, mica-dusted and rough-cast finewares, class F fabrics, are reasonably common at Nantwich at 6.9%. The principal supplier of these would seem to be the kiln site at Wilderspool which probably produced at least 4.6%. Wilderspool fabrics include red-slipped ware (F03 and F031), brown-slipped ware (F08), roughcast ware (F11/F12), mica-dusted ware (F41) and a little red-painted ware (F61).

Other supplies included Nene Valley colour-coated ware (F01/F02) which was a much minor source, reflecting its geographical position and the presence of local competition. There is also a relatively large quantity (0.5%) of a red-slipped ware with a white core (F05), which is probably from a fairly local source. Later Roman finewares were restricted to a little Nene Valley ware and a couple of Oxfordshire ware vessels.

Fineware levels are uncommon compared to most of the sites tabulated by Leary, with only Barton Street, Manchester, approaching the Nantwich level. Their sources appear to be fairly typical, with Wilderspool and Cheshire Plain sources seeming to be dominant throughout the area in the 2nd century. These sources died out in the 3rd century and the market was then left to much more distant producers.

Gritted wares, class G, are not common, at 0.4%, in line with their occurrence at Middlewich, Warrington and Manchester, and with the lack of any source for these in the region.

The mortarium assemblage, class M, from the site is reasonably large, at 2.4%, slightly higher than at Middlewich and Warrington and comparable with that from urban and military sites. The dominant 2nd-century supplier is Wilderspool, despite its distance, and the nearer source at Holt is not well represented nor is Wroxeter, although it lies at little greater distance than Wilderspool. However, in the 3rd century supply becomes the monopoly of the much more distant Mancetter-Hartshill industry, with the only other late source being Oxfordshire.

At Middlewich early (Flavian-Trajanic) mortaria seem to mainly consist of Verulamium region vessels along with a couple of Rhone Valley vessels. The Hadrianic-early Antonine mortaria were mainly oxidised wares, mostly Wilderspool products with an occasional Holt piece, and very occasional Mancetter products. Leary (2008) also notes a Cumbrian piece, showing that occasional vessels travelled, presumably as personal possessions, on the reverse of the west-coast trade. As is general in the region, Mancetter products probably became dominant in the Antonine period, and a number of 3rd- to 4th-century Mancetter reeded hammerhead mortaria are also found along with a couple of lower Nene Valley vessels.

In contrast, at Plas Coch, Wrexham, nearly all the 2nd-century mortaria were from Holt, as might be expected given the close proximity of that site and its possible military connections. Thus whilst Holt seems to have had a major role in supplying Chester and its garrison, Wilderspool seems to have been dominant over a much wider area of Cheshire and many of its products seem to have travelled south of the kiln site, in contrast with the

Fabric	Middlewich King Street	Warrington Gifford	Manchester Barton Street	Manchester Castlefield	Tarbock	Mellor	Tarporley	Rocester	Plas Coch
A	8.3%	7.6%	6.6%	6.6%	1.3%	-	3.2%	4.3%	5.2%
B	10.3%	10.5%	14.5%	7.9%	9.9%	10.3%	31.2%	14.3%	19.3%
C	0.2%	0.1%	0.2%	1.1%	-	2.3%	-	-	0.1%
F	2.0%	1.7%	6.4%	0.7%	1.0%	-	-	1.4%	1.5%
G?	0.3%	0.4%	3.2%	-	-	16.0%	-	8.8%	-
M	2.2%	1.9%	3.4%	2.2%	1.1%	6.2%	-	2.5%	1.1%
O	19.3%	26.6%	22.1%	29.7%	73.0%	27.1%	60.0%	4.3%	56.3%
Q& W	12.3%	8.6%	16.4%	13.8%	0.1%	1.6%	3.9%	9.3%	1.7%
R	31.8%	37.4%	16.4%	24.7%	12.5%	31.6%	0.6%	35.5%	10.0%
S	13.2%	5.2%	14.0%	11.3%	1.0%	7.6%	-	10.0%	4.0%

Table 4.4: Comparison of ware proportions in the region, after Leary (2008, 100 table 13) and, for Plas Coch, Evans forthcoming b.

usual pattern of northern mortaria kilns. As with the North-East of England, from the late Antonine period onwards Mancetter-Hartshill seems to have become a near monopoly supplier for at least the first two thirds of the 3rd century.

Oxidised wares, class O, dominate the Nantwich assemblage at 38.8%. Oxidised wares are much commoner than at Middlewich, where they comprised only 19.3%, and the high levels are only matched by those from the latest phases at Walton-le-Dale (Evans & Rátkai forthcoming a). Levels, however, are higher at Plas Coch, Wrexham, at some 56%, and to the south at Wroxeter. At Wroxeter 4th-century oxidised levels amount to 35-38% (White 1997, figs 366-7) and there is no reason to believe they were lower earlier, the site being well within the Severn Valley ware tradition. This seems also to have been the case at Nantwich. Firstly there is a major element of Severn Valley ware (fabrics O03, O06, O07, O08 and O011), amounting to 13% of the assemblage. Then there are the putative Wilderspool sandy fabrics O01 and O02 at 18%, the forms of very many of which are Severn Valley ware forms. These may be Wilderspool fabrics but if so their purchasers seem to have selected the Severn Valley ware copies from Wilderspool amongst the form range produced there. Either way, Severn Valley ware style forms are the predominant oxidised wares on the site. This is not the case to the same extent at Middlewich and Wilderspool, although Plas Coch also had quite strong Severn Valley connections. Neither does it seem to be the case at Whitchurch (Jones & Webster 1968), although that site was south of Nantwich. It does seem likely that the predominance of Severn Valley ware forms at Nantwich reflects the cultural traditions of the resident group.

White-slipped wares, class Q, at 3%, are very low, compared to Walton-le-Dale where they are at levels of 10-21% by phase (Evans & Rátkai forthcoming a), but the latter is a production site. They are also low compared to Middlewich, Manchester, Wilderspool and Rocester, which probably reflects the predominantly Antonine date of the Nantwich collection and perhaps also its Severn Valley cultural connections.

Reduced wares, class R, are an insignificant element of the assemblage at Nantwich, at 3.8%. They include a number of polished greywares, not seen on the other sites, which may be of quite local origin or possibly relate to Holt. The level of greywares here is a marked contrast to the near 32% from Middlewich and 37% at Warrington. Further north levels are still higher than at Nantwich at Walton-le-Dale, where they fall from 29% in phase 2 to 14% in phase 4 (Evans & Rátkai forthcoming a). Levels at Plas Coch are also low at 10% and this site also shares some of the Severn Valley connections of Nantwich. There are no quantified data from Whitchurch (Jones & Webster 1968) but the illustrated material suggests that greywares may well have been commoner there than at Nantwich.

Whitewares, class W, are also a very minor element at Nantwich, at 1.3%. The low proportion reflects the late date of the assemblage.

By Zone

The breakdown of ware types by zone is shown in Appendices 4.11-4.14 and the difference between them represented in Figure 4.10. In many ways the differences in the Road Zone (the presence of class C, the smaller proportion of class B and the higher proportion of class F) are a result of the later date range of the pottery from this zone.

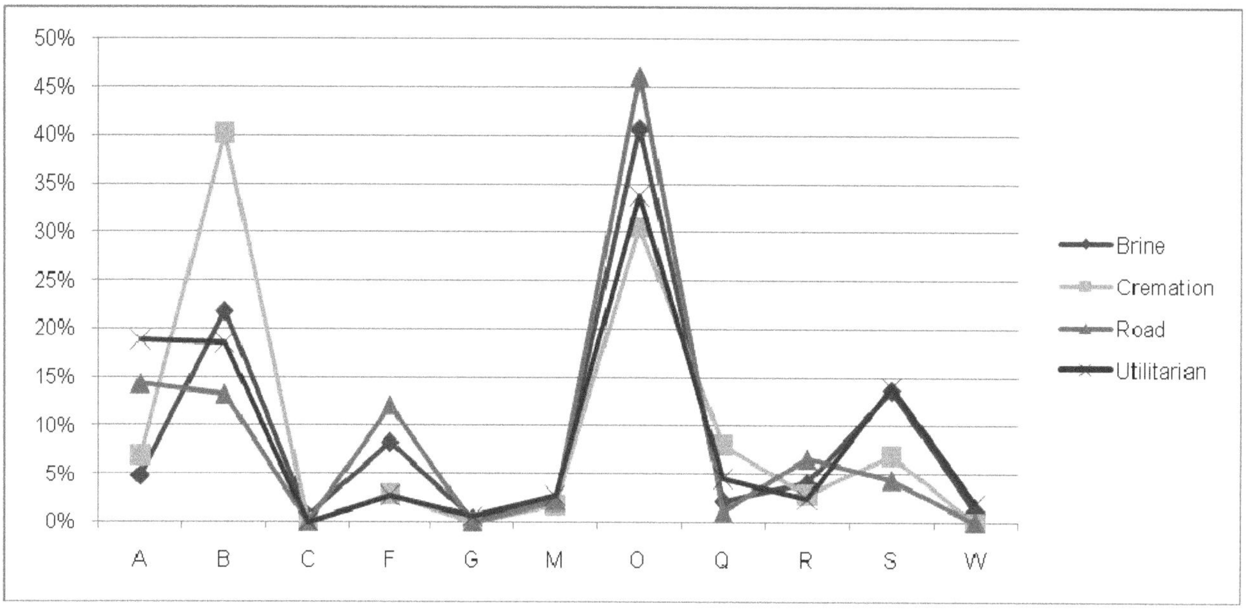

Figure 4.10: Ware by zone by NoSh%.

Fabrics

The fabrics are described in Appendix 4.2. Fabric quantities for the site are given in Appendix 4.3, and fabric occurrence by zone in Appendix 4.4. In the following catalogues, entries marked with an asterisk are not illustrated.

Class A Amphora (Fig 4.22)

Further evidence of the military supply at Nantwich comes from its amphora assemblage. In common with Walton-le-Dale (Evans & Rátkai forthcoming a), Lancaster (Evans & Rátkai forthcoming b) and Middlewich (Evans 2002), the amphora supply falls into a north-western pattern with a low level of Gallic wine amphorae (eg 7.9% of A01 to 0.2% of A02 at Middlewich (Evans 2002)). This is in strong contrast to the supply at Segontium and Chester where Gallic and other wine amphorae represent their usual 15% or so. It is further evidence of how the north-western sites are united in a single supply network, which operates to all military sites in the region, and which is different from those supplying military sites in north Wales (Evans & Rátkai forthcoming b; Evans forthcoming a).

Dressel 20 Amphora (8%)

This is a Spanish amphora, used to bring olive oil into the province. The large quantity of sherds and rims suggests that this was utilised on site and its presence in the Utilitarian Zone is highly suggestive of use there. The forms are late Antonine rim types illustrated by Martin-Kilcher (1987) and the dated stamp also seems to fit with this period. The pattern nicely matches the peak of deposition occurring in the mid to late Antonine period overall on the site.

A01.1	A Dressel 20 amphora, with a ring rim with a rounded triangular section and internal bead. Martin-Kilcher 1987, no 80. (1285) [1182].	AD 110-50
A01.2	A Dressel 20 amphora with a triangular section wedge rim and slight lid seating. Martin-Kilcher 1987, nos 93-6. (1283) [1182].	AD 150-210
A01.3	A Dressel 20 amphora with a rounded triangular section rim and internal lip. Martin-Kilcher 1987, nos 82-9. (1208) (1225) [1207].	AD 110-50
A21.1	An amphora lid. (1281) [1182].	C1-C3

Markings

Two stamped, as well as some white paint on a body sherd.

St20 A01.3, retrograde F.C. (1225) [1207].

St21 A01.1, LC ANT. Cf Funari 1996, no 52 from Colchester on a round handle with Martin-Kilcher 1987, no 91 type rim, AD 150-90. (1285) [1182].

P1 White paint (dipinto?) on body sherd A01. (1285) [1182].

A02 and A11 Gauloise Amphora (0.3%)

A02.1	A Pelichet 47/Gauloise 4 amphora. (1073). [1072].	C2

Class B Black-burnished Ware (21%) (Figs 4.22 & 4.23)

Amongst the black-burnished ware (BB1), 65 sherds have acute lattice decoration of Hadrianic-Antonine date, compared with 145 sherds with obtuse lattice decoration. Among the obtuse lattice decorated sherds, five have a groove above the lattice zone, suggesting a date after *c* AD 240. The functional breakdown of black-burnished ware pottery from the site shows a relatively weak jar presence, with a strong presence of bowls and dishes and a relatively high level of beakers (Table 4.5).

The functional breakdown by zone (Table 4.5) shows an interesting bias towards jars, and an absence of bowls, in the Cremation Zone. In the partly ritual depositions of the Brine Industrial Zone jars are a little commoner than tablewares. In the Utilitarian Zone jars are only slightly more common than the combined bowl and dish presence but just about maintain an absolute majority, unlike at Walton-le-Dale (Evans & Rátkai forthcoming a) and Binchester (Evans & Rátkai forthcoming c), although clearly belonging to the same trend.

		J	Bk	B	D	No
Site	MnR	50.9%	6.1%	15.8%	27.3%	165
	RE	57.4%	3.6%	16.8%	21.6%	2476
Brine Industrial Zone	MnR	50.4%	5.3%	17.7%	26.5%	113
	RE	57.0%	3.0%	19.4%	20.6%	1927
Cremation Zone	MnR	73.3%	6.7%	-	20.0%	15
	RE	86.5%	1.7%	-	11.8%	237
Utilitarian Zone	MnR	46.7%	10.0%	10.0%	33.3%	30
	RE	45.5%	10.2%	6.4%	37.9%	264

Table 4.5: Functional breakdown of class B pottery by site and zone.

Changes in the functional breakdown of the ware can be examined by utilising the date distribution data. Figure 4.11 shows a stacked bar chart of date distribution broken down by vessel function. Figure 14.12 shows the same data as a proportional stacked bar chart, using 50-year intervals.

It is interesting to note that jars make up a much more important part of the black-burnished ware assemblage after the late 2nd century, reflecting the distribution suggested by the proportions of sherds of different lattice types.

Black-burnished ware bowls paint a different picture, with a minor peak for flanged bowls in the early mid 2nd century and a much stronger one in the later 2nd century. There is a small presence of later bowls indicated by incipient beaded and flanged bowls dated to *c* AD 200-70 and developed beaded and flanged bowls dated to *c* AD 270-350. Incipient beaded and flanged BB1 bowls are always much rarer than flange rimmed bowls and developed beaded and flanged bowls, and the quantity reflects the weakness of pottery deposition after *c* AD 270.

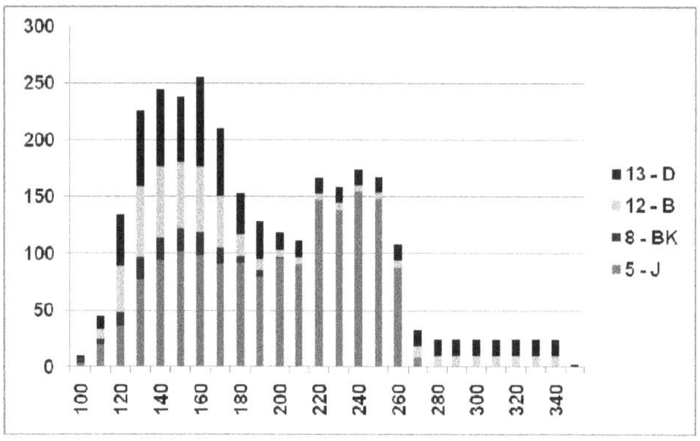

Figure 4.11: Functional breakdown of class B pottery by date distribution by RE.

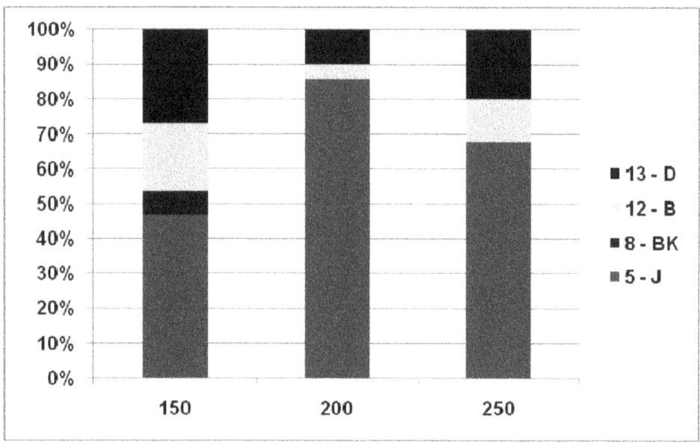

Figure 4.12: Proportional stacked bar chart showing functional breakdown of class B pottery, by RE, over 50-year intervals.

B01.1	A jar with an outcurving splaying everted rim with no bead. Gillam 1976, no 31.	Mid C2
B01.2	A necked jar with a strongly everted bead rim. Gillam 1976, no 3. (1285) [1182].	Mid-late C2
B01.3	A globular jar with a bead rim and acute lattice decoration. Cf Gillam 1976, 30. (1285) [1182].	Hadrianic-Antonine
*B01.4	A small jar with an everted rim. Gillam 1976, no 3.	Mid-late Antonine
*B01.5	A jar with an everted rim. Gillam 1976, no 5.	Late C2-early C3
B01.6	A jar with a strongly everted rim and acute lattice. Gillam 1976, no 1. (3041) [3020].	Early C2-mid C2
B01.7	A jar with an everted rim. Gillam 1976, no 30. (1337) [1207].	Hadrianic-Antonine
B01.8	A jar with a strongly everted thickening, almost bead, rim and obtuse lattice decoration. Gillam 1970, no 32. (1158) [1157].	Early C3
*B01.9	A jar with a straight everted rim and acute lattice. Gillam 1970, no 5.	Late C2-early C3
B01.10	A jar with an outcurving splayed everted rim and obtuse lattice decoration. Gillam 1976, no 7. (1279) [1182].	AD 200-40
B01.11	A jar with an outcurving everted rim with no bead. Gillam 1976, no 8. (1239) [1207].	AD 240-70
B01.12	A jar with a strongly everted rim and obtuse lattice. Gillam 1976, no 7. (1240) [1207].	AD 220-60
*B01.13	A jar with an everted rim. Cf Gillam 1976, no 22.	Mid-late C2
B01.14	A cooking pot with a cavetto rim with obtuse lattice decoration. Gillam 1976, no 6. (1283) [1182].	AD 200-20
B01.15	A cooking pot with a slightly everted rim. Gillam 1976, no 7. (1208) [1207].	AD 200-30
B01.16	A beaker with an everted rim and step below. Gillam 1976, no 25. u/s [1182].	Mid C2
B01.17	A beaker with a stubby rim. Gillam 1976, no 20. (2151) [2136].	C2?
B01.18	A handled beaker with an everted rim. Gillam 1976, no 27. (1239) [1207].	Mid C2
B01.19	A handled beaker with an everted stubby rim. Gillam 1976, no 28. (1285) [1182].	Mid-late C2
B01.20	A chamfered base flange rim bowl with a pointed arc decoration. (1285) [1182].	Mid C2
B01.21A	A flange rim bowl/dish, with no decoration. u/s [1182].	Hadrianic-Antonine
*B01.21B	A flange rim bowl/dish with acute lattice decoration.	Hadrianic-early Antonine
*B01.21C	A flange rim bowl/dish with pointed arc burnished lines.	Mid C2
*B01.21D	A flange rim bowl/dish with rounded arc burnished line decoration.	Late Antonine
*B01.21E	A flange rim bowl/dish with vertical burnished lines.	AD 200-70
*B01.22	A bowl with an incipient bead and flange rim. Gillam 1976, no 61.	AD 200-70
*B01.23	A developed beaded and flanged bowl. Gillam 1976, no 45.	AD 270-350
B01.24	A flanged rim bowl with a chamfered base. Gillam 1976, no 35. (1281) [1182].	Hadrianic-Antonine
*B01.25A	A flanged rim dish undecorated.	Hadrianic-Antonine
*B01.25B	A flanged rim dish with acute lattice decoration.	Hadrianic-early Antonine
B01.25C	A flanged rim dish with pointed arc decoration. Gillam 1976, no 65. (1284) [1182].	Mid C2
*B01.25D	A flange rim dish with intersecting arc decoration.	AD 160/80-200
*B01.26A	A dish with a grooved rim and chamfered base with acute lattice decoration.	Early-Mid C2
*B01.26B	A dish with a grooved rim with no decoration.	Hadrianic-Antonine
B01.26C	A dish with a grooved rim with pointed arc decoration. Cf Gillam 1976, no 52. (1285) [1182].	Mid C2
*B01.27A	A simple rim dish with intersecting arc decoration. Gillam 1976, no 75.	AD 160/80-350
*B01.27B	A simple rim dish with pointed arc burnished line decoration. Gillam 1976, no 76	AD 160/80-350
*B01.27C	A simple rim dish with acute lattice decoration. Gillam 1976, no 76.	Mid C2
B01.27D	A simple rim dish with indeterminate or no decoration. Gillam 1976, no 76. (1073) [1072].	Hadrianic-mid C4
*B01.28	A dish with a squared rim. Gillam 1976, no 79.	C3-mid C4
*B01.29A	A curving walled dish with a groove under the rim with	AD 160/80-200

	intersecting arcs. Gillam 1976, no 76.	
*B01.29B	A curving walled dish with a groove under the rim.	AD 120-350
B01.30	A jar base reworked as a simple rim dish. (1239) [1207].	AD 120-350

Class C Shell-tempered and Calcareous Tempered Wares (Fig 4.23)

Three shelly wares were recovered from the site and together make up a very small proportion of the overall assemblage. C01 is southern shell-tempered ware, from Harrold, Bedfordshire (Brown 1994), and is represented by two jars, one of 3rd- to 4th-century date and the other of late 3rd-century date or later. C02 appears to be a variation of Southern Shell-tempered ware and is also present as a 3rd- to 4th-century jar form. These forms perhaps indicate that southern shell-tempered ware was reaching the North-West somewhat earlier than the North-East where it seems to occur after AD 360 (Evans & Mills 2009; Evans & Rátkai forthcoming c).

There is also a heavily burnt example of a Dales ware rim (Loughlin 1977; Gillam 1951) dating from the later 3rd to mid 4th century.

C01 Southern Shell-tempered Ware (0.4%)

C01.1	A jar with an everted hooked rim. (1006) [1005].	C3-C4
C01.2	A jar with an everted rim. (1080) [1079].	Late C3-C4

C04 Dales Ware (<0.1%)

C04.1	A Dales ware rim. (1115) [1113].	Late C3-C4

Class F Finewares (Figs 4.23 & 4.24)

Wilderspool red-slipped wares (F03 and F031) are among the most common finewares (Figure 4.13), at 2% of the total assemblage, comprising mainly beakers but also including a bowl. A very small quantity of probable Wilderspool brown-slipped ware (F08) is present as a beaker.

A small number of later colour coats are present. There is a relatively large quantity (0.5%) of a regional red-slipped ware (F05) with a white core, which is present as a bowl and a 'Castor box'. A north-western source seems likely and Holt might be a possibility. Nene Valley colour-coated wares (F01) are very rare because of the geographical position of the site. They are represented by two beakers, a late 4th-century pierced lid and a 'Castor box'. Central Gaulish 'Rhenish' ware (F101) is represented by a single sherd but Trier 'Rhenish ware' is absent. A few sherds of Oxfordshire colour-coated ware (F21) are present in the form of two bowls.

There is a relatively large quantity of gold mica-dusted wares, perhaps from Wilderspool (F41), at 2%, comprising a jar, a bowl and a face pot. The face pot F41.3 fits into Braithwaite's (2007) type RB13, face jars with handles, within her late 1st- to early 3rd-century group of British face pots. The example here is a typical early 3rd-century bifid rimmed constricted-necked jar. Braithwaite's type RB13 is described as being made up of regional groups in local fabrics, with no obvious 'parent' group. The example from Nantwich has most morphological characterises in common with RB type 13A which was centred on Colchester. Face pots of this type are reported from Little Chester and Caerwent, and there are fragments of face pot reported in oxidised, probably Severn Valley ware, fabrics to the south of Nantwich. The use of face pots is very strongly associated with the Roman military in the 1st century, and the link is still strong between military and veteran presence and face pot usage in the later 2nd to early 3rd century (Braithwaite 2007). There is an additional gold mica-dusted fabric, F43.

There are a number of roughcast beakers from Wilderspool (F11 and F12) present at 0.6% and so much rarer than at Walton-le-Dale (Evans & Rátkai forthcoming a). Wilderspool is also represented by a red-painted ware bowl (F61).

There are several sherds in a black polished greyware (F32) not found at Middlewich or Wilderspool, which again might possibly be of Holt origin. A Dr 36 copy bowl in black polished ware F36.1 is also present. A number of other polished greywares were classified as too coarse for this group and are discussed with the greywares, class R (fabrics R21, R41, R42 and R43).

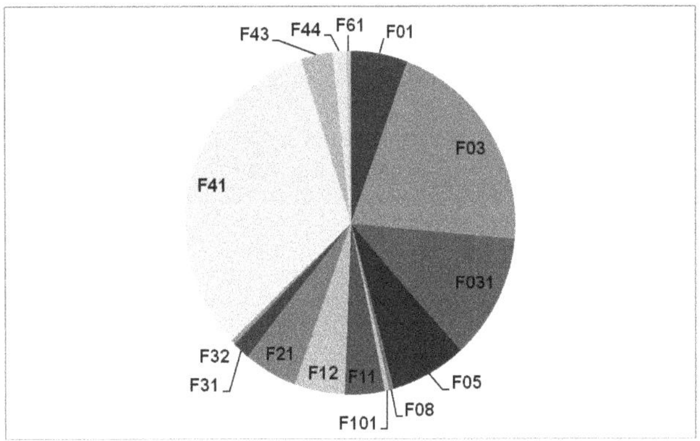

Figure 4.13: Pie chart showing proportions of class F fabrics.

Wilderspool Red-slipped Wares (2%): F03 and F031

F03.1	A beaker with a slightly everted thinning rim and rouletted decoration. Hartley & Webster 1973, no 35. (1285) [1182].	C2
F031.1	A narrow mouthed beaker. (3213) [3212].	C2
F031.2	A beaker with a pedestal base and a cornice rim. Hartley & Webster 1973, no 23. u/s [1207].	C2
F031.3	A roughcast beaker with a sub-cornice rim. Hartley & Webster 1973, no 23. (1073) [1072].	C2
F031.4	A beaker with a beaded rim. Hartley & Webster 1973, no 29. (1065) [1064].	C2

F031.5	A beaker with a wedge slightly undercut rim. Hartley & Webster 1973, no 23. (3479) [3470].	C2
F031.6	A bowl with a hooked rim. Hartley & Webster 1973, no 54. (1061) [1052].	C2

Wilderspool? Brown Colour-coated F08 (>0.1%)

F08.1	A beaker with a cornice rim. Hartley & Webster 1973, no 34. (1336) [1207].	C2

Red-slipped Ware F05 (0.5%)

F05.1	A bowl with two grooves in wall. (5018) [5017].	C2
F05.2	A 'Castor box'. (1205) [1182].	Late C2+

Wilderspool Red-painted Ware F61 (0.1%)

F61.1	A bead rim beaker with red paint traces on internal surface. Hartley & Webster 1973, no 54. (3093) [3092].	C2?

Wilderspool Roughcast Ware (0.6%): F11 (0.3%) and F12 (0.3%)

F11.1	A roughcast beaker with an everted sub-corniced? bead rim. Hartley & Webster 1973, no 26. (3118) [3117].	C2
F12.1	A roughcast beaker with a brown slip and sub-cornice rim. Hartley & Webster 1973, no 28. (1285) [1182].	C2
F12.2	A roughcast beaker with a strongly everted thinning rim. Perhaps Hartley & Webster 1973, no 30. Severn Valley ware copy, cf Webster 1976, no 15. (1285) [1182].	C2

Wilderspool? Mica-dusted Ware F41 (2%)

F41.1	A jar with an everted rounded triangular wedge rim. Hartley & Webster 1973, no 46. (1158) [1157].	C2
F41.2	A mica-dusted bowl with a slightly everted rim. Hartley & Webster 1973, no 49. (3284) [3195].	Flavian-Trajanic
F41.3	A face pot with three handles and a frilled everted bifid rim. The face comprises round eyes and mouth with a stubby nose and no eyebrows. Cf Braithwaite 2007, type RB13. (1272) [1264].	Late C2-early C3

Other Mica-dusted Ware F44 (0.1%)

F44.1	A bowl with an everted outcurving rim. Cf Hartley & Webster 1973, no 49. (1220) [1182].	Late C1-C2

Nene Valley Colour-coated Ware F01 (0.4%)

F01.1	A beaker with slightly thickening rim. (2002)	AD 160-400+
F01.2	An indented scale beaker. Howe *et al* 1980, nos 38-9. (4232) [4237].	Mid C3-late C3
F01.3	A flange rim lid with perforated knob. Howe *et al* 1980, no 72. (1106) [1105].	Late C4
F01.4	A 'Castor box'. Howe *et al* 1980, no 89. (1183) [1182].	AD 160-400+

Oxford Colour-coated Ware F21 (0.4%)

F21.1	A bowl with a bead rim. Young 1977, type C45, a Dr 31 copy bowl. (1209) [1207].	AD 240-400+
F21.2	A bowl with a grooved bead rim. Young 1977, type C81? (1061) [1052].	AD 300-400

Black Polished Greyware F32 (>0.1%)

F32.1	A Dr 36 copy dish. (1285) [1182].	C2

Glass G Gritted Wares (Fig 4.24)

There is only a small quantity of gritted wares, comprising an Antonine or later Derbyshire ware jar and a late 3rd- to 4th-century pink grog-tempered ware storage jar (Taylor 2004; Booth & Green 1989). The similarity to the supply of class C jars is of note and would indicate some sporadic contacts to the south-east in the 3rd and 4th centuries. Both of these fabrics occur sporadically in the later Middlewich assemblages (Leary 2008). Notably, Malvernian metamorphic tempered ware is absent, as is Warwickshire Museum fabric G12, both of which reached Middlewich in minute quantities from the Malvern Hills and Northamptonshire respectively.

Derbyshire Ware G108 (>0.1%)

G108.1	A jar with an everted incurving rim. Gillam 1939, no 1. (1284) [1182].	Antonine+

Pink Grog-tempered Ware G11 (0.2%)

G11.1	A necked storage jar with an everted hooked rim. (4002).	Late C3+

Class M Mortaria (Fig 4.25)

The breakdown of mortaria by fabric is shown in Table 4.6 and the date distribution, by MnR, in Figure 4.14. In the 2nd century supply is dominated by Wilderspool Raetian types. Oxidised Holt mortaria maintain a small presence in the 2nd to early 3rd century but whiteware Mancetter-Hartshill products become increasingly important from the mid 2nd century and peak at a higher level in the mid 3rd. This is quite different from the general date distribution for the site and shows mortaria making up an increased proportion of the assemblage in the mid 3rd century. Part of this increase may be explained by the presence of a large number of painted unreeded hammerhead mortaria (twenty, which have been rather arbitrarily assigned a range AD 220-70) and much fewer reeded hammerheads (which have been assigned to the period AD 220-300). It is not clear that the painted unreeded hammerheads represent a chronological horizon and it may be that the unusual number here indicates the arrival of a particular consignment of material from Mancetter. There is a small quantity of white-slipped Wilderspool/Walton/Cheshire Plain mortaria present from the late 3rd to early 4th century. There is also an additional whiteware mortarium, M11, possibly an outlying variant of Mancetter-Hartshill.

There are only occasional pieces of later 3rd- to 4th-century date; these include a small quantity of Oxfordshire mortaria with a 4th-century date. Also present are an unstratified body sherd of possible

Oxidised Wroxeter mortarium M07 and body sherds of possible Wroxeter whiteware mortaria M15 (0.1%).

Fabric	NoSh%	Wt%	MnR%	RE%	BE%
M01	14.10%	19.63%	23.81%	28.85%	-
M02	15.38%	14.65%	-	-	13.91%
M03	1.28%	0.08%	-	-	2.17%
M04	1.28%	10.55%	4.76%	8.70%	-
M06	1.28%	0.33%	-	-	10.87%
M11	2.56%	1.58%	4.76%	2.37%	-
M13	58.97%	41.78%	61.90%	52.17%	51.30%
M15	3.85%	7.05%	-	-	21.74%
M21	1.28%	4.38%	4.76%	7.91%	-
Total	**78**	**4000g**	**21**	**253**	**230**

Table 4.6: Mortaria by fabric.

Wilderspool Red-slipped Raetian Tradition Mortaria M01 (0.3%)

M01.1	A Wilderspool mortarium with a flange rim, having white quartz trituration grits. (1285) [1182].	Antonine
M01.2	A flange mortarium with a groove on lower lip of rim. (1295) [1207].	Antonine
M01.3	A mortarium with a bead and thickening flange rim. (4001).	Antonine
M01.4	A mortarium with a flange rim. (1111) [1096].	Antonine

Holt Oxidised Mortaria M04 (>0.1%)

M04.1	A mortarium with a bead and flange rim 'Raetian' red-slipped. (2002).	Mid-late C2
M04.2	A mortarium with a bead and slightly hooked thickening flange. (3067) [3066].	Romano-British

Mancetter-Hartshill Whiteware Mortaria M13 (1.7%)

M13.1	A flanged mortarium. u/s [1182]	AD 170-200
M13.2	A flanged mortarium with inner lip. u/s.	AD 140-80?
M13.3	A mortarium with an undercut flanged rim. u/s.	AD 150-80?
M13.4	An unreeded hammerhead mortarium rim. (i) (6018)[6017]; (ii) u/s.	AD 220-70
M13.5	A painted, reeded hammerhead mortarium. (1285) [1182].	AD 220+
M13.6	A reeded hammerhead mortarium rim, with a slight bead. (i) u/s; (ii) u/s.	AD 220+
M13.7	A hammerhead mortarium with cordons at top and base. (i) u/s [1182]; (ii) (6002).	AD 200-20
M13.8	An unreeded panted concave hammerhead mortarium. (1080) [1079].	AD 220-70
*M13.9	An unreeded straight hammerhead mortarium.	AD 220-70
M13.10	A bead and flanged mortarium. (1337) [1336].	AD 160-200

Mancetter-Hartshill Variant? Whiteware Mortaria M11 (0.1%)

M11.1	A reedless hammerhead mortarium. (1209) [1207].	AD 220+

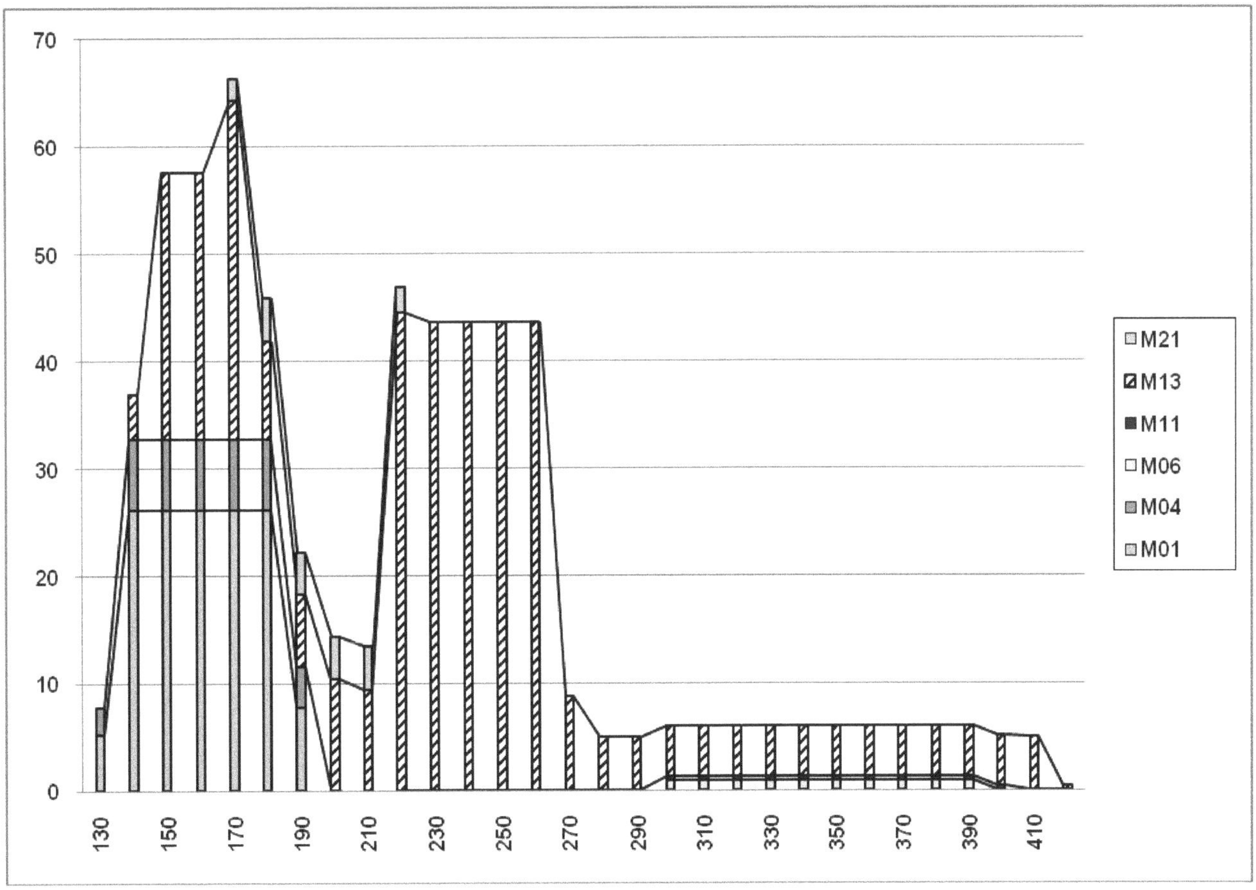

Figure 4.14: Stacked bar chart showing date distribution of mortaria fabrics by RE.

Wilderspool/Walton/Cheshire Plain White-slipped Mortaria M21 (>0.1%)

M21.1	A white-slipped reeded hammerhead mortarium with quartz and brown stone trituration grits. (1205) [1182].	Late C2-early C3

Oxford Mortaria M06 (0.1%)

*M06.1	Oxford mortarium. Young 1977, C100.8/10.	AD 300-400

Class O Oxidised Wares (Figs 4.26-4.29)

The majority, 18%, of class O wares seem to come from Wilderspool. There is a substantial element of Severn Valley tradition forms, some 13%, probably from sources within the Severn Valley region and a further 7% from a source which may well be related to Wilderspool (O04) but which carries a wide range of Severn Valley ware tradition forms. As at Walton-le-Dale, most forms are tablewares, along with some constricted-necked jars and quite a number of Severn Valley ware type wide-mouthed jars. There is also a small quantity of a micaceous fabric.

Figure 4.15 shows a date distribution, by RE, of the class O pottery broken down into wares. The Wilderspool component begins in the 2nd century, increases in the mid 2nd and peaks in the 3rd century. Interestingly, by RE Wilderspool fabrics are dominated by Severn Valley ware fabric O03 in the later 2nd century and to a lesser extent in the 3rd century. O04 appears to be about equivalent to the Wilderspool fabrics during this period. The discrepancy between the NoSh figures and calculations based on RE can perhaps be explained by the large number of wide-mouthed jars present in the Severn Valley ware tradition and not found in the Wilderspool fabrics.

Figure 4.16 shows a proportional stacked bar chart, divided into 50-year periods, of the functional breakdown of all class O wares. Constricted-necked jars gradually rise in importance from the mid 2nd century. Jars have a slight peak in the early 3rd century but are always less common than tablewares, although they are close to parity in the early 3rd. Wide-mouthed jars suffer a dramatic reduction in presence from the 2nd century to

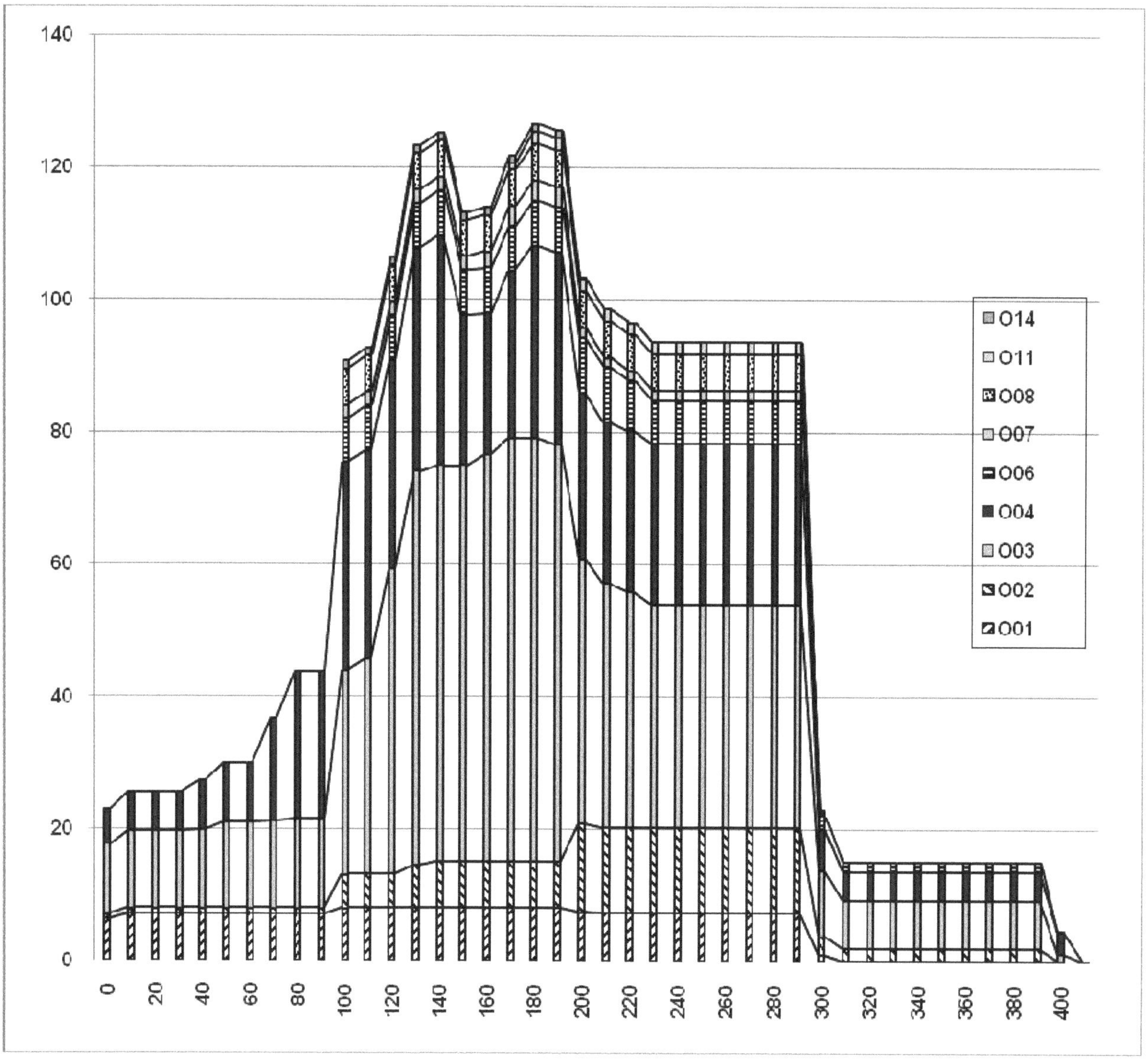

Figure 4.15: Date distribution of class O fabrics by RE.

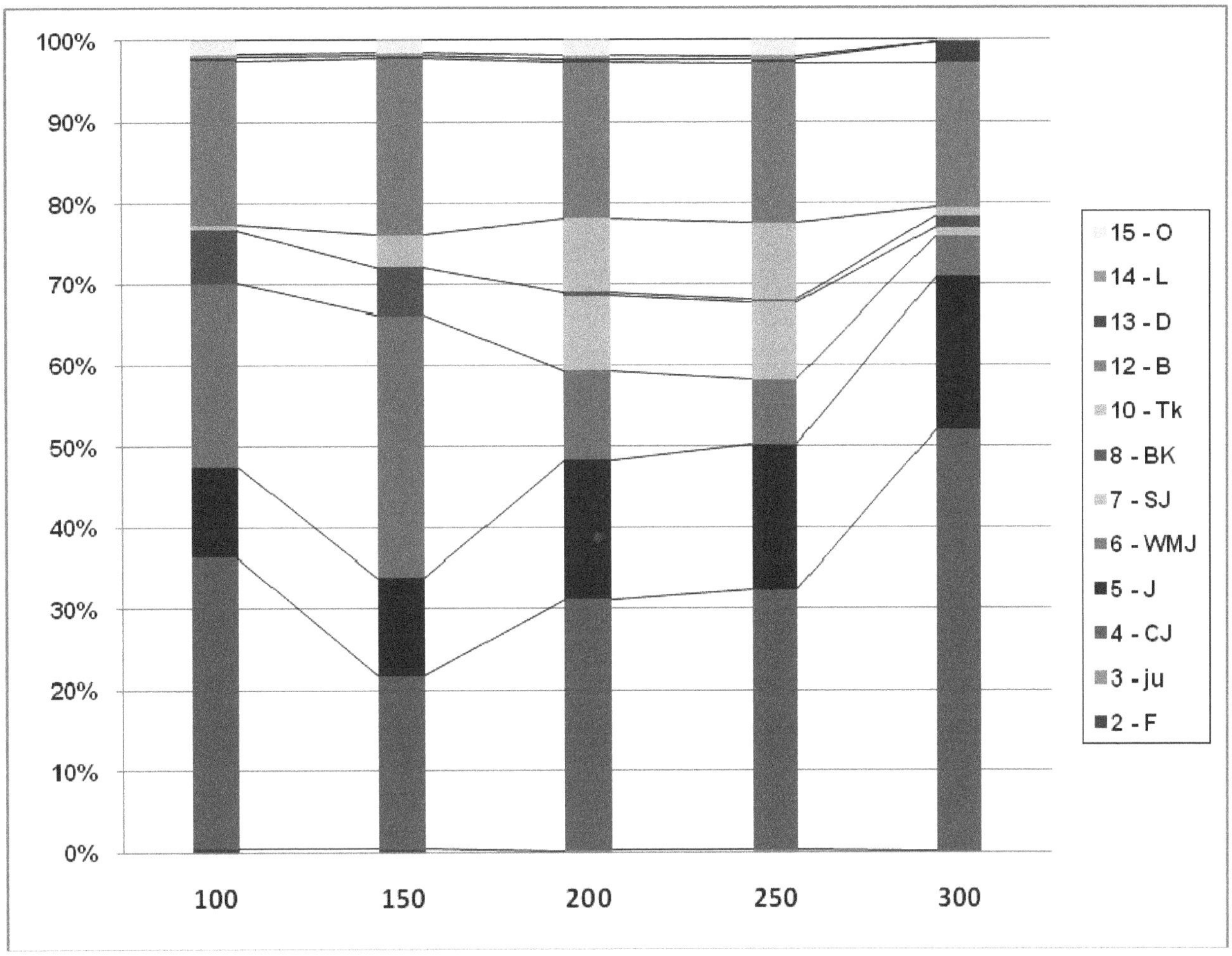

Figure 4.16: Proportional stacked bar chart showing functional breakdown of class O pottery, by RE, over 50-year intervals.

the 3rd. Storage jars seem confined somewhat to the early 3rd. Beakers and tankards begin to appear in the mid 2nd century, presumably substituting for black-burnished ware beakers which are dwindling in numbers by this stage.

The functional breakdown for fabric O02 shows a collection which is heavily dominated by bowls, with only 31% of jars by MnR (28% by RE) and a small quantity of storage jars and beakers (Table 4.7). The number of vessels in O01 is too small to give a reliable function figure but the forms recognised were a constricted-necked jar, a jar, two bowls and a lid, also suggesting an importance of tablewares at the expense of jars.

	J	SJ	Bk	Tk	B	No
MnR	30.80%	7.70%	7.70%	7.70%	46.10%	13
RE	27.90%	37.60%	1.30%	4.60%	28.70%	237

Table 4.7: Function breakdown of O02 vessels.

O03	F	Ju	CJ	J	WMJ	Bk	B	O	No
MnR	1.90%	1.90%	15.10%	30.20%	18.90%	5.60%	24.50%	1.90%	53
RE	0.50%	0.80%	40.70%	14.30%	20.40%	3.50%	16.70%	3.20%	1069

Table 4.8: Functional breakdown of O03 vessels.

O03		Ju	CJ	J	WMJ	Bk	B	No
Brine	MnR	2.50%	17.50%	30.00%	17.50%	10.00%	22.50%	40
Industrial	RE	0.90%	44.60%	11.30%	20.30%	7.70%	15.30%	941
Utilitarian	MnR	-	10.00%	20.00%	30.00%	-	40.00%	10
Zone	RE	-	16.70%	15.60%	30.00%	-	37.80%	90

Table 4.9: Functional breakdown of O03 vessels by zone.

O04	Ju	CJ	J	WMJ	Bk	Tk	B	D	No
MnR	2.90%	25.70%	11.40%	14.30%	8.60%	8.60%	20.00%	8.60%	35
RE	0.30%	53.70%	7.70%	10.10%	4.50%	13.40%	7.80%	2.60%	626

Table 4.10: Functional breakdown of O04 vessels.

Bowls are much less important in fabric O03 (Table 4.8), although they are only slightly less commonly represented than other jars. There are, however, an important group of water containers (F, Ju and CJ), at some 19% by MnR (42% by RE), and a large presence of wide-mouthed jars. Table 4.9 shows a comparison of the functional presence of vessels in O03 between the Utilitarian Zone and the Brine Industrial Zone. Water containers and jars are more strongly represented in the Brine Industrial Zone, and beakers are only present there. Bowls and wide-mouthed jars are much more important within the vessels deposited in the Utilitarian Zone.

Water containers (constricted-necked jars and jugs) are the most important group in fabric O4 (Table 4.10), at 29% by MnR and 54% by RE, the discrepancy between the two figures being a result of the 'chunkiness' shown by RE measures for small rims (Orton 1989). Beakers and tankards make up 18% by MnR (18% by RE). Wide-mouthed jars are also relatively strongly present, at 14% by MnR (10% by RE). Beakers and tankards make up 18% by MnR (18% by RE), a high figure but one appropriate to assemblages in the Severn Valley. Jars are very weakly represented in this fabric at 11% by MnR (8% by RE), again a feature typical of Severn Valley ware assemblages in the Severn Valley, with tablewares more common at 29% by MnR (11% by RE).

Wilderspool Wares O01 and O02 (18%)

O01 (7%)

O01.1	A constricted-necked jar. Severn Valley ware copy? Cf Hartley & Webster 1973, no 9; Webster 1976, no 3. (1285) [1182].	C2?
O01.2	An everted jar rim with a slightly thickening end. Hartley & Webster 1973, no 10; Webster 1976, no 1. (1183) [1182].	Mid C1-C4
*O01.3	A bowl with a bead rim. Hartley & Webster 1973, no 56?; Webster 1976, no 74.	C2?
O01.4	A curving walled bowl with a grooved rim. Webster 1976, no 34. (4140) [4139].	C2-C4
O01.5	A lid with a squared simple rim. (1110) [1109].	C1-C3

O02 (11%)

O02.1	A jar with an everted triangular, slightly undercut rim. Cf Hartley & Webster 1973, no 10; Webster 1976, no 21. (1209) [1207].	Mid-late C2
O02.2	A necked jar with a strongly everted thickening rim. Hartley & Webster 1973, no 29; Webster 1976, no 16 or 18. (3331) [3330].	C2-C4
O02.3	A tankard with a slightly everted rim with internal groove. Cf Webster 1976, no 42. (1073) [1072].	C2-C3
O02.4	A jar with a triangular sectioned rim. (4140) [4138].	C2-(C3)
O02.5	A bowl with a wedge rim. Hartley & Webster 1973, no 55; Webster 1976, no 35. (1192) [1191].	C2-C4
O02.6	A storage jar with a strongly everted hooked rounded triangular rim. Hartley & Webster 1973, no 10; cf Webster 1976, no 8? (1285) [1182].	C3
*O02.7	A hemispherical bowl. Wilderspool copy of Severn Valley ware form. Webster 1976, no 61.	Mid C2-C4
O02.8	A hemispherical bowl with a thickening rim. Webster 1976, no 72. (1239) [1207].	C1-C2
O02.9	A segmental bowl with a bead and flange rim. (3331) [3330].	C2
O02.10	A bowl with a flange rim. Webster 1976, no 57. (3002).	C3
O02.11	A jar with a perforated base. (3002).	C1-C3
O2.12	A counter reworked from a jar base. (3370) [3359].	C1-C3

Severn Valley Wares O03, O06, O07, O08, O11 (13%)

O03 (12%)

O03.1	A flagon with a flange ring neck. Hartley & Webster 1973, no 12. (5089) [5088].	C2
O03.2	A constricted-necked jar with a triangular, slightly undercut, rim. Cf Webster 1976, no 3; Hartley & Webster 1973, no 9. (i) (1277) [1182]; (ii) (1285) [1182].	C1/2-C3
O03.3	A constricted-necked jar with a horizontal thickening slightly undercut rim. Webster 1976, no 1. (1239) [1207].	Mid C1-C4
*O03.4	A constricted-necked jar with an everted thickening rim Webster 1976, no 1. (1094) [1081].	Mid C1-C4
O03.5	A constricted-necked jar with a bifid rim. Cf Webster 1976, no 9. (1208) [1207].	C3-C4
O03.6	A jar with a hooked rim. Webster 1976, no 3? (3402) [3401].	C2-C3
O03.7	A jar with a horizontal undercut rim. Cf Webster 1976, no 3. u/s [1182].	C2-C3
O03.8	A handled shouldered jar with an everted thickening rim. Webster 1976, no 1. u/s [2039].	Mid C1-C4
*O03.9	A jar with an outcurving, everted frilled rim. Cf Webster 1976, no 13; Hartley & Webster 1973, no 16.	C3+
*O03.10	A jar with an everted rim. BB copy? Webster 1976, no 2. u/s [1182].	Hadrianic-Antonine
O03.11	A jar with a triangular wedge rim with slight lid seating. Webster 1976, no 3? (1239) [1207].	Mid C1-C2
O03.12	A wide-mouthed jar with a rounded undercut rim. Cf Webster 1976, no 21. (1239) [1207].	Mid-late C2
O03.13	A necked wide-mouthed jar with a thickening everted rim. Cf Webster 1976, nos 20-1. (2160) [2177].	C2-early C3
*O03.14	A wide-mouthed jar with a strongly everted, thickening rim. Cf Webster 1976, no 22.	C2-early C3
O03.15	A wide-mouthed jar with an everted triangular wedge rim. Cf Webster 1976, no 21? (1265) [1264].	C3
O03.16	A necked beaker with a cupped everted rim. Cf Hartley & Webster 1973, no 35. (1073) [1072].	C2
O03.17	A beaker with an everted almost horizontal hooked rim. Cf Hartley & Webster 1973, no 41. (1188) [1182].	C2
O03.18	A beaker with an everted wedge-shaped rim. Cf Hartley & Webster 1973, no 36? (1062) [1053].	C2?

O03.19	A Severn Valley ware bowl with a small everted rim. Cf Webster 1976, no 35. (1278) [1182].	C2-C4
O03.20	A Dr 37 copy bowl. Cf Webster 1976, no 34. (4002).	C2
O03.21	A bowl with a hooked rim. Cf Webster 1976, no 50. (1100) [1099].	Late C2-late C3
O03.22	A flange rim bowl. Cf Webster 1976, no 50? (1220) [1182].	Late C2-late C3
O03.23	A curving walled (hemispherical) bowl with a grooved rim. Cf Webster 1976, no 61, Dr 37 copy. (1285) [1182].	Mid-late C2-C4
*O03.24	A flanged rim bowl with an inturned lip. Cf Webster 1976, no 47.	C2-C3
O03.25	A curved flange rim bowl. Cf Webster 1976, no 48? u/s [1182].	C2-C3
*O03.26	A segmental bowl with a beaded and flanged rim.	Late C1-C2
O03.27	A bowl with a flange rim and grooves below. Cf Webster 1976, no 50? (1225) [1207].	Late C2-late C3
O03.28	A colander with a developed bead and flange rim. (4002).	Romano-British
O03.29	A triple vase jar rim and wall sherd with joining pipe to next vessel. (1220) (1188) [1182].	Romano-British

O06 (1%)

*O06.1	A jar with a thickening everted rim. Webster 1976, no 5.	C2-C3
O06.2	A necked jar with a strongly everted frilled rim. Webster 1976, no 13. (1229) [1207].	C3-C4
O06.3	A wide-mouthed jar with an everted triangularly-sectioned, slightly undercut, rim. Cf Webster 1976, no 22. (1279) [1182].	C2-early C3
*O06.4	A wide-mouthed jar with an everted slightly thickening rim. Cf Webster 1976, no 22.	C2-early C3
O06.5	A bowl with a grooved rim. Cf Webster 1976, no 74. (1183) [1182].	Mid C2-C4
O06.6	A hemispherical bowl with a flange rim. Cf Webster 1976, no 47. (1225) [1207].	C2-C3

O07 (1%)

*O07.1	A jar with a hooked rim. Cf Webster 1976, no 5.	C2-C3
O07.2	A bowl with a flange rim and lid seating. Cf Webster 1976, no 50? Hartley & Webster 1973, no 63. (1158) [1157].	Late C2-late C3
O07.3	A segmental? bowl with a flange beaded rim. Cf Webster 1976, no 65. (1214) [1227].	C2

O08 (1%)

O08.1	A wide-mouthed jar with an everted, hooked rim. Cf Webster 1976, no 25. (1188) [1182].	C2-C3
*O08.2	A hemispherical bowl with a flange rim. Cf Webster 1976, no 74.	C2-C3
*O08.3	A flanged bowl with an internal bead rim. Cf Webster 1976, no 52.	C2-C3

Malvernian Severn Valley Ware O11 (0.2%)

O11.1	Malvernian Severn Valley ware tankard. Cf Webster 1976, no 4. (1285) [1182].	C2-C3

O04 Possible Severn Valley Type (7%)

This fabric is perhaps closer to the Wilderspool types than the classic Severn Valley ware definition. However, the range of vessels does cover the classic Severn Valley types.

O04.1	A constricted-necked jar with an everted thickening rim. Cf Webster 1976, no 3. (1183) [1182].	C2-C4
O04.2	A Severn Valley ware constricted-necked jar. Cf Webster 1976, no 3. (1238) [1207].	C1/2-C3
O04.3	A constricted-necked jar with a bead rim. Cf Webster 1976, no 1. (1097) [1096].	MC1-C4
O04.4	A constricted-necked jar with a strongly everted with triangularly-sectioned, slightly undercut, rim. Cf Webster 1976, no 3. (1225) [1207].	C1/2-C3
O04.5	A constricted-necked jar with an everted rim. Cf Webster 1976, no 2. (1142) [1096].	Late C1-mid C2
O04.6	A constricted-necked jar with an everted, triangularly-sectioned, undercut, wedge-shaped rim with lid seating. Cf Webster 1976, no 5. (1039) [1038].	C2-C3
*O04.7	A constricted-necked jar with a frilled triangularly-sectioned wedge-shaped rim. Cf Webster 1976, no 13; Hartley & Webster 1973, no 18.	C3+?
O04.8	A jar with an outcurving everted rim. (1239) [1207].	Hadrianic-Antonine?
O04.9	A Severn Valley ware jug with a beaded rim and burnished on neck. (1240) [1207].	C3+
*O04.10	A necked jar with a hooked rim. Cf Webster 1976, no 5.	C3
O04.11	A jar with a slightly everted rim. Cf Webster 1976, no 14. (1285) [1182].	Late C2-C3
O04.12	A wide-mouthed jar with a wedge-shaped rim. Cf Webster 1976, no 21. (1236) [1207].	Mid-late C2
O04.13	A wide-mouthed jar with a strongly everted hooked rim. Cf Webster 1976, no 25. (5096) [5098].	C3
*O04.14	A wide-mouthed jar with an everted thickening rim. Cf Webster 1976, no 22.	C2-C3?
O04.15	A wide-mouthed jar with a rounded triangularly-sectioned rim and a narrowing neck. (1208) [1207].	C2-C3
*O04.16	A wide-mouthed jar with an everted outcurving, thickening rim. Cf Webster 1976, no 16.	C3-C4
O04.17	A beaker with a wedge rim. Cf Webster 1976, no 15; Hartley & Webster 1973, no 36. (1239) [1207].	C2-C4
O04.18	A beaker with a stubby everted rim. Cf Hartley & Webster 1973, no 35. (3058) [3057].	C1-C2
O04.19	A beaker with a horizontal undercut rim. (3021) [3144].	C2
O04.20	A tankard with a simple rim. (1183) [1182].	C2-C3
O04.21	A Severn Valley ware tankard with an everted rim. Webster 1976, no 43. (i) (1285) [1182]; (ii) (1291) [1182].	Late C2-C3
O04.22	A bowl with an everted rim. (1239) [1207].	Romano-British
O04.23	A bowl with a triangularly-sectioned rim with slight groove below and	C2

	chamfered base. Cf Hartley & Webster 1973, no 56. (3139) [3138].	
O04.24	A bowl with a rounded triangular section wedge rim with wide groove below. Hartley & Webster 1973, no 24. (3219) [3218].	C2
O04.25	A globular jar with an everted thinning rim. (1225) [1207].	Romano-British
*O04.26	A curving walled bowl with a thinned outcurving everted rim. Webster 1976, no 66.	C2-C3
O04.27	A Dr 29 bowl copy. Cf Hartley & Webster 1973, nos 53-4. (3230) [3206].	C2
*O04.28	A straight-sided dish with a simple rim.	Romano-British
O04.29	A dish with a simple rim. Webster 1976, no 74. (4002).	Romano-British
O04.30	A cheese press. (i) u/s [1182]; (ii) (1183) [1182].	Romano-British

Micaceous Oxidised Ware O14 (0.04%)

*O14.1	A beaker with an everted thickening rim. Cf Hartley & Webster 1973, no 26.	C2

Class Q White-slipped Wares (Fig 4.29)

There are small quantities of white-slipped wares. The most abundant supply is from Wilderspool, at 1%, and is well represented by the typical flagons of this ware, alongside a beaker and a flagon base which has been reused as a dish. The other main white-slipped fabric, Q03, is also probably related to the Wilderspool industries and is represented by a flagon.

Wilderspool White-slipped Wares Q01 (1%)

Q01.1	A flagon with a pulley wheel rim. Cf Hartley & Webster 1973, no 1. (1239) [1207].	C2
Q01.2	A flagon with a cupped mouth. Hartley & Webster 1973, no 2. u/s.	Late C1-early C2?
Q01.3	A flagon with a bead rim. Cf Hartley & Webster 1973, no 5. (1281) [1182].	C2
Q01.4	A beaker with a slightly everted rim. Cf Hartley & Webster 1973, no 36. (3217) [3216].	C2
Q01.5	A costrel end, reworked into a dish with a simple rim. (3445) [3440].	C1-C2
Q01.6	A triple vase. Cf Hartley & Webster 1973, no 69; Gillam 1970, type 344. (1277) [1182].	AD 110-50

Possible Wilderspool White-slipped Fabric Q03 (1%)

Q03.1	A ring necked flagon. (1351) [1347].	C2

Class R Reduced wares (Fig 4.30)

There are very few greywares present. The largest group is made up of sandy greyware (R02) whose products include a possible reduced Severn Valley ware constricted-necked jar and a wide-mouthed jar form, and black-burnished ware copies of a late 2nd- to mid 3rd-century jar and a later 3rd- to 4th-century developed beaded and flanged bowl. Also present in this fabric are an ungentarium and a handmade crucible fragment. Another sandy greyware present is R01 but only a crucible fragment was identified in this fabric.

One of the next largest groups, at 0.6%, are micaceous fabrics F31 and F36, a group which comprises a number of everted jar rims. Also making up 0.6% of this class are polished greywares R21, R41, R42 and R43. These were considered not really of good enough quality to include with the class F wares. Forms include a Dr 30 copy bowl (R21.1), a ring necked flagon (F43.1) and a narrow-mouthed beaker (R42.1).

R01 (0.4%)

R01.1	A crucible/briquetage. (1188) [1182].	Romano-British

R02 (2%)

R02.1	A constricted-necked jar with an everted bifid rim with notches on the lower lip. A Severn Valley ware form. Cf Webster 1976, no 15; Hartley & Webster, 1973, no 14. u/s [1182].	C3-C4
R02.2	A jar within an everted outcurving rim. BB copy? Cf Gillam 1970, type 143. Cf Webster 1976, no 21. (1188) [1182].	AD 180-280
R02.3	A necked wide-mouthed jar with an everted thickening outcurving rim. Severn Valley ware copy? Cf Webster 1976, no 22. (6018) [6017].	C2-C3
R02.4	A developed beaded and flanged bowl, a BB copy. (1069) [1068].	AD 270+
R02.5	An ungentarium with an everted lid seated cupped rim. (1239) [1207].	Romano-British
R02.6	A crucible. (1220) [1182].	Romano-British

Micaceous Wares (0.6%)

R31 (0.5%)

R31.1	A jar with a slightly everted rim. (4062) [4059].	Romano-British
R31.2	A jar with a horizontal slightly undercut rim. (1006) [1105].	Romano-British

R36 (0.1%)

R36.1	A jar with an everted outcurving rim. (2019) [2018].	Romano-British

Polished Greywares (0.6%)

R21 (0.1%)

R21.1	A fine surfaced carinated bowl, a Dr 30 copy. (1239) (1236) [1207].	C2

R41 (0.3%)

R41.1	A necked jar with an everted outcurving slightly thickening rim. (1073) [1072].	Romano-British

R42 (0.1%)

R42.1	A narrow-mouthed beaker. (1057) [1055].	C1-early C2?

R43 (0.1%)

R43.1	A disc-necked flagon. (2115) [2114].	Romano-British

Class S Samian Ware

The samian ware is discussed by Felicity Wild (see Chapter 5).

The proportion of decorated samian ware is moderate at 24% compared to the remarkable 52% at Walton-le-Dale (Evans & Rátkai forthcoming a) but is more closely comparable with the 30.3% from Mitchell's Brewery, Lancaster (Evans & Rátkai forthcoming b), and 30.1% of CG material from Middlewich by minimum numbers of vessels (Evans 2002). These measures tend to suggest a 'military' type of supply to Nantwich, as to the other sites.

Class W Whitewares (Fig 4.30)

There is a small quantity of whitewares present, as is typical for the region. The most abundant fabric is W01, represented by a flagon, a beaker and a Dr 36 copy bowl. Also present is a 3rd-century grooved rim flagon in W02 and an Oxfordshire parchment ware bowl, fabric W03, of type P24, dated AD 240-400+.

W01 (1%)

W01.1	A flagon rim with lid seating. (3188) [3187].	Mid-late C2
W01.2	A beaker with a bead rim. (1006) [1105].	Romano-British
W01.3	A bowl with a grooved rim. Dr36 copy? (1069) [1068].	Late C2?

W02 (0.3%)

W02.1	A flagon with a grooved rim and flange below. (1225) [1207].	C3+?

W03 (0.2%) Oxfordshire Parchment Ware

W03.1	A reed rim bowl. Young 1977, type P24. (2068) [2067].	AD 240-400+

Function

The functional breakdown for the site (Table 4.11) indicates a very Romanised assemblage. 'Other' jars are low and even combined with wide-mouthed jars give a value of just 27% by MnR (26% by RE), compared to tablewares at 46% by MnR (30% by RE). Beakers and cups are strongly represented at 13% by MnR (12% by RE), with tankards making up a further MnR of 1% (1% by RE). Liquid containers are present with a MnR of 2% (7% by RE) for jugs and flagons, and a further 4% (12% by RE) for constricted-necked jars. Lids are quite low at 1% by MnR (2% by RE). Mortaria are reasonably high at 4% by MnR (3% by RE). Amphorae are relatively very strongly represented at 2% by MnR (4% by RE).

There can be no doubt that this site falls within the more urbanised end of sites of contemporary date in terms of its jar levels and tablewares (cf Evans 2001). It is also well supplied with drinking vessels and especially liquid containers. Leary provides comparative data from other sites in the North-West (Table 4.13). Nantwich groups well with Middlewich and Manchester, although the Warrington Gifford site would appear to be a marginal one with barely urban characteristics.

Further comparative data can be added from Plas Coch, Wrexham, and Walton-le-Dale (Tables 4.14 & 4.15). The overall level of jars at Plas Coch, 42.1%, is at the higher end of the scale for the other sites here but is still fairly low, and the level of dishes and bowls is quite high at 34.6%. Drinking vessels are modestly represented; there is a surprising lack of tankards, owing to their poor representation amongst the Severn Valley wares. This is something that Plas Coch shares with Nantwich, where again the large collection of Severn Valley ware types might suggest that tankards ought to be much commoner, as they are in the Severn Valley itself. Flagons are reasonably represented and constricted-necked jars well so, resulting in an unusually high level of possible liquid storage and serving vessels. These levels are much higher than those found in northern England (Evans 1993) and higher than those on most sites in the Midlands (Evans 1996). Evans (2001, fig 6) compares the jar and tableware ratio from Plas Coch with those from a series of sites in the West Midlands and the South-West; Plas Coch phases 3 and 4 fall within the urban range, well separated from villa and rural sites. However, Phase 5 (by which time the military-style coin supply has ceased) falls between the rural and urban sites and suggests a change in the settlement to a more rural pattern.

	A	F	Ju	CJ	J	WMJ	Bk	Cu	Tk	M	B	D	L	O	No
MnR	1.7%	1.9%	0.4%	3.8%	23.7%	3.6%	8.0%	4.6%	1.0%	4.0%	30.0%	14.9%	0.6%	1.5%	523
RE	4.0%	7.3%	0.1%	10.9%	22.2%	4.7%	7.0%	4.7%	1.3%	2.9%	21.7%	8.2%	1.8%	2.0%	8876

Table 4.11: Functional breakdown.

		A	F	Ju	CJ	J	WMJ	SJ	Bk	Cu	Tk	M	B	D	L	O	No
Road	MnR	-	6%	-	6%	19%	13%	-	-	-	-	13%	31%	13%	-	-	16
	RE	-	4%	-	6%	12%	18%	-	-	-	-	18%	34%	9%	-	-	124
Cremation	MnR	4%	-	-	-	54%	-	-	8%	4%	-	4%	8%	19%	-	-	26
	RE	7%	-	-	-	67%	-	-	6%	2%	-	3%	4%	12%	-	-	351
Utilitarian	MnR	1%	3%	-	1%	19%	3%	-	10%	1%	-	5%	30%	27%	-	-	100
	RE	4%	12%	-	2%	19%	3%	-	12%	0%	-	7%	23%	19%	-	-	1000
Brine Industrial	MnR	2%	2%	1%	5%	23%	3%	0%	8%	6%	1%	3%	32%	12%	1%	1%	378
	RE	4%	7%	0%	13%	21%	5%	1%	7%	6%	1%	2%	22%	7%	2%	2%	7340

Table 4.12: Functional breakdown by zone.

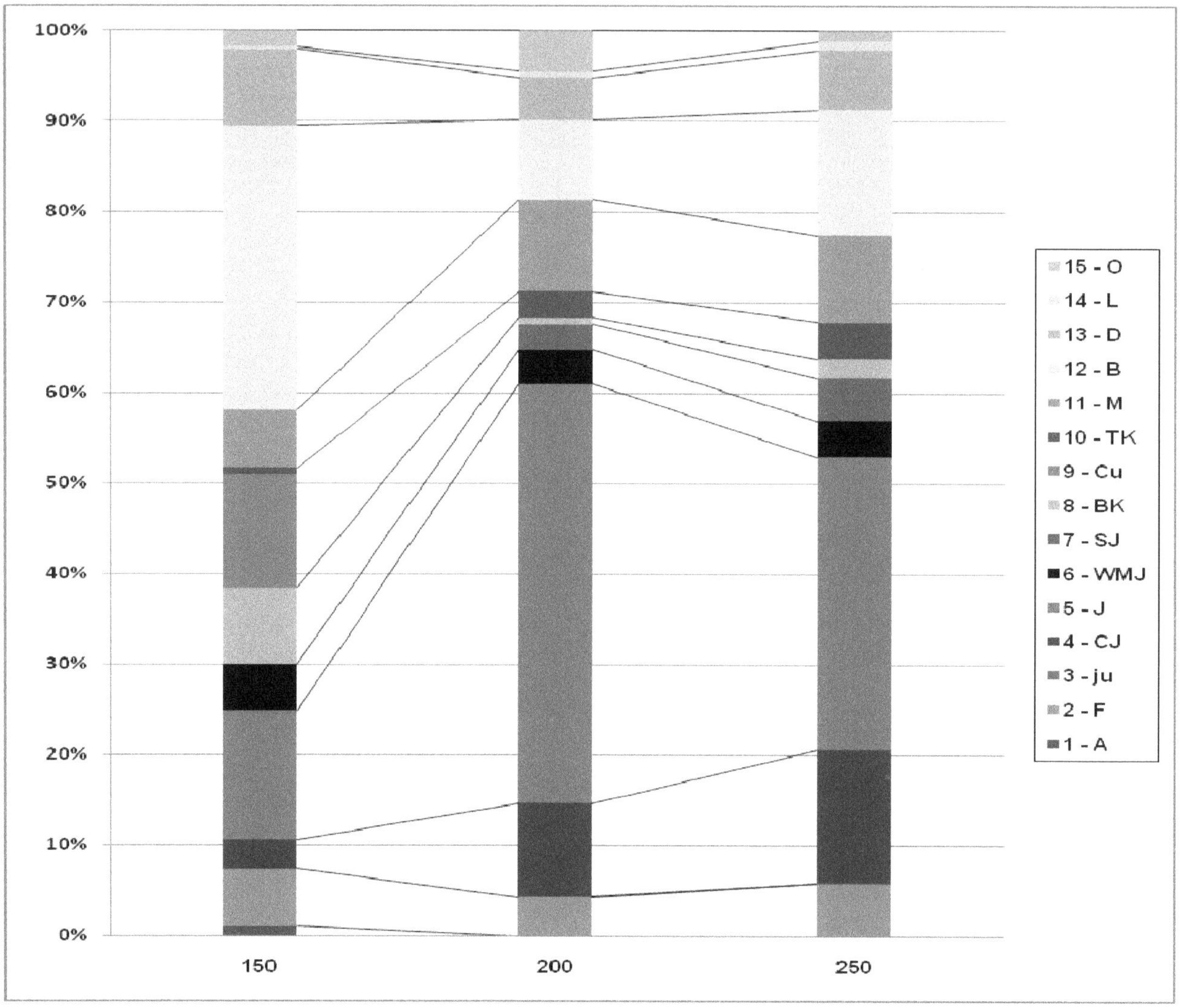

Figure 4.17: Function by approximate 50-year periods for the whole site.

Jar levels from Walton-le-Dale are even lower than those from most of the other sites and tableware levels, at generally over 40%, a little higher. Liquid containers are well represented again, a particular feature of these sites, and drinking vessels are also well represented at urban levels. The strong representation of mortaria at Walton-le-Dale is a site-specific feature, representing their production here, a site to which many of the Wilderspool potters migrated.

Overall, Nantwich groups well with the other Lancashire and Cheshire Plain sites, from Walton-le-Dale to Middlewich, in having an assemblage which would broadly fit within the range found on early northern military sites or on quite well-urbanised sites.

A functional analysis of the Nantwich site by the excavators' zones reveals some differences between the areas (Table 4.12). The most striking is the very high proportion of jars from the Cremation Zone, although whether this is a result of disturbance of burial vessels or from visitors coming to prepare a funerary memorial meal or from other activity here is not clear. The first suggestion can probably be ruled out because material from this area is not particularly complete and disturbance of burial urns might have been expected to yield larger than average sherds. Tablewares are also significantly under-represented in this zone, but not drinking vessels, although liquid containers are absent. In the Brine Industrial Zone the only oddity is the presence of the 'cheese presses', a rare type, possibly associated with making hard cheese from ewe's milk. The Road Zone seems surprisingly well supplied with mortaria and the Utilitarian Zone seems to be very poorly supplied with liquid containers.

The change in the functional composition over 50-year periods is shown for the whole site in Figure 4.17. It indicates a sharp increase in jars at the expense of tablewares, especially bowls. Drinking vessels also sharply decline. This can be compared to the change in the functional make-up of the brine tanks over the same period (Fig 4.18) which shows a much sharper increase in all jar types and a reduction of drinking vessel types to exclusively tankards. There is a slight increase in the types of 'other' vessels deposited, although this is small in all the date groups.

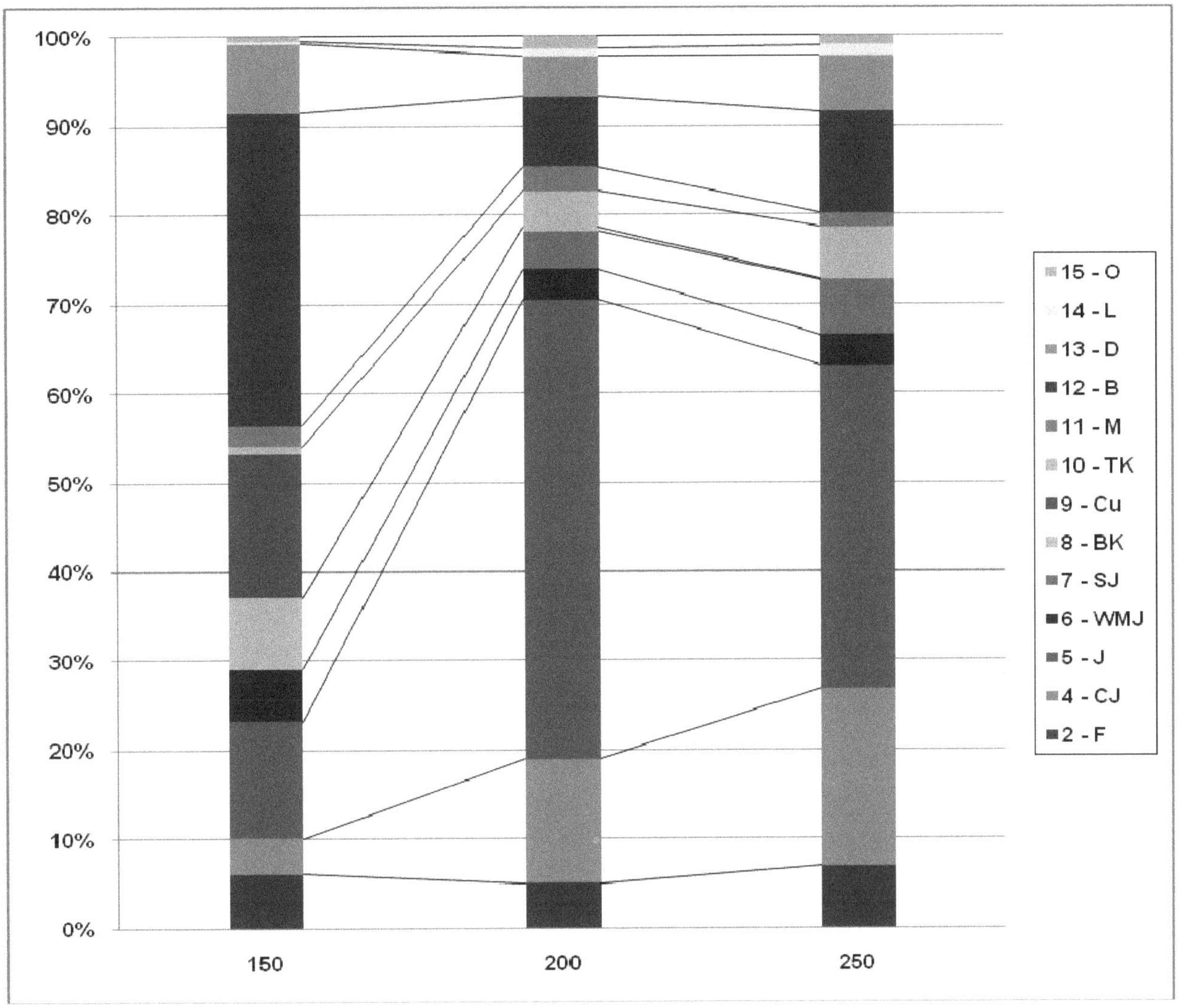

Figure 4.18: Function by approximate 50-year periods from brine tanks [1182] and [1207].

	A	F/Ju	CJ	J	WMJ	Bk/Cu	M	B/D	L	O
Middlewich King Street	1.0%	6.5%	3.7%	27.4%	2.7%	10.9%	4.2%	37.7%	5.3%	0.3%
Warrington Gifford	1.3%	9.3%	3.1%	46.4%	1.7%	8.7%	9.3%	18.3%	1.8%	-
Manchester Barton Street	1.4%	9.3%	10.2%	21.7%	1.8%	15.7%	5.4%	29.6%	3.8%	1.2%
Manchester Castlefield	1.5%	11.3%	-	21.1%	1.5%	9.8%	7.5%	39.9%	1.5%	3.0%
Tarbock	2.0%	4.1%	-	36.7%	-	14.3%	12.2%	30.6%	-	-
Mellor	-	-	4.1%	39.9%	3.37%	1.9%	9.4%	41.4%	-	-
Tarporley	-	1.7%	-	33.9%	-	3.4%	5.1%	40.6%	1.7%	13.6%

Table 4.13: Relative quantities of vessel types at other sites in the region, after Leary (2008, 99 table 11).

	F	CJ	J	WMJ	SJ	Bk/Cu	Tk	M	B	D	D/B	O	No
All site	4.0%	9.5%	37.5%	4.3%	0.3%	5.6%	0.7%	3.6%	18.1%	15.8%	0.7%	0.7%	304
Phase 3	12.5%	-	37.5%	-	-	-	-	6.3%	25.0%	18.8%	-	-	16
Phase 4	4.6%	8.4%	28.2%	3.1%	0.8%	6.9%	1.5%	3.1%	20.6%	22.1%	-	0.8%	131
Phase 5	2.2%	10.8%	46.8%	5.8%	-	4.3%	-	3.6%	15.8%	10.1%	-	0.7%	139

Table 4.14: Functional analysis of the Plas Coch assemblage by MnR (Evans forthcoming b).

Phase	A	F/Ju	CJ	J	WMJ	SJ	Bk	Tk	M	B	D	D/B	L	O	Indet	No
2	-	4.7%	1.6%	32.8%	3.1%	1.6%	-	-	9.4%	21.9%	23.4%	-	1.6%	-	-	64
3A	-	1.7%	3.5%	17.2%	-	-	10.3%	-	10.3%	27.6%	22.4%	-	-	6.9%	-	58
3B	2.9%	3.7%	3.7%	23.5%	1.5%	-	12.5%	-	8.8%	22.1%	15.4%	1.5%	0.7%	1.4%	2.2%	136
2-3	5.4%	8.1%	8.1%	5.4%	2.7%	-	2.7%	-	27%	16.2%	24.3%	-	-	-	-	37
3	1%	4.8%	1.9%	16.9%	1.3%	-	11.5%	-	13.1%	24%	20.1%	0.3%	1%	1.2%	2.9%	313
3-4	-	1.3%	3.3%	19.1%	2%	-	14.5%	-	11.8%	25%	22.4%	-	-	0.7%	-	152
4	2.2%	2.8%	3.7%	18.9%	3.3%	0.2%	13.9%	0.9%	10.9%	22.6%	15.2%	0.2%	2%	1.4%	1.7%	460
4-5	0.5%	1.1%	10.3%	15.1%	1.6%	-	14.1%	1.1%	11.4%	26%	17.3%	-	0.5%	-	-	182
5	0.9%	1%	16.9%	15.5%	2%	0.1%	12.5%	1%	9.7%	18.3%	17.6%	0.4%	-	0.9%	1.7%	704

Table 4.15: Functional analysis of the Walton-le-Dale assemblage by MnR (Evans & Rátkai forthcoming a).

Sooting and Other Deposits

The proportion of sooting and other deposits, recorded by ware, is shown in Table 4.16. Not surprisingly the fabric for which cooking vessels are dominant, black-burnished ware, shows the highest level of burning and other deposits, with over 50% of all sherds displaying some form of deposit. Somewhat surprisingly, the samian ware has the next highest level of burning, with just over 20% of the body sherds counted as burnt. Amphora and shelly wares show relatively high levels of burning but this pattern may be slightly distorted by the small number of sherds present in each ware. Sooting in mortaria is common and comparable with Antonine levels of mortarium burning at Piercebridge (Evans & Mills 2009, fig 9.21). Reduced wares, white slips and whitewares all have rather higher levels of sooting than may normally be expected.

Table 4.17 presents the data for sooting by vessel function. Dishes show a higher level of sooting than jars, a result which does not seem to have much changed over time. Similarly there is a high number of bowls with sooting. Of note are the drinking vessels (Bk and Cu) which exhibit sooting, as well as a reasonably high proportion of constricted-necked jars. This latter phenomenon has also been observed at Hayton where the sooting on such vessels is 23% (Mills forthcoming). Figure 4.19 presents the date distribution of sooting on mortaria by RE as a stacked bar chart and shows a change in the nature of sooting deposits after the late 2nd century lacuna, and an increase in the number of sooted rims (about a doubling) in the early 3rd. Only two types of mortarium are sooted, M01 in the late 2nd century and M13 in the mid 3rd. Figure 4.20 shows the variation in sooting on the whole pottery assemblage by zone. The largest group of sooted sherds comes from the Utilitarian Zone, presumably reflecting a focus of cooking activities here. The lowest levels are from the Road Zone and probably reflect the more itinerant nature of later pottery deposition in this zone.

Ware	Sooting type	No%	Wt%	MnR%	RE%
A	Burning - general	0.7%	1.5%	-	-
	Sooting - external	17.0%	3.5%	10.0%	9.9%
B	Burning - general	2.1%	1.4%	2.4%	0.9%
	Burning - post depositional	0.1%	-	0.6%	0.2%
	Sooting - external	44.0%	53.7%	51.5%	62.6%
	Burnt food residue - internal	0.4%	0.6%	0.6%	0.8%
	Lime scale and sooting	4.3%	5.0%	1.2%	4.4%
	Lime scale	0.8%	0.5%	0.6%	0.9%
	Sooting - internal and external	0.3%	0.5%	0.6%	0.2%
C	Sooting - external	35.7%	16.8%	50.0%	29.6%
F	Burning - general	0.9%	2.8%	2.6%	2.5%
	Sooting - external	1.3%	0.8%	2.6%	1.3%
M	Burning - general	2.6%	6.8%	-	-
	Sooting - external	3.8%	3.7%	4.8%	6.7%
	Composite	5.1%	4.8%	-	-
	Sooting - internal and external	2.6%	8.2%	4.8%	9.5%
O	Burning - general	0.8%	0.8%	1.6%	1.4%
	Sooting - external	3.0%	8.5%	6.5%	11.5%
	Lime scale	0.2%	0.6%	-	-
Q	Burning - general	12.0%	12.4%	-	-
R	Burning - general	6.4%	4.4%	-	-
	Sooting - external	7.2%	7.2%	7.1%	9.2%
S	Burning - general	20.9%	11.1%	16.5%	9.8%
W	Burning - general	2.3%	5.6%	-	-
	Sooting - external	9.3%	11.8%	20.0%	7.3%

Table 4.16: Deposit type by ware.

Form	Sooting type	MnR%	RE%
A	Sooting - external	11.1%	11.4%
CJ	Sooting - external	15.0%	23.2%
J	Burning - general	1.6%	0.7%
	Sooting - external	33.1%	44.5%
	Lime scale and sooting	1.6%	5.5%
	Sooting - internal and external	0.8%	0.3%
Bk	Burning - general	2.4%	0.8%
	Sooting - external	11.9%	7.3%
	Burnt food residue - internal	2.4%	3.1%
Cu	Burning - general	12.5%	5.0%
M	Sooting - external	4.8%	6.7%
	Sooting - internal and external	4.8%	9.5%
B	Burning - general	7.6%	6.7%
	Burning - post-depositional	0.6%	0.3%
	Sooting - external	14.0%	17.5%
D	Burning - general	14.1%	7.7%
	Sooting - external	34.6%	56.4%
	Lime scale	1.3%	3.1%

Table 4.17: Deposit type by vessel function.

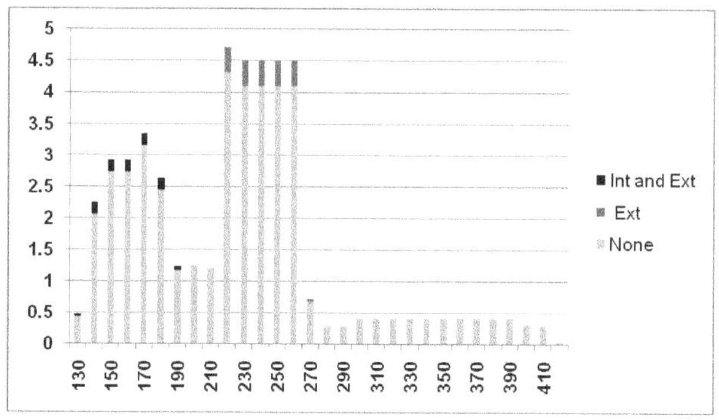

Figure 4.19: Sooting date distribution on mortaria by RE.

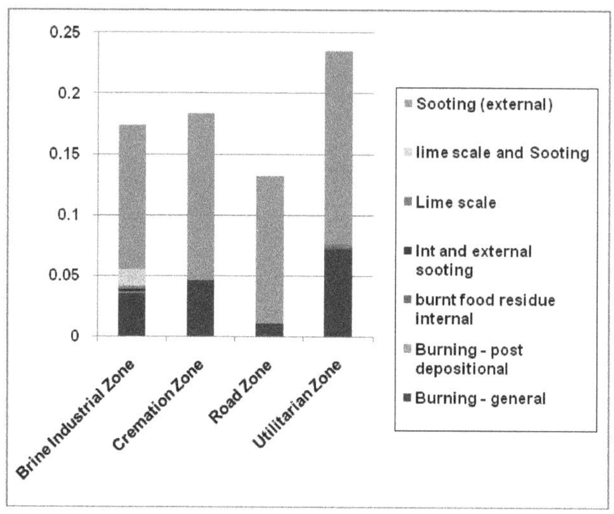

Figure 4.20: Sooting of all pottery by zone.

Feature/Context		Comments	Period	Ware	Rim type
Subsoil	(3002)	With rivet hole	Hadrianic-Antonine	S21	37
Subsoil	(3002)	Rim with rivet hole	Antonine	S21	Bowl
(1225)	[1207]	Rim fragment with rivet holes	Antonine	S21	31
(1206)	[1182]	5 rivet holes	Second half C2	S21	31R
(1240)	[1207]	Burnt on section which shows 6 drill holes for rivets?	Antonine	S21	38 or 44
(1281)	[1182]	1 rivet hole	Antonine	S21	18/31R or 31?
Topsoil	(1002)	With rivet hole	Antonine	S21	37

Table 4.18: Repairs.

Graffiti

There is one sherd with a graffito. This is a decorated samian Dr 33, D9, dated *c* AD 160-85, from context (1225) of brine tank [1207]. It is illiterate, a six-ray star, and may have been a mark of ownership.

Repairs

Repairs were only noted on samian vessels. Some 1.1% of the entire samian assemblage showed evidence of repairs (Table 4.18). They are all CG, mainly of Antonine date, although one example is given a suggested date range of Hadrianic-Antonine.

Complete Vessels and Profiles

There are a large number of vessels described as complete, almost complete or with complete profiles (Table 4.19). Three unstratified vessels with complete profiles were recovered from the subsoil and comprise bowl B01.23, dish O043.29 and cheese press O04.30. These functional types are the most likely to be recovered with complete profiles from normal deposits. Similarly the two dishes with complete profiles, B01.27A and B01.27D, recovered from the Utilitarian Zone would be expected from normal depositional activities. There are four vessels with complete profiles from the Cremation Zone, including dish B01.27A which could be a product of normal depositional activities, but also almost complete jar B01.2, which was used as a cremation urn, and a CG Dr 31R bowl. There is also dish Q01.5, whose presence in the Cremation Zone is of note.

The majority of the complete vessels and profiles were recovered from the Brine Industrial Zone, and the majority of them from the brine tanks. Table 4.20 shows the functional distribution of the vessels retrieved from brine tank [1182] and tank [1207] and from both tanks together. The brine tanks have relatively many more drinking vessels (BK, Cu and TK) than the site overall, as well as more constricted-necked jars, tablewares and lids. There are no amphorae, flagons, jugs or mortaria in the brine tank group. The absence of flagons and jugs is interesting but does not detract from the clear dominance of vessels associated with drinking, echoing the ritual drinking discussed by Leary (2008) at Barton Street, Manchester. There is also a difference in functional composition between the two tanks, with many vessels associated with drinking, constricted-necked jars, beakers, tankards and lids, only appearing in tank [1182], and jars, wide-mouthed jars and dishes only appearing in tank [1207], implying perhaps a different or parallel ritual centred on food. Figure 4.21 shows the date distribution of the complete vessels from the tanks. This contrasts with the overall date distribution of the zone and gives rather more prominence to the complete vessels of the early to mid 3rd-century depositions. It may possibly reflect small special depositions in the later period rather than emphasising the normal survivals of smaller vessels such as beakers, as well as dishes and bowls.

Cross-joins

Neither the samian ware nor the coarse pottery was presented marked and it was not possible to examine contexts for cross-joins as much as the assemblages warrant. Most of the cross-joins are between mid Antonine forms from the brine tanks (Table 4.21). They suggest some degree of disturbance of complete vessels from these deposits in the mid 3rd century and later. There is also a cross-join of a stamped (S3) samian Dr 31 between the Cremation Zone pit [4231] and context (1285) in brine tank [1182].

Discussion

The date distribution of the Nantwich pottery shows a marked contrast with that from many other sites in the region in lacking their major Hadrianic peak. However, the tail-off of activity here after the early decades of the 3rd century is fairly typical of virtually all nucleated sites in the North-West south of Hadrian's Wall.

The earliest pottery on the site is a single example of a Flavian-Trajanic mica-dusted jar, F41.2. This was mixed with other later material and would appear to have been an heirloom. Similarly the slightly larger group of Hadrianic-early Antonine vessels would seem to have arrived with the community who first developed the site.

Use of the site seems to have peaked in the Antonine, to which period brine tank [1207], Buildings 1 and 2, the large enclosure and the hearths can be attributed. This activity significantly declined in the last decades of the 2nd century before rising to a lower plateau in the first half of the 3rd century. It then declined again to reflect little more than sporadic deposition of material in the later 3rd and 4th centuries. The chronology of this site is quite different from that of Walton-le-Dale, Middlewich and other sites on King Street in terms of its origins. All those sites start in the 1st century and have a highly characteristic Hadrianic-early Antonine peak to their activity, declining from that point onwards. However, in terms of their 3rd- and 4th-century history Nantwich shares a pattern with Middlewich (Evans 2002; Leary 2008), Wilderspool (Lucas 2007), Walton-le-Dale (Evans & Rátkai forthcoming a) and Lancaster (Evans & Rátkai forthcoming b).

None of these sites shows any evidence of extensive occupation dating to later than the mid 3rd century but all show evidence of sporadic pottery use, presumably related to persons passing by on the road network which they sit on. With the exception of Middlewich, none of these sites south of Lancaster is known to have an early Roman fort and it is unlikely that most would have one. The function of the sites has been discussed in detail elsewhere (Evans forthcoming a). The consistent evidence of timber-built production halls, the recurrent furnaces and pottery kilns, the absence of strip buildings and usual civilian domestic buildings, the presence of a military pottery and amphora supply, but the low quality diet and frequent circulation of low-denomination coinage, all point towards centres, leased to contractors, for production for the army and probably largely staffed by a servile workforce. These centres are not towns. They have little relationship to the surrounding countryside, which shows no sign of having purchased their produce in any real quantity.

The Nantwich site and Middlewich were clearly involved in salt production. Wilderspool and Walton-le-Dale clearly both produced pottery, although this should certainly not be seen as the principal activity in either place. At these sites, some process involving heat and water is the only certainty.

Given that all these Lancashire and Cheshire sites were in the business of military supply for the frontier, their history becomes more rational. The construction of King Street, according to one dating, in *c* AD 90 (Wild 2002; Evans & Rátkai forthcoming a) saw the development of sites on that route. Most of these, including Middlewich, Wilderspool and Walton-le-Dale, have samian ware lists with a start date in the late Flavian period, followed by a peak in the Hadrianic-early Antonine period and relative declines in the mid to late Antonine period. The coarse pottery on these sites shows a major decline in the early 3rd century to a very minor level by the later 3rd. The Hadrianic peak would seem to reflect the demand for supplies and material for the construction of Hadrian's Wall. The lower samian levels in the mid to late Antoine period runs counter to the usual pattern for CG samian across the province (Willis 2005). The obvious reason for this would be the lower levels of investment needed for the construction of the Antonine Wall and the subsequent refurbishment of Hadrian's Wall.

Although the presence of a fort at Nantwich cannot be ruled out from the ceramic evidence, it equally does not require the presence of one. There is no evidence of a fort at Walton-le-Dale, for example, and there probably will not be. It is relatively unlikely that a fort would be established at Nantwich at so late a date. The 'military type' pottery supply is a feature common to other Cheshire Plain and Lancashire sites discussed here but this is because they were sites involved in production for the military not because they were military establishments.

The last major known military campaigns in the north, under Severus, may be reflected in the continuing early 3rd-century occupation on these sites, before they peter out in the mid 3rd century. There is no revival for these sites in the later 3rd or 4th centuries.

	Feature	Context	Fabric	Form	Period	NoSh
Cremation Zone	Pit [3440]	(3445)	Q01	Q01.5	Romano-British	1
	Cremation burial pit [4004]	(4003)	B01	B01.2	Mid-late C2	18
	Pit [4231]	(4230)	S21	Dr 31R	Antonine	2
	Pit [4237]	(4232)	B01	B01.27A	AD 160/80+	2
Utilitarian Zone	Pit [3078]	(3079)	B01	B01.27A	AD 160/80+	1
	Posthole [3094], Building 2	u/s	B01	B01.27C	Hadrianic-mid Antonine	4
	Subsoil	(2002)	O04	O04.30	C1/2-C3	2
Brine Industrial Zone	Brine tank [1182]	u/s	B01	B01.27A	AD 160/80+	1
		u/s	B01	B01.27B	Mid C2?	1
		u/s	O04	O04.30	C1/2-C3	1
		(1188)	B01	B01.27A	AD 160/80+	1
		(1188)	O04	O04.30	C1/2-C3	2
		(1215)	S21	Dr 33	Antonine	1
		(1220)	B01	B01.27D	C3?	1
		(1220)	O06	O06.6	C2-C3	2
		(1273)	S21	Dr 38	Antonine	1
		(1277)	B01	B01.27A	AD 160/80+	1
		(1277)	O03	O03.2	C1/2-C3	1
		(1279)	S21	Dr 31	Antonine	2
		(1281)	A21	A21.1	C1-C3	1
		(1281)	B01	B01.24	Hadrianic-Antonine	7
		(1281)	B01	B01.25A	Hadrianic-early Antonine	1
		(1281)	B01	B01.25B	Mid C2	1
		(1282)	S21	Dr 33	Antonine	2
		(1284)	B01	B01.25C	AD 160/80-200	15
		(1285)	B01	B01.26C	Mid C2?	1
		(1285)	S21	Curle15	Hadrianic-Antonine	1
		(1285)	S21	Dr 33	Antonine	1
		(1285)	F03	F03.1	C2	5
		(1285)	O03	O03.2	C1/2-C3	1
		(1285)	O11	O11.1	LC2-C3	5
	Brine tank [1207]	u/s	B01	B01.27A	AD 160/80+	1
		u/s	S21	Dr 30	Hadrianic-Antonine	1
		(1225)	S21	Dr 37	Antonine	5
		(1225)	S21	Dr 37	Antonine	5
		(1225)	S21	Dr 31R	Antonine	2
		(1225)	S21	Dr 33	Antonine	4
		(1239)	B01	B01.11	AD 240-70	7
		(1239)	B01	B01.25A	Hadrianic-early Antonine	1
		(1239)	B01	B01.27D	C3?	3
		(1239)	O03	O03.12	Mid C2-late C2	17
		(1239)	R02	R02.5	Romano-British	1
		(1239)	R21	R21.1	C2?	1
		(1240)	B01	B01.12	AD 220-60	15
		(1240)	B01	B01.20	Mid C2	2
		(1240)	O03	O03.24	C2-C3	5
	Chamber [1347]	(1349)	B01	B01.20	Mid C2	1
		(1351)	Q03	Q03.1	C2	15
	Pit [1096]	(1098)	S21	Dr 38	Antonine	1
		(1112)	B01	B01.29A	AD 160/80-200	1
	Pit [1105]	(1106)	F01	F01.3	C4	1
	Trough [1264]	(1272)	F41	F41.3	Early C3	44
Area C/D	Subsoil	(4002)	O04	O04.29	Romano-British	1
Area G	Subsoil	(7002)	B01	B01.23	Hadrianic-Antonine	3

Table 4.19: Complete vessels and profiles.

		A	F/Ju	CJ	J	WMJ	Bk	Cu	Tk	M	B	D	L	O	No
[1182]	MnR	-	-	14.3%	-	-	7.1%	21.4%	7.1%	-	28.6%	-	7.1%	14.3%	14
	RE	-	-	33.7%	-	-	11.1%	6.2%	4.2%	-	32.1%	-	9.5%	3.3%	549
[1207]	MnR	-	-	-	20.0%	6.7%	-	6.7%	-	-	40.0%	26.7%	-	-	15
	RE	-	-	-	30.8%	12.5%	-	10.7%	-	-	32.0%	14.0%	-	-	798
Both	MnR	-	-	6.9%	10.3%	3.4%	3.4%	13.8%	3.4%	-	34.5%	13.8%	3.4%	6.9%	29
	RE	-	-	13.7%	18.3%	7.4%	4.5%	8.8%	1.7%	-	32.0%	8.3%	3.9%	1.3%	1347

Table 4.20: Functional breakdown of complete vessels and profiles from brine tanks [1182] and [1207].

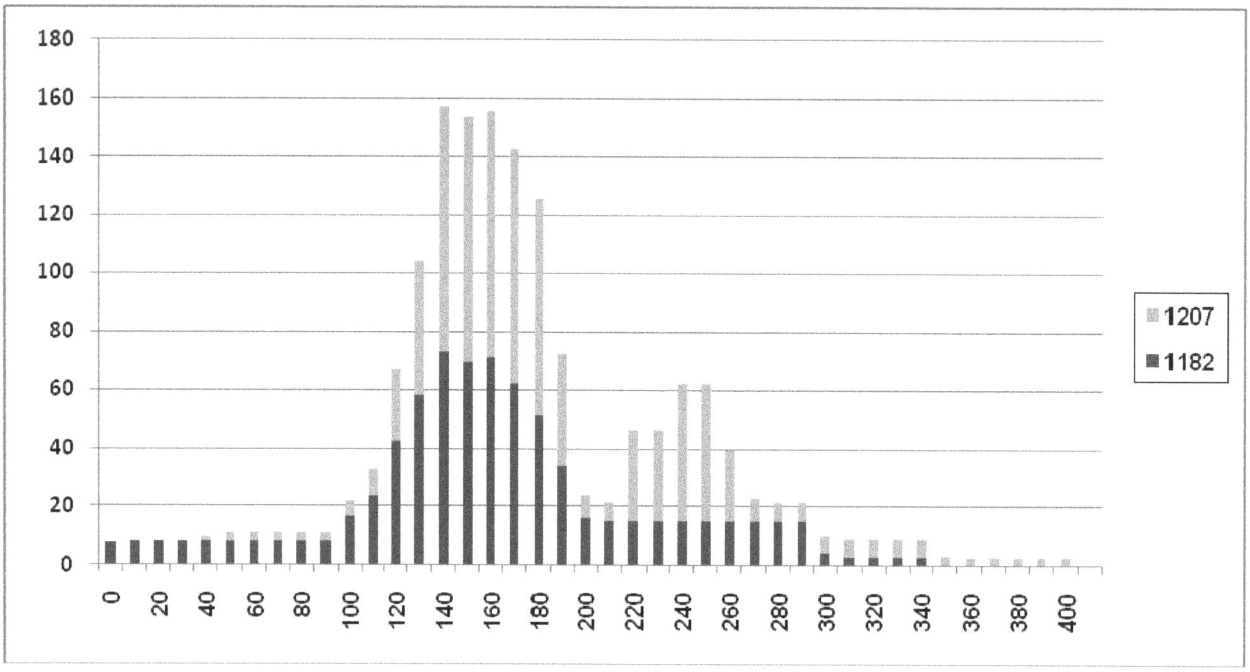

Figure 4.21: Date distribution of complete vessels from brine tanks [1182] and [1207] by RE.

	Feature	Context	Form	Joins	Stamped/ Decorated
Cremation Zone	Pit [4237]	(4232)	F01.2	(4236) [4237]	
Utilitarian Zone	Posthole [2156], Building 1	(2157)	CG	(2157) [2156] (2159) [2158]?	
	Posthole [2177], Building 1	(2160)	CG	(2159) [2158]	
Brine Industrial Zone	Brine tank [1182]	(1183)	CG Dr 33	(1188) [1182]	
		(1188)	CG Dr 33	(1183) [1182]	
		(1188)	O04.30	(1183) [1182]	
		(1205)	CG Dr 31	(1188) (1205) [1182]	S7
		(1277)	O01.1	(1285) [1182]	
		(1279)	CG Dr 31	(1277) [1182]	S5
		(1281)	CG Dr 37	(1282) (1285) [1182]	D18
		(1282)	CG	(1281) (1282) (1285) [1182]	S2, D18
		(1285)	CG Dr 31R	(1281) (1282) (1285) [1182], (4230) [4231]	S3
		u/s	O04.30	(1188) [1182]	
		u/s	EG Dr 31R	(1208) (1239) (1240) [1207]	D12
	Brine tank [1207]	(1208)	A01.3	(1225) [1207]	
		(1208)	CG Dr 37	(1208) [1207] (1285) [1182]	D11
		(1225)	CG Dr 37	(1209) (1225) (1239) [1207]	D9
		(1239)	CG Dr 37	(1209) (1225) (1239) [1207]	S10, D9
		(1240)	SG Dr 37	(1239) [1207]	D4
	Chamber [1347]	u/s	R21.1	(1239) (1336) [1207]	
		(1351)	Q03.1	(1340) [1347], (1164) [1162]	
	Pit [1096]	1097	CG Dr 37	(1097) (1098) [1096]	D1
		1098	CG	(1277) [1182]	
	Trough [1264]	(1267)		(1267) (1270) (1272) [1264]	
		(1270)		(1267) (1270) (1272) [1264]	
		(1272)	F41.3	(1267) (1270) (1272) [1264]	

Table 4.21: Cross-joins.

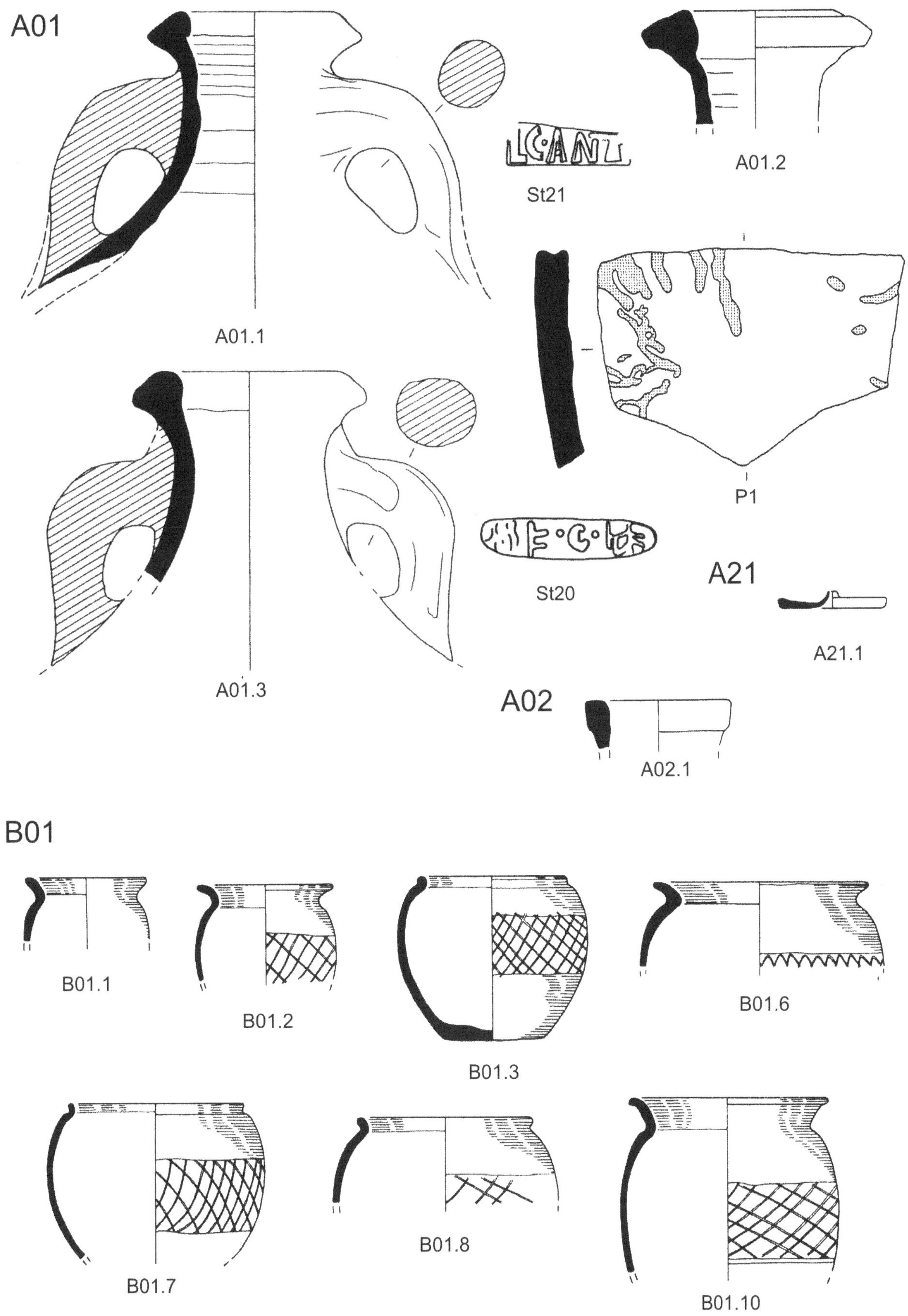

Figure 4.22: Coarse ware. Scale 1:4.

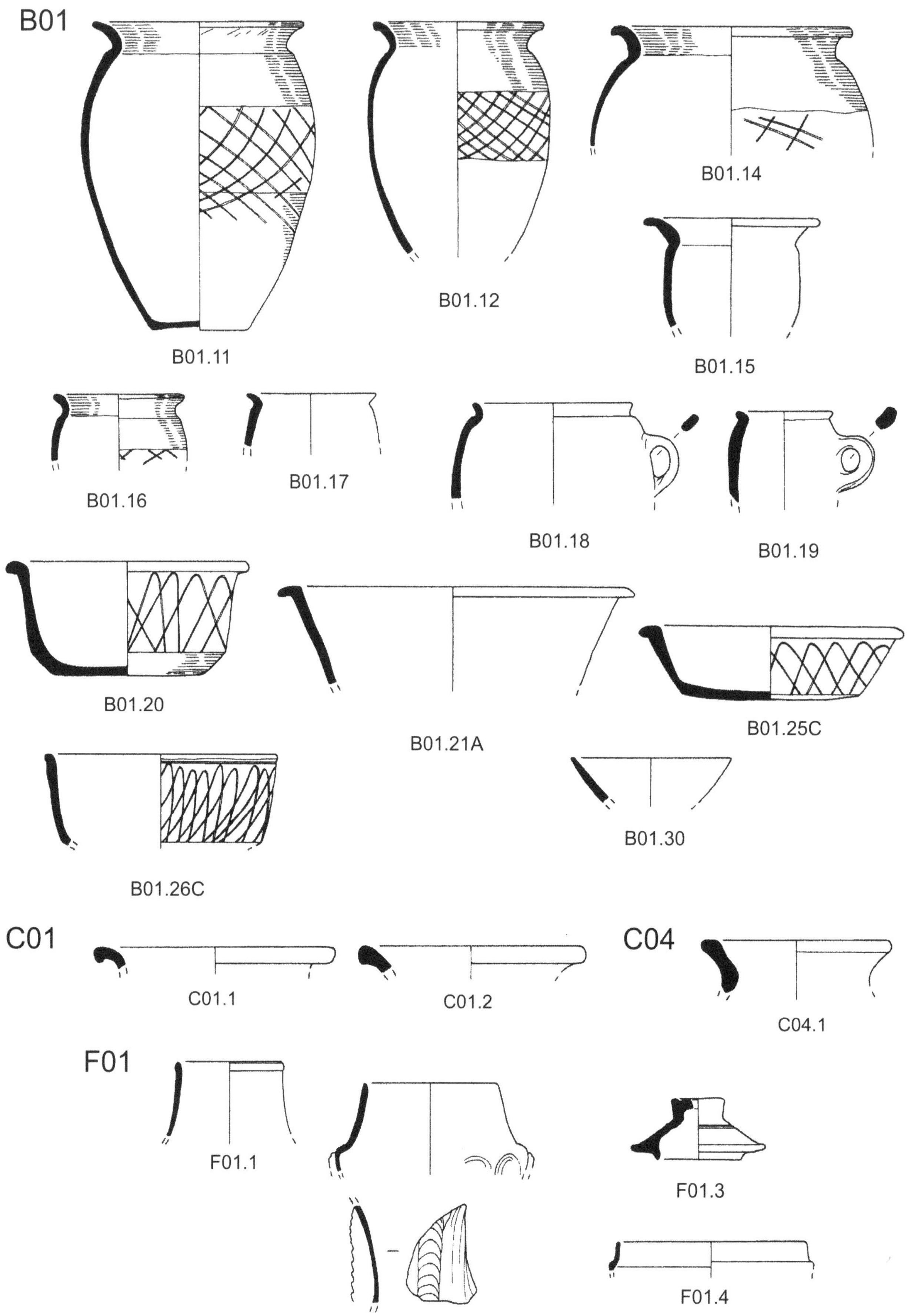

Figure 4.23: Coarse ware. Scale 1:4.

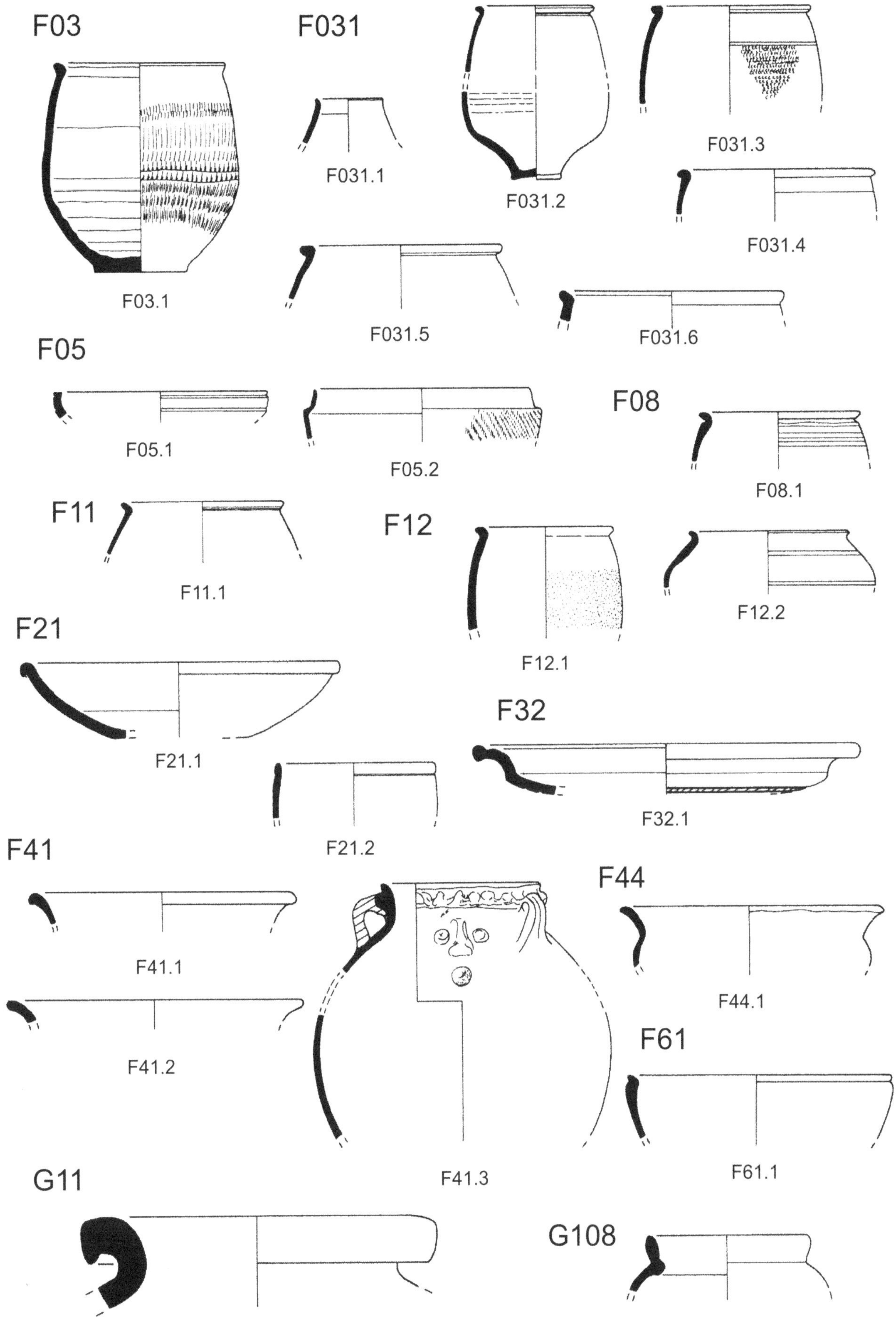

Figure 4.24: Coarse ware. Scale 1:4.

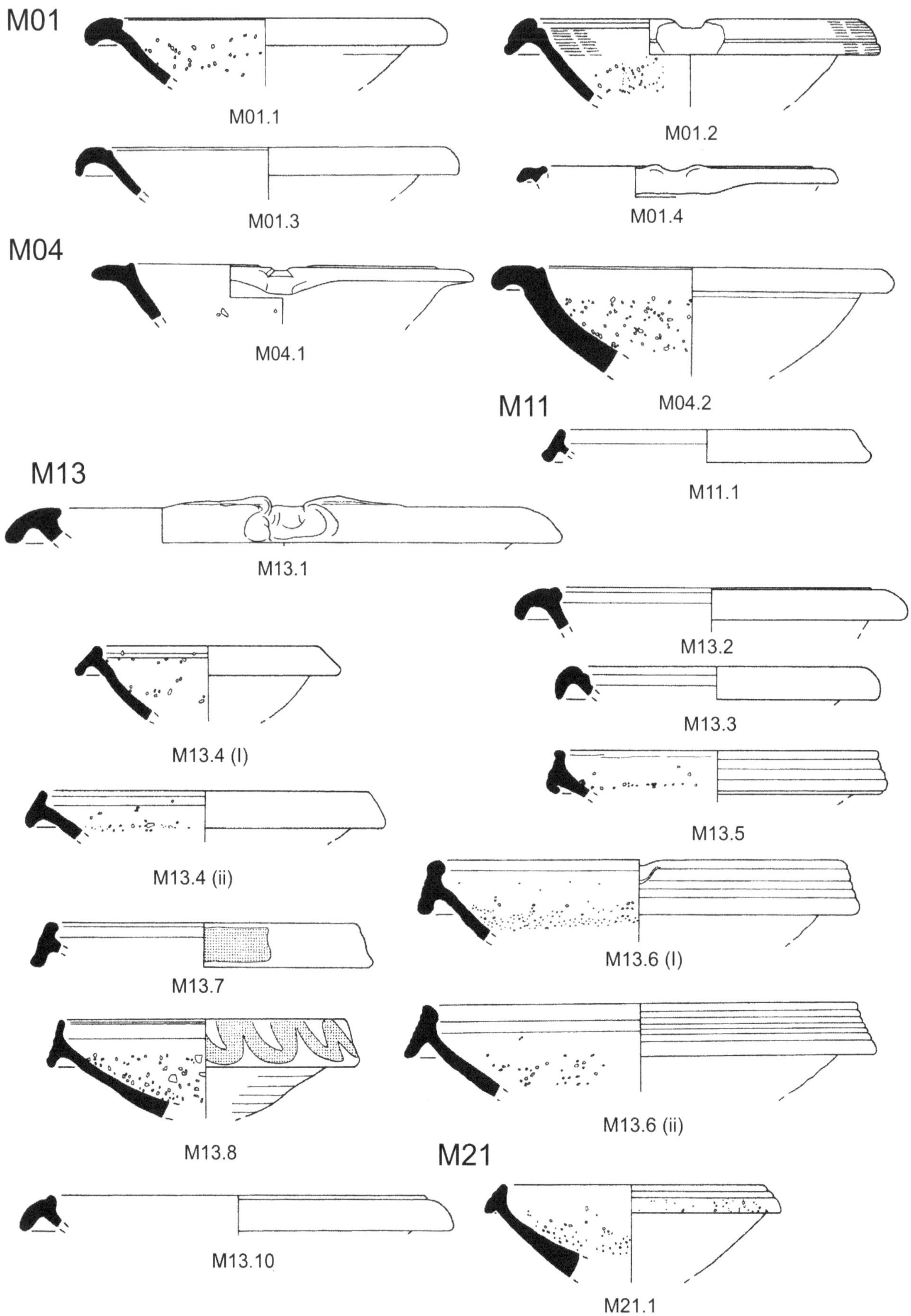

Figure 4.25: Coarse ware. Scale 1:4.

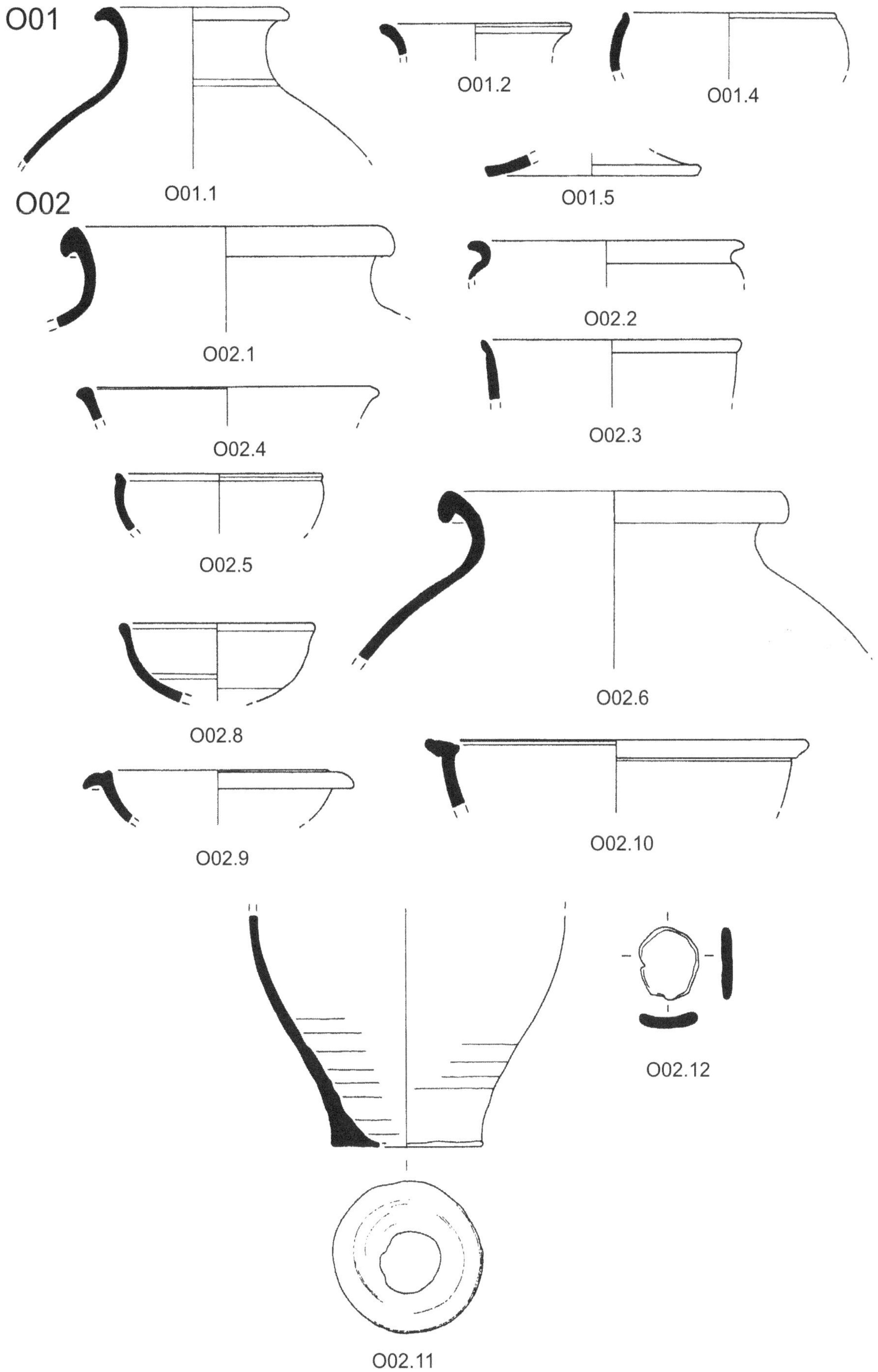

Figure 4.26: Coarse ware. Scale 1:4.

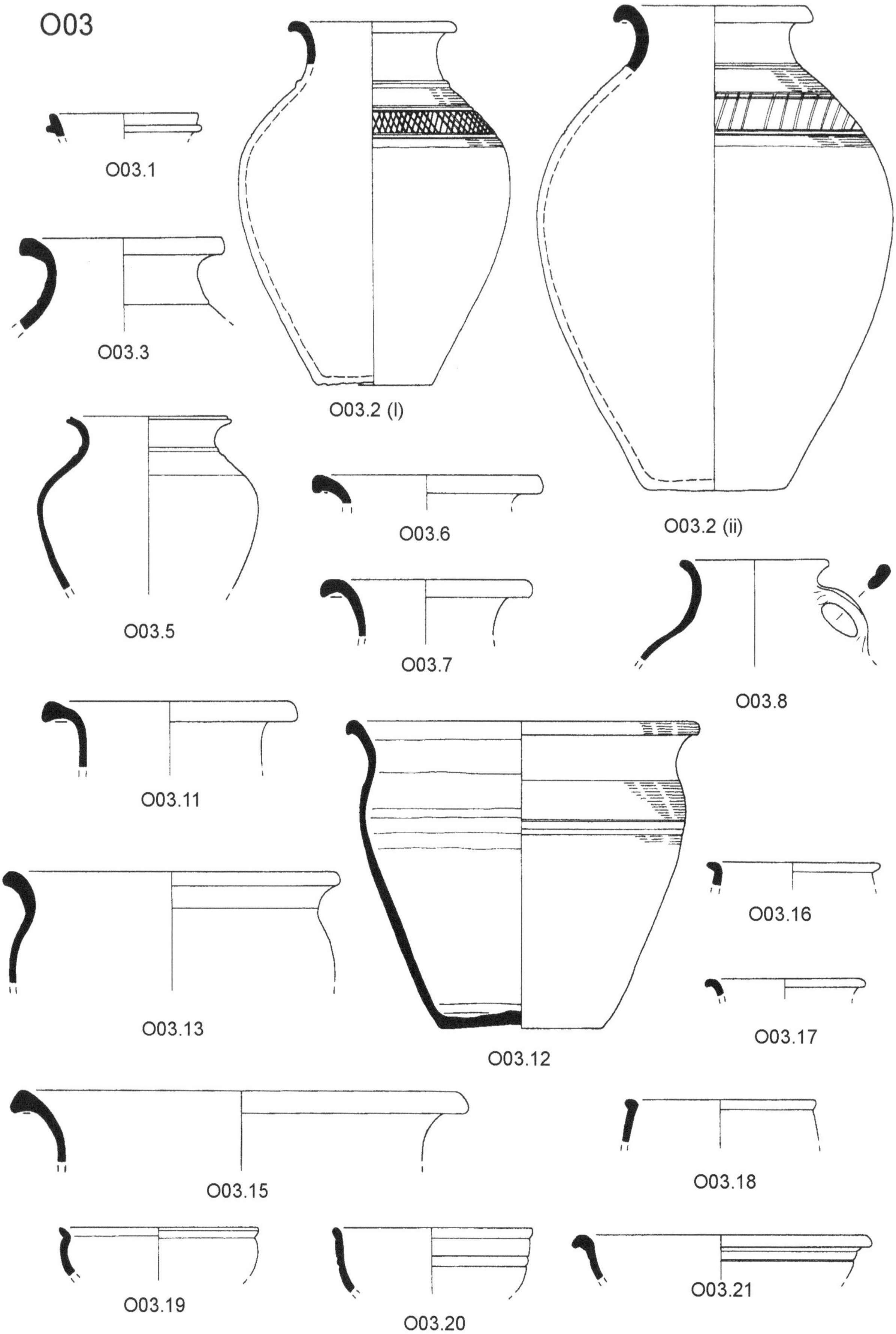

Figure 4.27: Coarse ware. Scale 1:4.

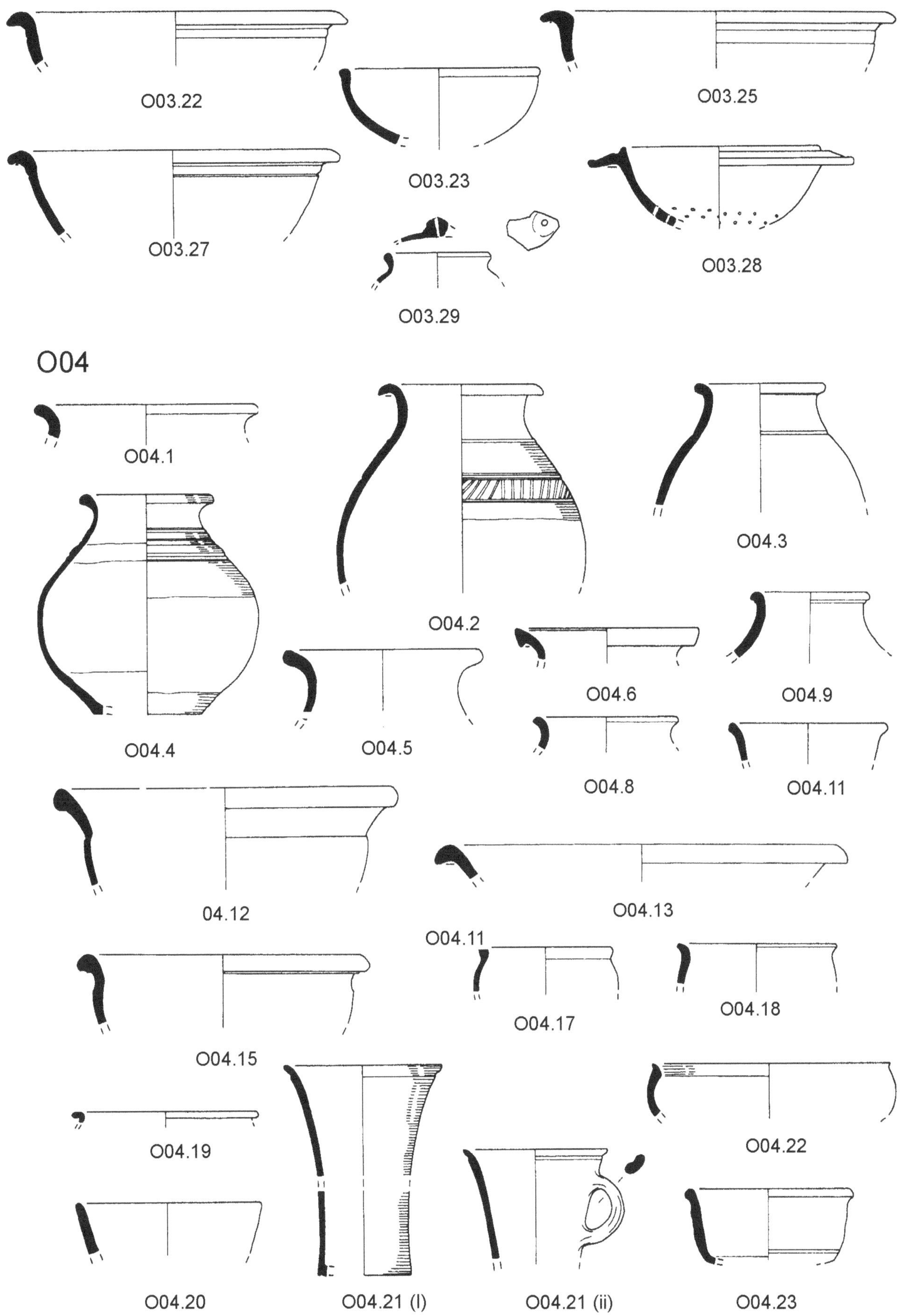

Figure 4.28: Coarse ware. Scale 1:4.

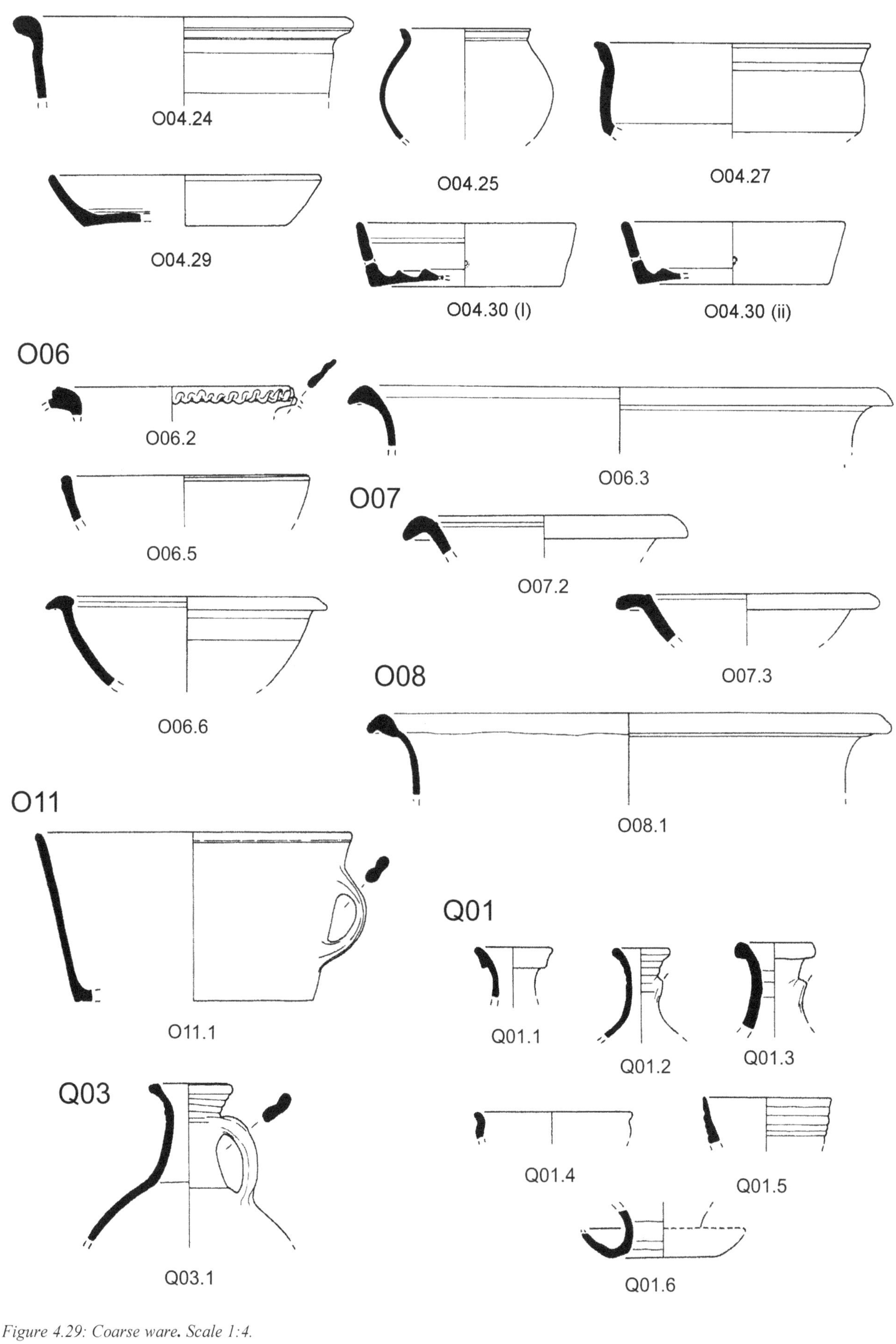

Figure 4.29: Coarse ware. Scale 1:4.

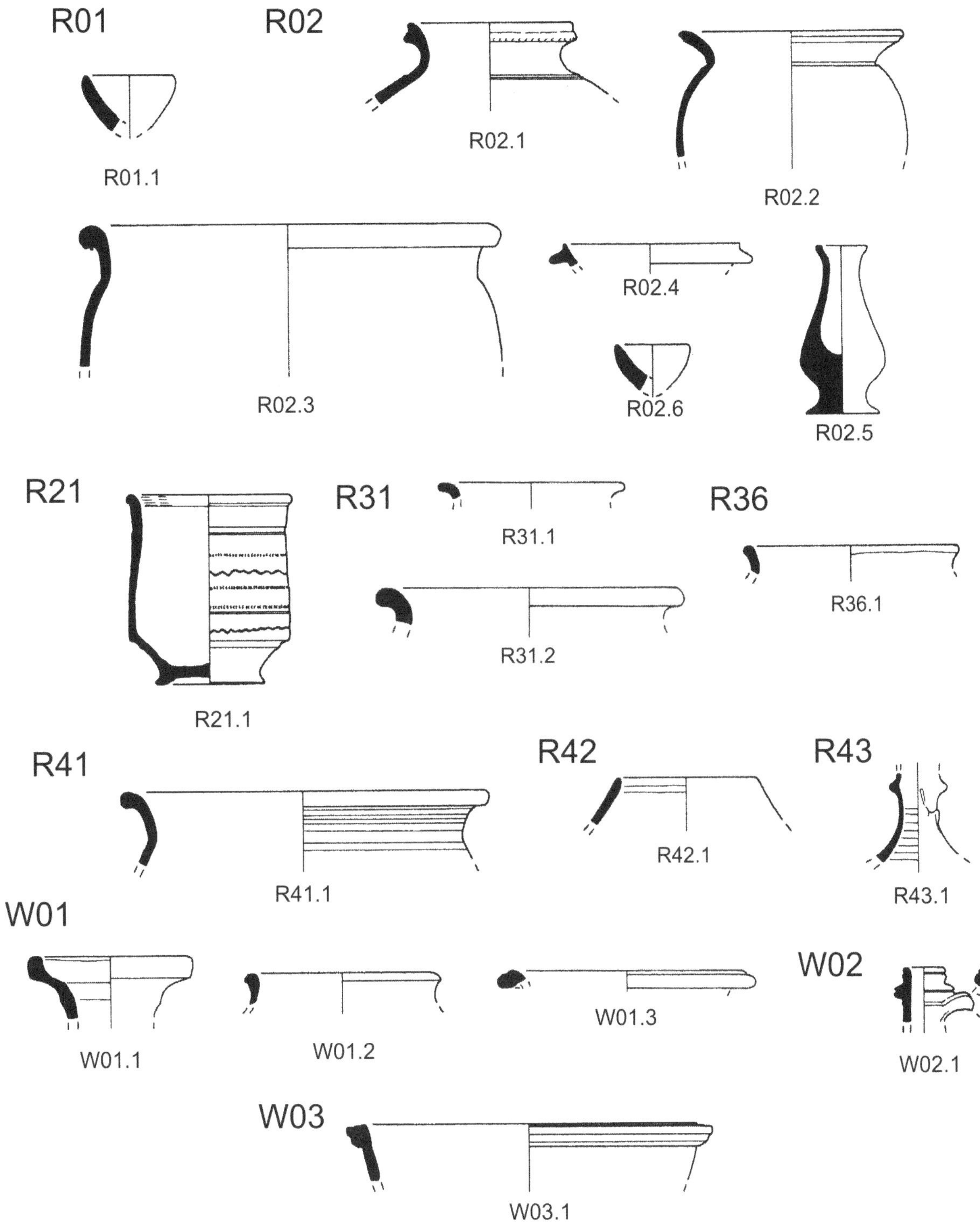

Figure 4.30: Coarse ware. Scale 1:4.

5. Samian Ware

Felicity C Wild

The site produced 588 sherds of samian ware from at least 376 vessels of recognisable form. Joins between sherds from different contexts were frequent, making a precise estimate of vessel numbers difficult, particularly in the case of plain forms. For this reason, and owing to the large number of small or indeterminate sherds, the statistics given below should only be taken as approximations. Decorated ware amounted to at least 21% of the total, based on a sherd count. Apart from three sherds of form 30 (none showing decoration) and a few scraps of beaker, all the decorated ware was from form 37.

As is to be expected in an assemblage of 2nd-century date, the wares were almost entirely Central Gaulish and from Lezoux. There were, in all, only four sherds of South Gaulish ware, from two vessels. One, of form 15/17R, is likely to be of Flavian date and was doubtless an heirloom. The other, of form 37 (D4 below), was likely to have been produced at Montans and is of Hadrianic-early Antonine date. Equally scarce are wares attributable to Les Martres-de-Veyre: a mere three sherds from plain vessels. Of the two decorated bowls by potters known to have worked at Les Martres-de-Veyre (D5 and D6), both are in Lezoux fabric and are likely to have been manufactured there. In all, it would seem unlikely that samian ware was reaching the site before the Hadrianic period at the earliest.

Wares from East Gaul were present, although, as is only to be expected on sites on the west side of the country, not in great quantities. They amounted to 3-4% of the total. While plain sherds could tentatively be assigned to Rheinzabern and Trier as well as La Madeleine or the Argonne, there was only one East Gaulish decorated bowl (D12), from La Madeleine. In a group of late 2nd- to 3rd-century date, the absence of any decorated ware from Rheinzabern seems curious. None of the potter's stamps was East Gaulish.

The Central Gaulish ware was consistent with a mid to late 2nd-century date. 51% of the total assemblage came from the two brine tanks, [1207] and [1182]. A break-down of the forms, where identifiable, from each of the tanks and other contexts is set out in Table 5.1. Taking the assemblage as a whole, the Hadrianic-early Antonine forms 27 (10 examples) and 18/31 (7 examples) were heavily outnumbered by their later equivalents, 33 (49 examples) and 31 (63 examples). Forms typical of the second half of the 2nd century were present (38 (12 examples) and 31R (18 examples)), although there were only six examples of form 79 and/or 80 and one of form 79R. Samian mortaria were absent.

The range of forms identified was comparatively limited, the vast majority of sherds coming from the bowl form 37, the cup form 33 and the dish forms 31 and its rouletted variants 18/31R and 31R. Bowl forms (37, 30, 38 and/or 44) amounted to *c* 36% of the total, cups (27 and 33) to 16% and dish forms (18/31, 31 and rouletted variants) to 42%.

As noted above, just over half the samian ware from the site as a whole came from the two brine tanks [1207] and [1182]. Although [1207] contained some of the earlier decorated pieces (D4-D6), there was little real difference in date between the contents, and the stamps from both tanks are uniformly late 2nd-century.

Form	[1207]	1182]	Other contexts	Total
37	20	29	46	95
30	-	1	2	3
Beakers	3	1	4	8
27	2	2	6	10
33	14	15	20	49
18/31	1	2	4	7
18/31 or 31	2	-	15	17
31	21	28	14	63
18/31R	-	-	13	13
18/31R or 31R	6	6	6	18
31R	3	5	11	18
31 or R	-	7	12	19
38	4	3	5	12
bowl	3	6	16	25
79	-	-	4	4
79 or 80	-	-	2	2
79R	-	-	1	1
81	-	1	-	1
36	2	2	-	4
Curle 11	-	1	1	2
Curle 15	-	1	-	1
Curle 15 or 23	-	-	1	1
15/17R	-	1	-	1
15/31	-	-	1	1
Total	**81**	**111**	**184**	**376**

Table 5.1: Total of samian forms in brine tanks [1207] and [1182] and other contexts.

Decorated Ware

In the following report, figure types are quoted from Oswald 1936-7 (O.), with additional types from Rogers 1999 (R.); Central Gaulish decorative details from Rogers 1974 (Rogers) and parallels from Stanfield and Simpson 1958 (S&S). Lower-case Roman numerals after a potter's name denote homonyms in the system used in Hartley and Dickinson 2008-.

D1. Form 37, Central Gaulish (Fig 5.1). Two joining fragments showing ovolo (Rogers B233) over two zones

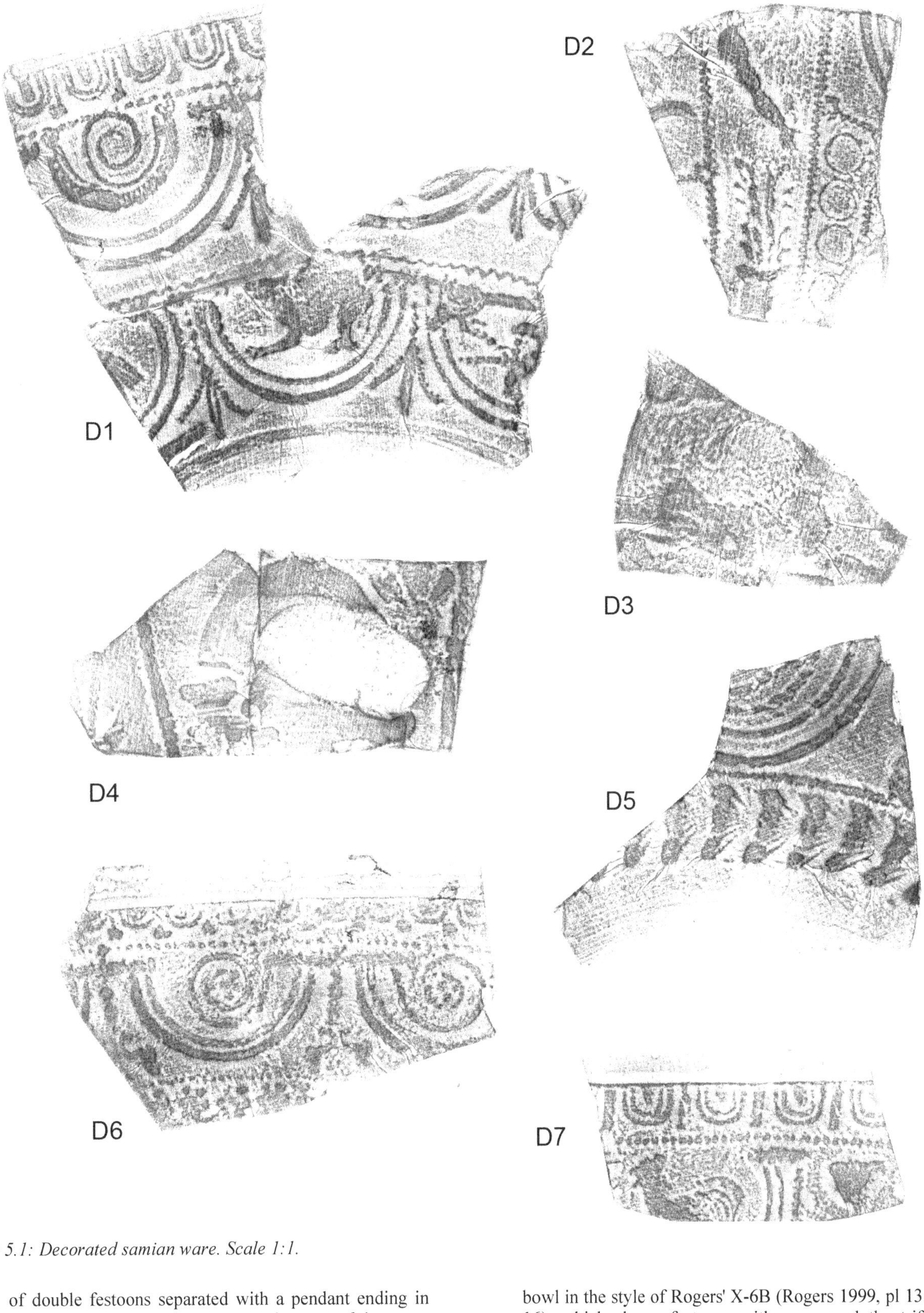

Figure 5.1: Decorated samian ware. Scale 1:1.

of double festoons separated with a pendant ending in the trifid motif (Rogers G32). The festoons of the upper zone contain a small festoon with spiral (Rogers F76), those of the lower zone contain the panther (O.1566) and a cross motif with goose to right (O.2314) and possibly another to left (O.2286A?). The style is that of Stanfield and Simpson's potter X.6, a style which embraces the work of a number of potters whose names are as yet largely unknown. The closest parallel is a bowl in the style of Rogers' X-6B (Rogers 1999, pl 135, 16), which shows festoons with geese and the trifid G32. X-6B also used the panther, though Rogers attributes the ovolo and F76 to X-6D. The present piece shows what appears to be a potter's plain-ware name stamp in the mould at the base of the decoration. Unfortunately it is not legible. *c* AD 125-50. Fills (1097) (1098) of pit [1096].

D2. Form 37, Central Gaulish (Fig 5.1), showing panel

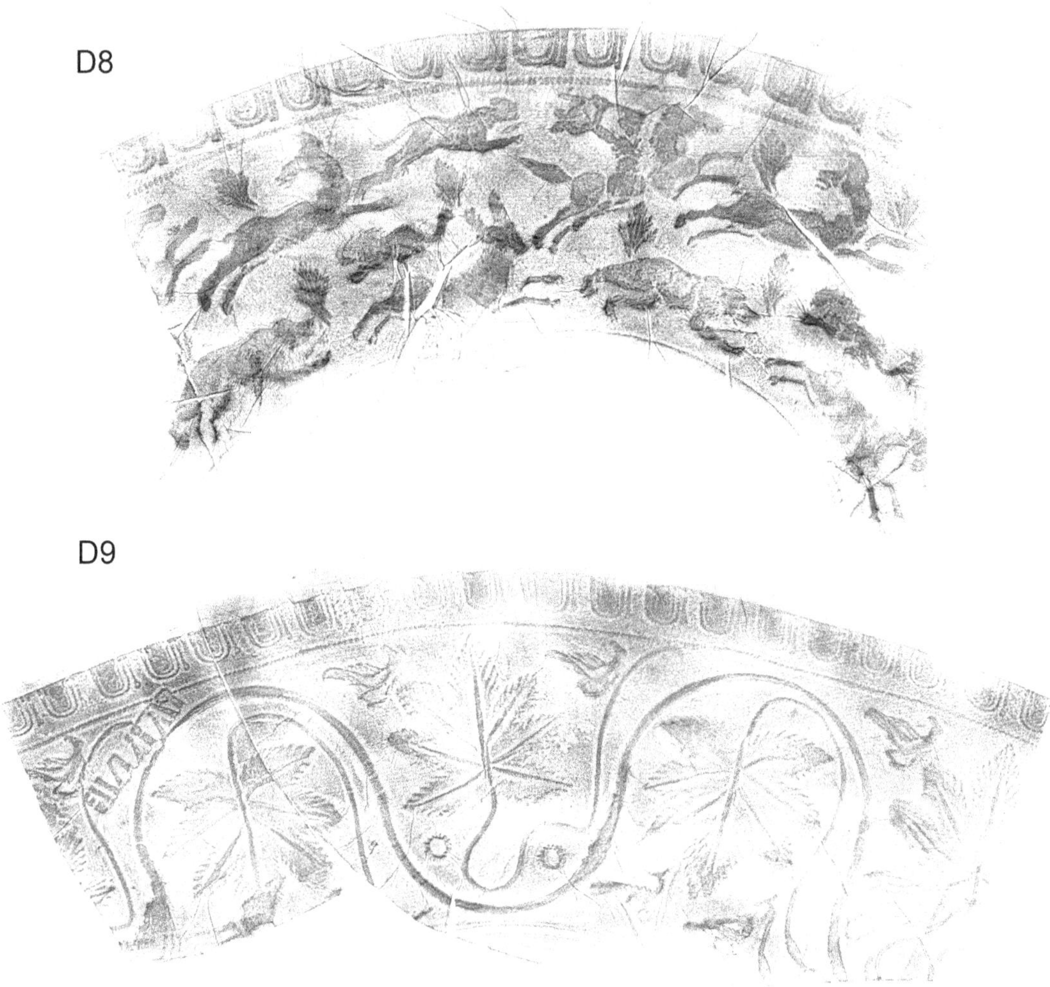

Figure 5.2: Decorated samian ware. Scale 1:2.

decoration with caryatid (O.1207A) and a double festoon over a vertical row of circles. The caryatid and junction rosette are typical of the work of Divixtus i and Criciro v. *c* AD 135-70. Fill (3400) of pit [3399].

D3. Form 37, Central Gaulish (Fig 5.1), burnt, showing freestyle hunting scene with stag (O.1720). The style is probably that of Butrio, who used the stag and the leaf (Rogers J160) hanging from the wavy-line border beneath the ovolo (S&S, pl 57, 653). The ovolo is too fragmentary and abraded to be identified with certainty, but may be his ovolo (Rogers B82). *c* AD 120-45. Fill (2159) of posthole [2158], Building 1.

D4. Form 37, South Gaulish (Fig 5.1). Two joining sherds showing crude and poorly impressed decoration, with a column and arcade in the central panel. A large (vegetable?) impurity in the clay has blown a lump off the surface of the bowl, removing much of the surviving decoration. The style is typical of the latest products of South Gaul and, with the fabric, suggests probable manufacture at Montans. A similar column and arcade occur on a bowl stamped VII (Mees 1995, Taf 247, 2), which also shows plant motifs in the adjoining panel of a generally similar type to what may occur here. A bowl stamped by Chresimus of Montans from recent excavations in Deansgate, Manchester (DGM04 (195)), as yet unpublished, shows a similar colonnade. Montans ware was reaching the North-West of Britain and as far as the Antonine Wall during the Hadrianic-early Antonine period, so this piece need be no earlier than the other pieces from the site. The arcade, or a closely similar one, occurs as a festoon on sherds from Balmuildy (Miller 1922, pl 34, 37, 39). *c* AD 120-45. Fills (1239) (1240) of brine tank [1207].

D5. Form 37, Central Gaulish (Fig 5.1), burnt, showing a basal wreath consisting of the upper part of a bud (Rogers G18) beneath scroll decoration (?). The basal wreath was used on a number of bowls by Rogers' potter P10 (1999, pl 126, 14, 15, 18, 19), though normally with panel decoration rather than a scroll, if that is indeed what the present piece shows. Although burnt, the fabric is that of Lezoux rather than Les Martres-de-Veyre, suggesting a Hadrianic rather than Trajanic date. Fill (1239) of brine tank [1207].

D6. Form 37, Central Gaulish (Fig 5.1), showing zonal decoration with double festoons containing spirals separated by a pendant ending in the pair of dolphins (O.2407A), above a row of dot rosettes (Rogers C280). The style is that of Drusus i, who used the various motifs and the ovolo (Rogers B28). However, where he used spirals inside his festoons, they generally seem to point in the same direction (Terrisse 1968, pl 9, 130: 11, 10129) rather than in opposing directions, as here, and although he used the dolphins, he generally used other motifs on his pendants. An identical bowl, which could

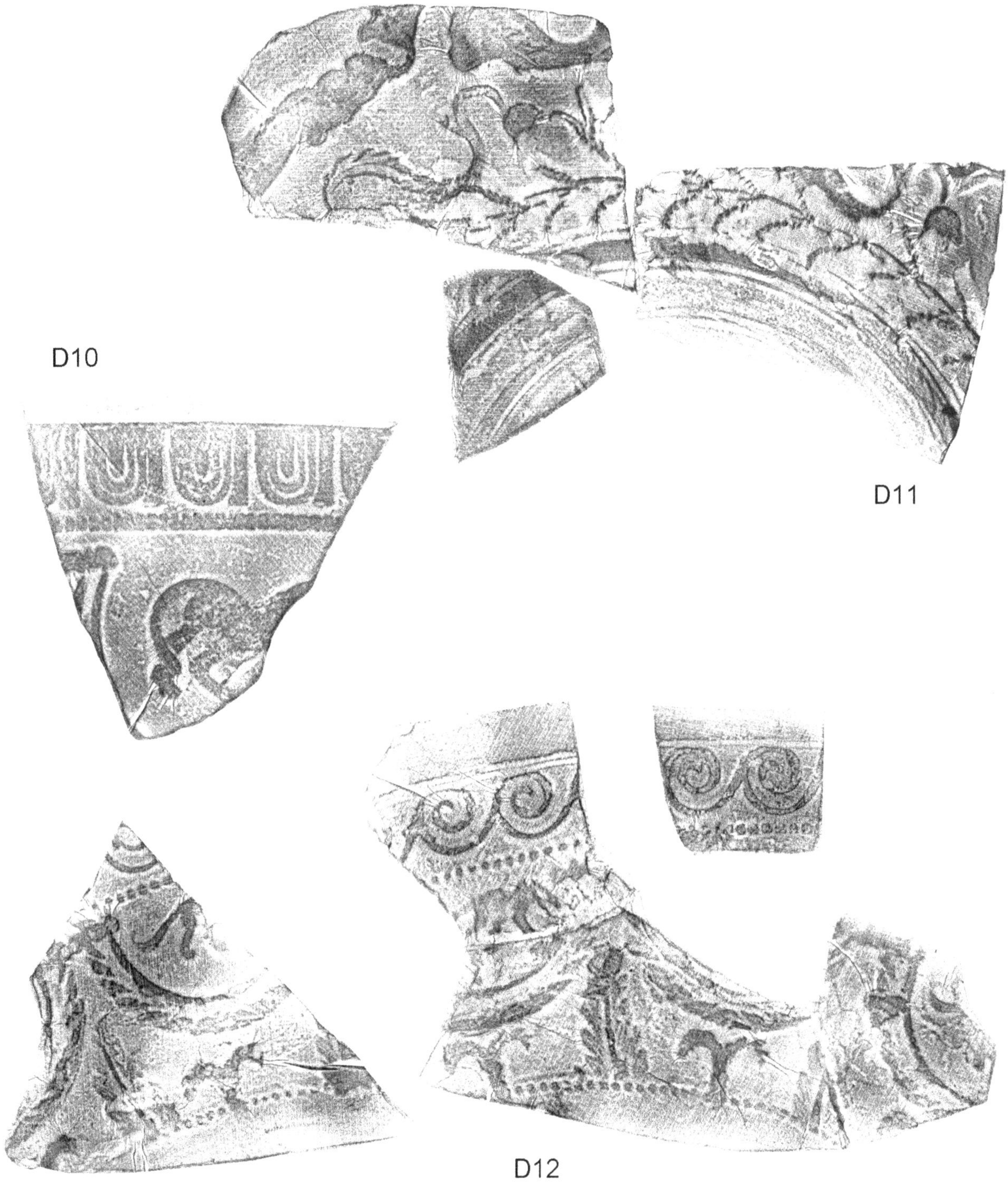

Figure 5.3: Decorated samian ware. Scale 1:1.

be from the same mould, was found in the excavations at Deansgate, Manchester (DGM04 (198) (215)). The fabric of the present piece appears typical of the micaceous fabric of Lezoux rather than that of Les Martres-de-Veyre, and the piece is likely to be Hadrianic. The Manchester bowl, on the other hand, is in a fabric that could indeed be from Les Martres. Fill (1208) of brine tank [1207].

D7. Form 37, Central Gaulish (Fig 5.1). Fragment from a small bowl. Although the definition is poor, the ovolo is probably Rogers B143, used on Cinnamus ii's mature style, rather than B144, the Cerialis-early Cinnamus ovolo. Decoration shows the small cock (O.2350) in a festoon and probably the head of Venus (O.331). The cock is not one of Cinnamus' common types. *c* AD 150-80. Fill (1241) of brine tank [1207].

D8. Form 37, Central Gaulish (Fig 5.2). Complete bowl, showing freestyle hunting scene in the style of Cinnamus ii, with his ovolo (Rogers B233) and bush space-filler (Rogers N15). Types are the horseman (O.245), panther (O.1507), hind (O.1822I), stag (O.1720), small lion (O.1421) and bear (closest to O.1633L). A stamped bowl from Lezoux (Rogers 1999, pl 32, 45) shows the same ovolo, bush, horseman, stag

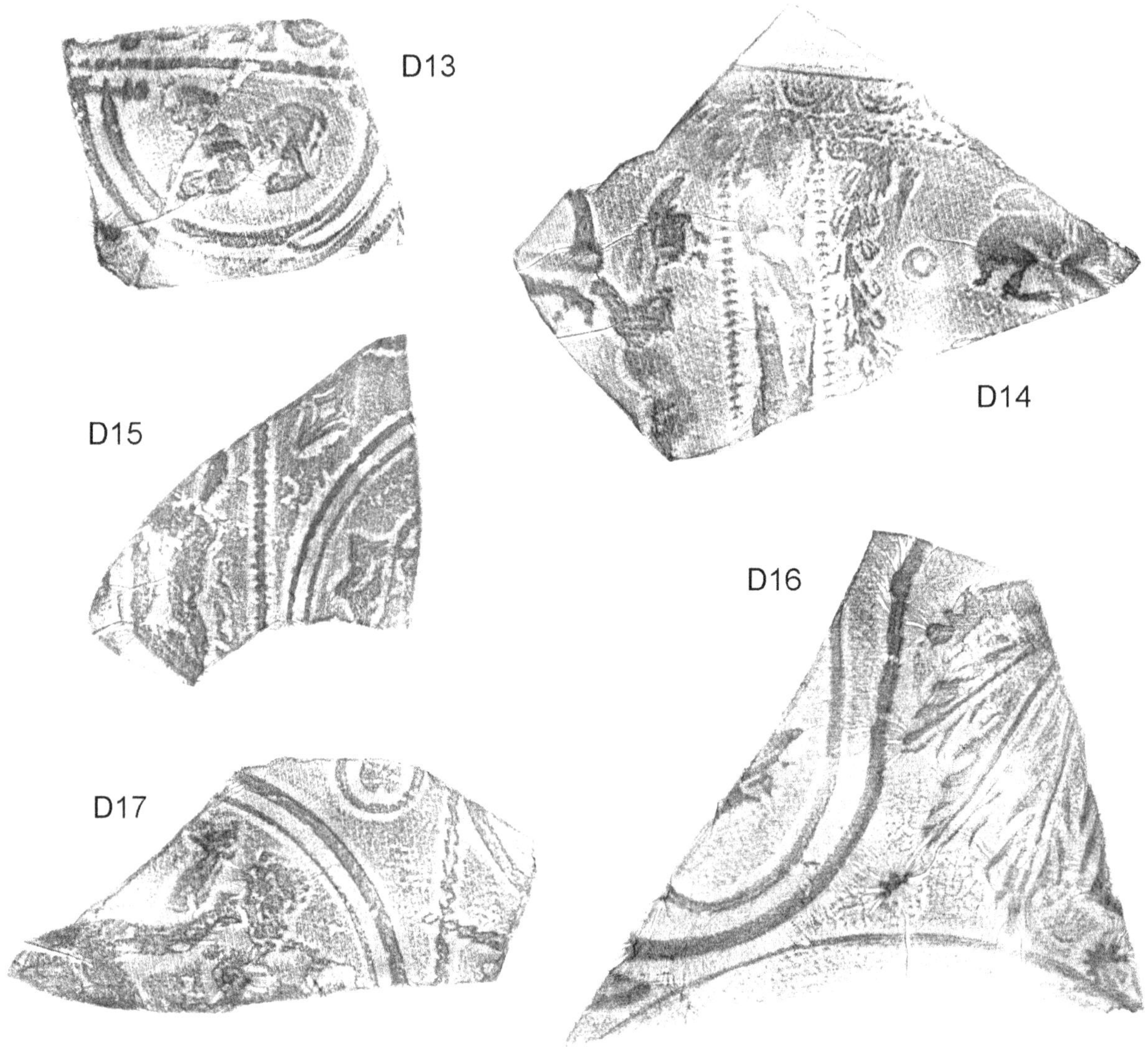

Figure 5.4: Decorated samian ware. Scale 1:1.

and hind. *c* AD 150-80. Fill (1225) of brine tank [1207].

D9. Form 37, Central Gaulish (Fig 5.2). Most of a bowl with leaf-scroll, showing the large label stamp (die 7a) of Paternus v and his ovolo (Rogers B105). The scroll contains his large leaf (Rogers H2) with the bird (O.2250A) and his serrated ring (Rogers E57) in the upper, and the peacock (O.2365) in the lower, concavities. *c* AD 160-85. Fills (1225) (1239) of brine tank [1207], u/s.

D10. Form 37, Central Gaulish (Fig 5.3), showing a deep festoon containing a panther (O.1521). The ovolo (Rogers B161) with large-beaded border is typical of the style of Do(v)eccus i, who also used the festoon (S&S, pl 147, 10). The panther is attested on work in his style. Although it fits O.1521 in size, the tail here is clearly wrapped around the panther's leg, as on the larger type O.1520. *c* AD 165-200. Fill (1224) of brine tank [1207].

D11. Form 37, Central Gaulish (Fig 5.3). Three fragments of bowl with worn footstand, showing a basal wreath of the trefoil (Rogers G32) above a thick basal ridge, a characteristic feature of the potter X.6 (cf Rogers 1999, pl 135 for the trefoil wreath; S&S, pl 75, 13 for both features). Freestyle decoration of sea creatures shows a sea horse or bull (not listed by O. or Rogers) and the dolphin (R.4015) used by X.5 and Rogers' X-6B (1999, pl 135, 1, which also shows a trefoil wreath). *c* AD 125-50. Fill (1208) of brine tank [1207], fill (1285) of brine tank [1182], u/s area B.

D12. Form 37, East Gaulish (Fig 5.3). Five fragments, in the style of La Madeleine, showing spirals (Ricken 1934, Taf 7, 33) used as an ovolo replacement above festoons containing the animal (Taf 7, 101), with leaf (Taf 7, 50) and bifid bud (Taf 7, 15) below. Although no very close parallel is forthcoming, the individual types were all common at La Madeleine. Ricken 1934, Taf 9, 1 shows the spirals and festoon, Taf 9, 2 the animal in a different festoon and Taf 9, 13 the leaf, bifid and bead-row border beneath the decoration. *c* AD 130-60. Fills (1208) (1239) (1240) of brine tank [1207], u/s [1182], subsoil (1002).

D13. Form 37, Central Gaulish (Fig 5.4). Two joining fragments showing the panther (O.1566) in a festoon with a pendant ending in the trifid (Rogers G32). The decoration is identical to the lower zone of the bowl in the style of X.6 (D1 above). The ovolo is probably the same (Rogers B233), though this is clearly a different bowl. *c* AD 125-50. Fill (1277) of brine tank [1182].

D14. Form 37, Central Gaulish (Fig 5.4), showing panels with the Diana and hind (O.106), caryatid (probably O.1206) and festoon (Rogers F16) with hare (O.2057).

Figure 5.5: Decorated samian ware. Scale 1:1.

The style suggests the work of Criciro v, who used the caryatid and distinctive circular junction motif. The ovolo is badly impressed but may be Rogers B52, which appears on a bowl in his style which also shows the small circle in the field (Rogers 1999, pl 38, 5). Another bowl in his style (Rogers 1999, pl 38, 4) shows the wavy-line horizontal border and caryatid. The festoon occurs on work in his style and on a bowl with his cursive signature from Bath (publication forthcoming). The Diana and hare do not appear to be among his signed types but both were used by Divixtus i, with whom he had close links. *c* AD 135-70. Fill (1281) of brine tank [1182].

D15. Form 37, Central Gaulish (Fig 5.4), with panels showing Bacchus (O.580) and double medallion with the diamond (Rogers U33) in the panel corner. The medallion, diamond and Bacchus were all used by Cinnamus ii, though the Bacchus normally attributed to him is O.581, whereas this is the larger version O.580. The leaf inside the medallion (Rogers J89) is not one of Cinnamus' usual types, though was used by his earlier associate Attianus ii. *c* AD 150-80. Fill (1188) of brine tank [1182].

D16. Form 37, Central Gaulish (Fig 5.4). Base, showing traces of wear, of a bowl with scroll decoration in the style of the Paternus v group. The leaf (Rogers H27) and rosette (Rogers C194) were used by both Laxtucissa (cf S&S, pl 97, 2) and Paternus himself (S&S, pl 107, 26). *c* AD 160-90. Fill (1206) of brine tank [1182].

D17. Form 37, Central Gaulish (Fig 5.4), showing panel decoration with corded borders and double medallion containing the erotic group (O.Y) and rosette (probably Rogers C123) inside a circle in the panel corner. The style is that of Paternus v. A bowl in his style (Rogers 1999, pl 78, 25) shows both the features seen here. *c* AD 160-90. Fill (1220) of brine tank [1182].

D18. Form 37, Central Gaulish (Fig 5.5). Five sherds, with weak joins, of a bowl with the mould stamp of Censorinus ii (die 1a). The ovolo (Rogers B105) is his, with characteristic astragalus border. Decoration, not strictly in panels, as it lacks regular vertical divisions, shows a large double medallion containing the stamp and probably the Pan mask (O.1214), Apollo (O.92), caryatid (O.1201) and satyr (O.627). There appears to be a vertical row of the column (Rogers P3) between the caryatid and the satyr, and possibly an incomplete row between the Apollo and the caryatid (cf S&S, pl 101, 4 for a similar 'semi-freestyle' arrangement, divided by a row of columns). The leaf in the medallion and hanging from the astragalus border is Rogers J153. All the types and motifs are already attested on Censorinus' signed work or work in his style. *c* AD 160-90. Fills (1281) (1282) (1285) of brine tank [1182].

D19. Form 37, Central Gaulish (Fig 5.6), in very hard, red fabric, which appears to be overfired or burnt, with a dull red slip. The ovolo is almost certainly Rogers B5, used at Vichy, Terre Franche by Cinnamus ii and Banvus. The untidy border is more likely to indicate Banvus. No parallels are as yet forthcoming for the ram's horns, which are not identical to any illustrated by Rogers. Antonine, presumably second half of the 2nd century AD. Fill (1277) of brine tank [1182].

D20. Form 37, Central Gaulish (Fig 5.6). Base fragment, showing wear, of small bowl with freestyle decoration. Although he rarely made freestyle bowls, the style is that of Do(v)eccus i. A stamped bowl from Ancaster (unpublished) shows a similar hunting scene with the same space-filler, consisting of partial impressions of his leaf (Rogers J189). The stag (O.1786) occurs on work in his style. Too little of the other animal types survives for identification. *c* AD 165-200. Fill (1220) of brine tank [1182].

D21. Beaker, Central Gaulish (Fig 5.6). Scrap, showing lower (?) border of decoration with the diamond motif (Rogers U28) used both horizontally and vertically. There are possible connections to the Quintilianus i group, who made beakers (Rogers 1999, pl 96, 67, 69) and who used G28 horizontally (ibid, pl 96, 70) as well as vertically, suggesting a date *c* AD 125-45. Fill (1158) of pit [1157].

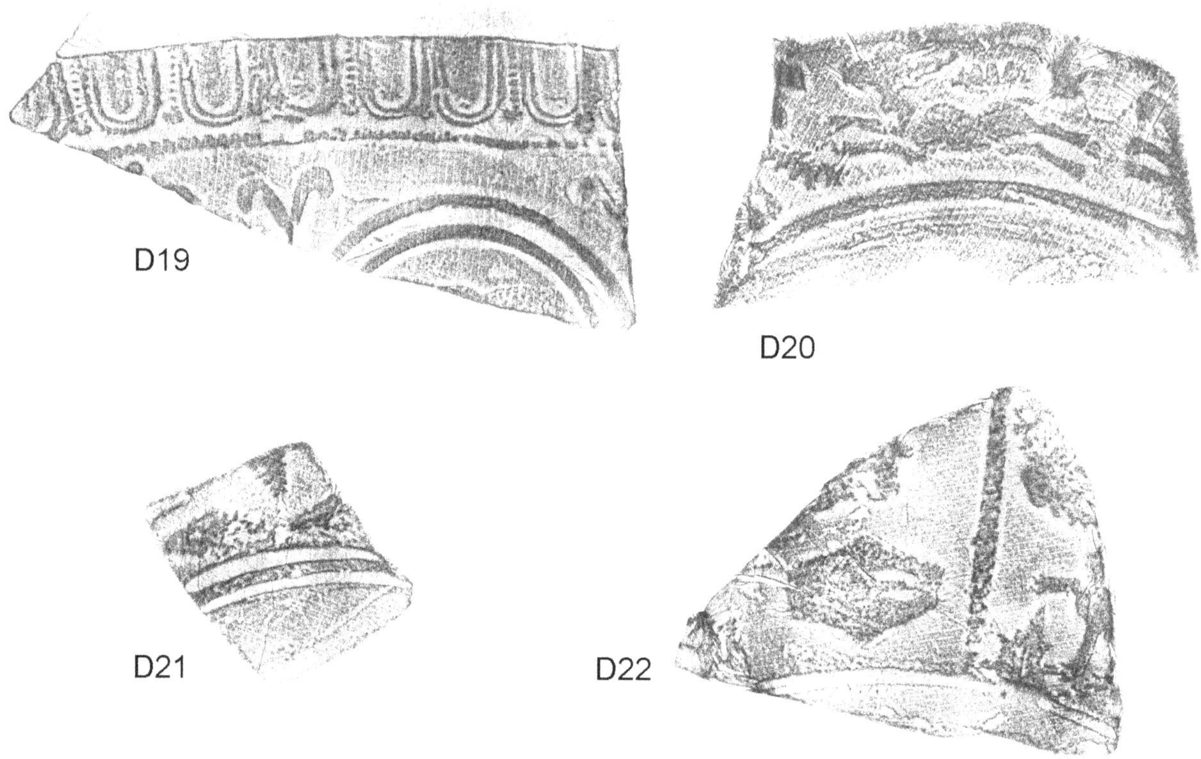

Figure 5.6: Decorated samian ware. Scale 1:1.

D22. Form 37, Central Gaulish (Fig 5.6), showing panels with the shield (Rogers U210) and the sphinx (O.857). The style is likely to be that of Do(v)eccus i, who used the types and large-beaded borders. The leaves are probably his (Rogers J149). *c* AD 165-200. Topsoil (3001).

The Potters' Stamps

I am grateful to Miss Brenda Dickinson for the stamp identifications and notes on which this report is based. The potter and die numbers are those currently appearing in Hartley and Dickinson 2008- . Each entry gives potter, die number, form of vessel, reading of stamp and pottery of origin. Superscript (a) after the name of the pottery denotes that the stamp is attested at the pottery in question, (b) that the stamp is not attested at the pottery in question, although the potter is known to have worked there, (c) that the stamp is assigned to the pottery on the evidence of fabric, distribution and/or form. Underlining of letters denotes ligatures.

S1. Catiolus 1a 31R CΛTIOLIM (blind A, lambda L) Lezoux[c]. *c* AD 155-95. Fill (1225) of brine tank [1207].

S2. Censorinus ii 1a 37 CENSORINI Lezoux[a] (D18 above). *c* AD 160-90. Fill (1282) of brine tank [1182].

S3. Cintusmus i 2b 31R CINTVSMIM Lezoux[a]. *c* AD 150-80. Fill (4230) of pit [4231].

S4. Cocuro 1a 31 [COCV]RO.F Lezoux[a] (cf Hartley 1972, fig. 82, 136). *c* AD 130-55. Fill (1214) of cut [1227].

S5. Doeccus i 13c 31 DOIICCVS (two lines at the end of the stamp are more likely to be a twig than letters) Lezoux[b]. *c* AD 170-200. Fill (1279) of brine tank [1182].

S6. Genitor ii 5a 31 G.E.N.I.T.O.R.F Lezoux[a]. An example of this stamp, with the final two letters missing, is illustrated from St Magnus House, London (Dickinson 1986, 189, 3.54). *c* AD 160-200. Fill (1208) of brine tank [1207].

S7. Marcellinus ii 2a 31 MARCELLINIF (N reversed and top bar of F missing) Lezoux[a] (cf Dickinson 1986, 191, 3.93). *c* AD 160-200. Fill (1188) of brine tank [1182].

S8. Maternianus i 3a 33 MΛTERIΛIIΛI Lezoux[a] (cf Dickinson 1986, 192, 3.113). *c* AD 170-200. Fill (1225) of brine tank [1207].

S9. Paterclinus 1a 33 PΛTERCLINIOF Lezoux[a]. *c* AD 155-80. Subsoil (1002).

S10. Paternus v 7a 37 PΛTERNFE Lezoux[a] (D9 above). *c* AD 160-85. Fill (1239) of brine tank [1207].

S11. Sacrillus 3a 79R SACRILL.I.M (A with a dot and lambda Ls) Lezoux[a] (cf Dickinson 1986, 195, 3.180). *c* AD 160-200. Fill (3017) of cut [3016].

S12. Sextus v 2b 33 SEXTIMΛN Lezoux[b]. *c* AD 160-90. Fill (1282) of brine tank [1182].

S13. Sextus v 4c 38 SEXTIMA Lezoux[b]. *c* AD 160-200. Fill (1273) of brine tank [1182].

Illiterate and Broken Stamps

S14. Illiterate 33 Λ.......TS (T upside down, S reversed) Lezoux[c]. Antonine. Fill (1279) of brine tank [1182].

S15. Form 31]I.M Central Gaulish, Antonine. Fill (1220) of brine tank [1182].

S16. Form 33]..I.MA Central Gaulish, Antonine. Fill (1285) of brine tank [1182].

S17. Form 33 illegible Central Gaulish, Antonine. Fill of ditch [3366], large enclosure.

S18. Form 31R illegible Central Gaulish, mid to late Antonine. Topsoil (4001).

6. Coins

David Shotter

A total of 52 coins were found in the course of the excavation, some from the excavation itself, others by metal detectorists working alongside the archaeological team. These latter finds are indicated in the following list by (MD). The coins, which comprised Roman issues, some British issues of the 18th century and at least one trade token, were examined at the time in a preliminary assessment but unfortunately have not been available for further study. Many were fragile and/or badly corroded. As a result, in the absence of further cleaning and treatment, it has not proved possible in most cases to make detailed identifications or assessments of wear.

In addition to the pieces which were obviously coins of various kinds, there were a number of 'bronze discs' which, though not badly corroded, were evidently completely featureless. Their sizes varied between the equivalents of *dupondii* and *sestertii*. It is possible that these were a forger's blanks, or perhaps pieces which were being prepared for votive use.

This report provides as much information as possible on individual coins, with separate lists of the Roman coins and post-medieval coins in chronological order. The report also takes account of other Roman coins which have been reported over the years from the Nantwich area.

The Coins

Trajan (AD 98-117): 5 coins

1. *Denarius* (fragmentary); moderately worn — AD 103-11
   IMP TRAIANO[
   ] COS V P P
2. (MD) *Sestertius*; little worn — 112-14
   [IMP CAES NERVAE TRAIANO AVG GER
   DAC P M] P M TR P COS VI P P
   CONSERVATORI PATRIS PATRIAE S C
   (*RIC* 2 (Trajan), 619)
3. *Sestertius*; area A; very worn — 103-17
4. *Sestertius*; area B; very worn — 103-17
5. *As*; area B; moderately worn — 103-17

Hadrian (AD 117-38): 1 coin

6. *Sestertius*; area B; moderately worn — 119-22
   IMP CAESAR TRAIANVS HADRIANVS AVG

Antoninus Pius (AD 138-61): 5 coins

7. *Denarius* (fragmentary); little worn — 145-61
   COS IIII Fortuna?
8. *Sestertius* (fragmentary); area A, subsoil (1002); little worn — 138-61
9. *Sestertius*; area C, subsoil (3002); very worn — 138-61
10. *Dupondius* or *as*; area B; very worn — 138-61
11. *Dupondius* or *as*; very worn — 138-61

Faustina I (AD 138-41+): 3 coins

12. (MD) *Sestertius*; very worn — 141+
13. *Dupondius* or *as*; moderately worn — 141+
14. *As*; area B; very worn — 141+

Marcus Aurelius (AD 161-80): 1 coin

15. (MD) *Sestertius*; very worn — 161-80

Lucius Verus (AD 161-9): 1 coin

16. *Sestertius*; moderately worn — 161-9

Faustina II (AD 145-76): 3 coins

17. *Dupondius*; area C, 3002; very worn — 161-76
   FAVSTINA AVGVSTA
   FECVNDITATI AVGVSTAE S C
   (*RIC* 3 (Marcus), 1641)
18. *Dupondius* (fragmentary); moderately worn — 145-76
19. *Sestertius*; moderately worn — 176+
   DIVA AVG [FAVSTINA]

Commodus (as Caesar) (AD 172-80): 2 coins

20. *Denarius*; area B; very worn — 172-80
21. (MD) *Sestertius*; little worn — 175-6
   L AVREL] COMMODO [CAES AVG
   FIL GERM SARM
   [IOVI CONSERVATORI] S C
   (*RIC* 3 (Marcus), 1524)

Commodus (AD 180-92): 2 coins

22. *Sestertius*; moderately worn — 183
   [M COMMODVS ANTONINVS AVG]
   [TR P VIII IMP VI] COS IIII P P [S C
   Minerva walking right
   (*RIC* 3 (Commodus), 368a)
23. *Dupondius* or *as*; area B; moderately worn — 180-92

Tetricus I (AD 271-3): 1 coin

24. Radiate Copy (fragmentary); little worn — 271-3

Unassignable Radiate Copy (*c* AD 280): 1 coin

25. (MD) Radiate Copy; very worn — *c* 280

House of Constantine: 1 coin

26. *Nummus*; area A; moderately worn — 341-6
   [VICTORIAE DD AVGGQ NN]
   Facing Victories

Illegible Coins (8)

27-34. These comprised one *sestertius*, one *dupondius* or *as* (fragmentary), and four *asses* (three fragmentary), all or which probably fell within a date range of *c* AD 70-200; one of the *asses* came from Trench 47. There were also two fragmentary Æ coins, of which no identification was possible, and nine featureless Æ 'discs'; all came from road (8057) in Trench 47.

Discussion

Although this group of coins is relatively small, it does

exhibit a few distinctive features. Firstly, the complete absence of any pre-Trajanic coins suggests that there was little or no activity on this site until, at least, the latest years of the 1st century. This is a dating feature which the present site has in common with a number of other 'industrial' sites in the North-West, most obviously Wilderspool (Shotter 2007, 69-70). In contrast, at first sight Holt (Shotter 2000, 91), Middlewich (Shotter 2000, 104) and Walton-le-Dale (Shotter 2000, 24-7) appear different in this respect. This, however, is due largely to the appearance of earlier coins at these sites, which is probably to be explained by a known or likely nearby military presence at an earlier point in the 1st century.

It appears likely that such 'infrastructural' sites saw considerable development following the decision in AD 87 to withdraw from what is now Scotland and, contrary to the implication of the Roman historian Cornelius Tacitus (*Histories* 1.2), to invest seriously in the consolidation of Rome's hold on northern England (Shotter 2004, 52-74). It will have been at this point that legion II *Adiutrix* was transferred to the Continent and replaced on a permanent basis at the fortress at Chester by XX *Valeria-Victrix*. This move must certainly have prompted major reorganisation in the field of military supply.

The Roman army's dynamic role in the development of frontier policy in Britain during the 2nd century will have placed a great emphasis on supplies, especially salt. It was essential not only for the preservation of foodstuffs but also for the preparation of hides for the manufacture and repair of tents, which will have been in great demand for a large campaigning army. This is reflected by the evident dominance on the present site of coinage of the 2nd century, which accounts for an overwhelming 88.46% of the legible coins in this sample. There are no sites in the North-West which share such a statistic with Kingsley Fields.

Two points of particular interest stand out in the profile of 2nd-century coin loss. Firstly, in both the sample from the excavation and the casually reported finds, the coins of Hadrian are very poorly represented (Table 6.1). Assuming that the coins of an emperor continued to circulate strongly in the reign of his successor, this phenomenon presumably reflects, as elsewhere (Shotter 1990, 118-19), the troop movements involved in the Antonine re-occupation of southern Scotland. Secondly, and of especial note, is the relatively high proportion of coins of Commodus. On virtually every other Roman site in the North-West, the coin loss of the 2nd century exhibits a steady downward trend from Trajan to Commodus, in which coins of Commodus are usually very few in number or completely absent. This is probably to be accounted for largely by the effects of growing inflation in rendering the smaller denominations of the coin system increasingly useless. Indeed, by the turn of the 2nd and 3rd centuries the system appears to have relied principally on the *denarius* and the *sestertius*, more valuable coins the loss of which, it is suggested, perhaps prompted a more diligent search. In addition the relatively large numbers of Commodan coins may point to reorganisation of some kind in the area, perhaps in connection with the campaigning of the contemporary governor, Ulpius Marcellus (Birley 2005, 162-70).

It should also be noted that all of the industrial sites in the region appear to have witnessed a decline in activity from the early 3rd century, from which there was no sign of a robust recovery, although it appears that the sites were not totally abandoned (Webster 1975, 91-2 for Wilderspool; Shotter 2008, 39 for Middlewich). In the case of the Nantwich excavation site, there are no recorded coins between those of Commodus (AD 180-92) and that of Tetricus I (AD 271-3), although there are two Severan *denarii* amongst the casual finds. It should be noted that the advanced state of wear exhibited by some of the coins of the 2nd century would not preclude their loss as late as the early 3rd century. Nevertheless, just two radiate copies were recorded from the excavation, together with one of the Constantinian period in the first half of the 4th century, periods which normally generate respectively more than 20% and 10% of site coin loss in the North-West. It may be, therefore, that the region saw a radical reorganisation in military supply in the 3rd century.

	Kingsley Fields 2002		Reports of casual finds	
Period	No	%	No	%
I (- AD 41)	-	-	-	-
II (41-54)	-	-	-	-
III (54-68)	-	-	-	-
IV (69-96)	-	-	5	17.24
V (96-117)	5	19.23	4	13.79
VI (117-38)	1	3.85	2	6.90
VII (138-61)	8	30.77	5	17.24
VIII (161-80)	7	26.92	4	13.79
IX (180-92)	2	7.69	3	10.34
X (192-222)	-	-	2	6.90
XI (222-35)	-	-	-	-
XII (235-59)	-	-	-	-
XIII (259-75)	2	7.69	1	3.45
XIV (275-94)	-	-	-	-
XV (294-324)	-	-	2	6.90
XVI (324-30)	-	-	-	-
XVII (330-46)	1	3.85	1	3.45
XVIII (346-64)	-	-	-	-
XIX (364-78)	-	-	-	-
XX (378-88)	-	-	-	-
XXI (388-)	-	-	-	-

Table 6.1: Chronological distribution of the legible coins from the excavation and casual finds reported from Nantwich.

Post-Roman Coins

These comprised four halfpenny pieces of George III (one of which was found by a metal detectorist, whilst one came from each of excavation areas A, B and D). There were also one illegible post-Roman coin, three illegible British coins of probable 18th-century date and one illegible trade token.

Other Roman Coin Finds from Nantwich

A considerable number of Roman, medieval and post-medieval finds made by metal detectorists in the wider area of the Kingsley Fields development (see Chapter 2) were recorded at the time of the archaeological programme of works. Amongst these were sixteen Roman coins, as follows:

Otho	1	AR (*RIC* 1^2 (Otho), 8)
Vespasian	1	AR (*RIC* 2^2 (Vespasian), 545?)
Domitian	1	AR (*RIC* 2^2 (Domitian), 451)
Nerva	1	AR (*RIC* 2 (Nerva), 25)
Trajan	1	Æ (*RIC* 2 (Trajan), 667)
Hadrian	1	AR (*RIC* 2 (Hadrian), 146ff)
Antoninus Pius	3	AR (*RIC* 3 (Antoninus), 282?); 2Æ (*Sestertii*; inc *RIC* 3 (Antoninus), 778)
Faustina II (under Antoninus Pius)	1	Æ (*RIC* 3 (Antoninus), 1377)
Marcus Aurelius	1	Æ (as *RIC* 3 (Marcus), 964)
Commodus	1	AR
Crispina	1	AR (*RIC* 3 (Commodus), 278)
Plautilla	1	AR (*RIC* 4 (Caracalla), 363a)
Unassignable Radiate Copy	1	Æ
House of Constantine	1	Æ (as *LRBC* I. 52)

W T Watkin (1886, 291-3) mentions, without giving very specific information, a small number of earlier finds of coins in Nantwich, including some that were evidently hoards (see below). In more recent years, a further fourteen finds have been reported (Shotter 2000, 108; 2011):

Domitian	2	(*Denarius*: *RIC* 2^2 (Domitian), 691; *Sestertius*)
Trajan	2	(*Denarii*)
Hadrian	1	(*Denarius*: *RIC* 2 (Hadrian), 146ff)
Antoninus Pius	1	(*Denarius*: *RIC* 3 (Antoninus), 137)
Marcus Aurelius	2	(*Denarius*: *RIC* 3 (Marcus), 1130; *Sestertius*)
Faustina II	1	(*Sestertius*: *RIC* 3 (Marcus), 1645)
Commodus	1	(*Sestertius*)
Julia Domna	1	(*Denarius*)
Claudius II	1	(Radiate Copy: *RIC* 5 (Claudius II), 53)
Constantine I	2	(*Nummi*: both of the SOLI INVICTO COMITI type; one minted in London)

These casual finds are broadly consistent in date range with those recovered in the excavation at Kingsley Fields (Table 6.1), although the denominational range is strikingly different, with many more *denarii* reported amongst the casual finds. This may point to a military involvement in the area. However, the presence of some coins earlier than those recovered from the excavation is not a matter for surprise, as coins of period IV (AD 69-96) would normally be expected to have remained in money circulating in the late 1st and early 2nd centuries. Indeed, in the cases of a number of Roman sites in north-west England we can distinguish between early Flavian and late Flavian/Trajanic establishments by the proportional relationship between coins of periods IV and V (Shotter 1993). Further, in the case of the coins from Nantwich, the reported condition of these earlier issues would suggest that they had been in circulation for some time prior to their loss.

In all, therefore, the coin evidence from the excavation, together with other finds, would confirm the presence from a date late in the 1st century or early in the 2nd of a substantial Romano-British settlement, presumably based principally upon the salt industry. A military presence of some kind seems also to be suggested, perhaps representing the market for Nantwich products or possibly in a 'policing' role. It should be emphasised that the coin evidence does not present a strong case, in the sampled areas at least, for activity earlier in the Roman period than the turn of the 1st and 2nd centuries AD.

Roman Coin Hoards from the Nantwich Area

Over the years a number of Roman hoards have been reported from Nantwich and the surrounding area, although mostly not in much detail (Shotter 2000, 167). Watkin (1886, 292) records the discovery 'many years ago in Marsh Lane' of twelve *aes*-coins, which appear to have been part of a hoard. He also reports the discovery in 1849 in 'Wall Lane tanyard' of fifteen *aes*-coins which were corroded into a 'hard lump'; all of the coins were said to have carried a helmeted head on the obverse, suggesting that they were probably Constantinian issues dating between *c* AD 317 and 335. In addition, it appears that a hoard of approximately 80 coins was found in Nantwich in *c* 1850. It is said that they were issues of Claudius and earlier emperors. Whilst it is possible that Claudius I was meant, it seems more likely, in view of the general coin profile, that this was a hoard of radiates and copies, containing issues of Claudius II (Thompson 1965, 106). As with the casual finds of Roman coins, the contents of these hoards appear to have had a similarity of dating with the coins excavated on the Kingsley Fields site.

7. Small Finds and Vessel Glass

H E M Cool

This report on the small finds from Kingsley Fields should be viewed as a preliminary rather than a definitive statement as a consequence of issues which arose during the analysis stage. The first relates to the condition of the material. Though some of the metalwork from the brine tanks is in very good condition, much of the rest of it is very poorly preserved. Under normal circumstances many items would have undergone investigative conservation to clarify details. This was not possible and many identifications have to remain possible rather than certain. In the circumstances it has been felt that the best way to illustrate much of the iron is to use Dr O'Connor's X-radiographs. The lack of conservation has also meant that some details about the pyre goods could not be checked. Secondly the wood and the stone would have benefited from expert species and geological identifications. Finally no stratigraphic site narrative was available whilst the report was being written. It has thus not been possible to comment on any issues relating to site formation processes nor to consider whether the date of the object is appropriate for the phase of the context it was found in. What is presented here is a basic typological account structured according to the part of the site the items were found in.

A considerable amount of metalwork was recovered during the machine clearance of the site and during a metal-detecting survey of the excavated spoil. Due to the circumstances of recovery much of this material could not be allocated to a specific area of the site. Some is clearly of Roman date but it also includes a decorated lead spindle whorl of late 13th- to 14th-century date (Brickstock *et al* 2007, 110 nos 62-3), musket balls and modern screw-threaded fittings. Unstratified and unlocated items that are clearly of Roman date are considered in a section at the end of the report.

Unstratified material that is not chronologically sensitive has not been generally included, even though some of it might well be of Roman date. An exception has been made in the case of the large amount of lead alloy runoff and sheet as it is felt that this may well relate to brine production on the site.

The evidence from the finds can be summarised as follows. They indicate that most of activity was taking place in the later 1st or 2nd century with an episode of cremation burial in the 3rd century. Various strands of the evidence point towards there being military involvement at Kingsley Fields during both phases. Various items found in the brine tanks strongly suggest that the decommissioning was associated with ritual structured deposition.

The Road Zone

The Road Zone produced a small assemblage of finds that belong to the later 1st and 2nd centuries where they can be dated.

Trumpet brooches such as 1.1 are common from the later 1st century to the middle of the 2nd century. Such features as can be discerned suggest it belongs to Bayley and Butcher's group A (Bayley & Butcher 2004, 160) like the unstratified, but better preserved 5.1 noted below.

This area also produced the handle of what I shall term here, though with some caution, a casserole (1.2) (Fig 7.7). These saucepan-shaped vessels have been given many different names in the modern literature. A popular one is *patera* but, as the Latin name was probably *trulla* (Koster 1997, 56), it appears best to avoid this. Den Boesterd, who was Dutch but who wrote her classic work in English, called them saucepans (Den Boesterd 1956, 1). Generally the modern term for saucepan seems accepted for these vessels in other languages (French *casserole*, German *Kasserolle*, Italian *casseruole*). This has the disadvantage that these words, like saucepan, have to modern eyes overwhelmingly culinary associations. In the ancient world the vessels were used for a variety of purposes and not merely culinary ones (Cool 2006, 47). The alternative is to use a descriptive term such as deep handled pan but that would rapidly become cumbersome. Casserole is preferred here to saucepan as it is similar to the word used in the other languages and is the English term preferred by Koster (1997, 56) in her continuation of Den Boesterd's work.

The need to establish what to call these vessels is important because this site is quite exceptional in having produced one complete and three fragmentary examples from a relatively small area (Fig 7.7). Two of these came from one of the brine tanks (4.42-3) but 1.2 and 5.2 appear to have been just ordinary site finds. Casseroles are found on all sorts of sites but overwhelmingly they are much commoner on military sites than on any other type of settlement (Cool 2006, table 15.3). The presence of so many in this area strongly points to a military presence.

Casseroles developed in the late Republican period and continued in use into the 3rd century. Dating normally depends on details such as the type of foot, the rim and wall form, and the decoration at the end of the handle. Fragments of handles such as 1.2 and 5.2 are not closely dateable on their own. Little further can be said about 1.2 but 5.2 retains a small punched circle. This, together with the size of the piece and the deeply incised grooves parallel to the edge, suggests that it might have come from a handle with a perforated disc end like the one from Corbridge (Eggers 1966, 108, Abb 13a). It is noticeable that the large number of disc-ended casserole handles which Tassinari publishes from Pompeii do not have this punched dot decoration (Tassinari 1993, G3000 series), and so it probably belongs to the late 1st century at the earliest. The small circles were often grouped in triangles and clearly represented bunches of grapes, reflecting the use of some of these vessels in the service of wine. Den Boesterd (1956, 10-11) notes that casseroles with this type of decoration were being made in Italy and Gaul from the late 1st century, so probably a 2nd-century date would be most likely for 5.2.

This area also produced the only glass vessel that could be identified as any form other than a blue/green bottle (1.3) (Fig 7.4). It is a funnel-mouthed indented jar of the late 1st and 2nd centuries (Price & Cottam 1998, 143-5). By normal standards a substantial amount of it is preserved.

Knobbed terrets (1.4) (Fig 7.5) are a widespread type coming into use in the 1st century and being commonest in the 2nd. The distribution mapped by MacGregor showed a large gap in the western Midlands with only a single example from Chester being shown (MacGregor 1976, 46-7, map 10). Interestingly the data from the Portable Antiquities Scheme does not appear to have modified this distribution as it has done for many classes of copper alloy items. The presence of this one at Nantwich is thus a welcome addition to the corpus.

The other items cannot be closely dated within the Roman period. 1.6 is a bell-shaped stud of the type discussed by Allason-Jones (1985) but is so encrusted that the variant cannot be identified. 1.7 is a pottery repair.

Personal Ornaments

1.1 Trumpet brooch, lacking spring, pin and catch plate. Copper alloy. Heavily corroded and details obscured. Cast loop on back of head for spring. Central button moulding appears to continue around the back of the brooch but precise form unclear. Brooch may have had a ribbed cylindrical footknob. Length 49mm, width of trumpet head 13mm. u/s Trench 47.

Household Equipment

1.2 Casserole; handle fragment (Fig 7.7). Copper alloy. Fragment from one side of handle close to junction with body of vessel; handle originally had a central constriction and two deep grooves on either side. Dimensions 31mm x 22mm, thickness 3mm. Road (8057) [8056].

1.3 Jar; seven non-joining rim, body and base fragments (Fig 7.4). Blue/green glass with many bubbles, some large. Funnel rim with fire-rounded edge; upper body sloping out, convex-curved body with elongated indentations; lower body sloping into edge of concave base. Rim diameter 70mm, base diameter 50mm, estimated height 140mm. Basal fill (5108) of well [5052].

Transport Equipment

1.4 Knobbed terret (Fig 7.5). Copper alloy. Oval-sectioned hoop with rectangular-sectioned bar; weakly biconical knobs with groove around widest part; traces of similar groove on the collar between the hoop and the bar. About three-quarters extant, lacking part of bar, one collar and adjacent part of hoop with knob. Depth 57mm, section of bar 15mm x 4mm, hoop section 6mm x 5mm. Road (8057) [8056].

Tools

1.5 Whetstone (Fig 7.5). Grey fine-grained and finely bedded stone. Rectangular section with rounded corners and damaged ends. Slightly waisted centrally through wear. Length 87mm, section 29mm x 26mm. Road (8057) [8056].

Fasteners

1.6 Bell-shaped stud. Copper alloy. All that is visible is the upper part; the rest is embedded in concretions and surrounded by stone. Slightly dished head with flat-topped conical central boss. Diameter of head 24mm. Road (8057) [8056].

1.7 Plug. Lead alloy. H-shaped; oval; retaining fragment of reduced pottery. Dimensions 28mm x 25mm. Weight 26g. Fill (5094) of pit [5093].

The Cremation Zone

Two of the burials produced pyre goods. In burial [4245] most of the items were made of iron and virtually all were recovered during the osteological analysis. Having been found late in the analysis programme it was not possible to submit them for X-radiography. When iron is burnt it can result in the object having a lack of corrosion which makes it easy to identify. This is the case for 2.4, 2.6 and some of the nails in 2.3. The other iron items do have corrosion crusts, which means that some of the identifications have to be tentative.

The items identified as nails (2.1 and 2.3) appear to be typical joinery nails which could have been used in the manufacture of a bier or a coffin, though the possibility also exists that old timbers with nails still in them were being used for the pyre. Mature oak of the type which formed the charcoal deposit (see Barnett, Chapter 8) would have been as suitable for building with, as for fuel. The presence of at least three upholstery nails (2.2, 2.7-8 and see also 2.9) would suggest that they were more likely to have been used to build a bier. It seems to have been upholstery nails of this sort (Manning 1985, 135 type 7) that were observed on the upholstered bier found in situ at Beckfoot (Bellhouse 1955, 52).

The deceased was also accompanied on the pyre by shoes, either worn or placed there. One hobnail can be identified with certainty (2.4), another group can be cautiously identified (2.5) and 2.8 consists of lumps that were either hobnails or upholstery nails. Hobnails are a regular inclusion in both cremation burials and in inhumations (Philpott 1991, 165-75). They were found both in this grave and in [4250]. No particular significance should be attached to the fact that none was recovered from the individual in the urned cremation burial (4004). As Mould has pointed out (2004a, 392), all that can be surmised from such an absence was that the individual was not accompanied by *nailed* shoes. They may, of course, have been accompanied by a pair of shoes in the numerous styles which did not use nails.

Half of a small blue square-sectioned bead was also found. It shows no obvious signs of having been melted by heat but the fact that it is split in two might indicate thermal shock. This was a common type in the 3rd-century cemetery at Brougham (Cool 2004, 386). Two examples were also found in the fortress baths drain deposit at Caerleon (Brewer 1986, 151 nos 71-2) dated to AD 160-230. It seems to be an introduction of the late 2nd or 3rd century, arriving with soldiers and their families who introduced the fashion for wearing strings of small beads. If this individual was accompanied by a bead string in the late 2nd or 3rd century, that would suggest military associations. The fashion was not to become widespread through the population as a whole until later. The individual in this burial has been sexed as a ??male. Bead string wearing was overwhelmingly a female fashion (Cool 2004, 389), so it is possible that this is a chance inclusion.

The individual in burial [4250] (a ??female) went to their pyre on a bier decorated with bone veneers or inlays (2.11-16) (Fig 7.1). This was a 3rd-century fashion. The largest group has come from Brougham (Greep 2004), but it was the group from a 3rd-century burial at Birdoswald which provided the information that they must have decorated biers rather than boxes (Wilmott *et al* 2009, 283-8). The Birdoswald ones were the subject of detailed conservation and were shown to have had red, blue and black colouring. It is possible that some of the Brougham ones had pink and black decoration. The fragments from burial [4250] have not been cleaned in any way and so it is not known whether they too were coloured, although one of the pieces in 2.11 shows discolouration which might indicate that they were.

The 3rd-century dating of the fashion seems secure as, in addition to the ones from Brougham and Birdoswald, they have been associated with 3rd-century cremation burials at Usk (Greep 1995) and at Low Borrowbridge where one was found but not identified for what it was (Cool 2004, 465). The only possibly earlier one came from Owlesbury in Hampshire (Collis 1977, fig 11). In the brief note about the cemetery this burial was stated to be of 2nd-century date but, as it has never been properly published, there is no way of verifying this. Where currently known, the associations of the burials with veneers are primarily military.

In addition to shoes, there appears to have been some form of iron fittings on the pyre of burial [4250] (2.18 and 2.25) (Fig 7.1) but what form the object they belonged to took cannot be suggested.

Amongst the other items from non-funerary contexts in this zone, only the frit melon bead 2.28 (Fig 7.3) can be independently dated. It would have belonged to the

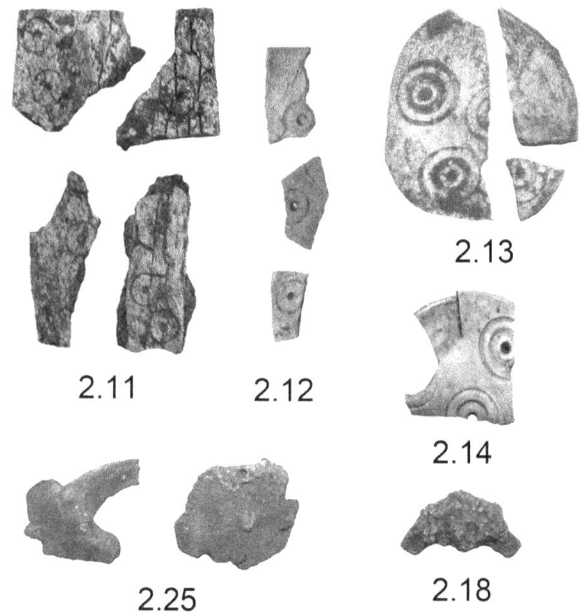

Figure 7.1: Small finds, pyre goods. Scale 1:1.

activity on the site before the cremation burials as it would have been in use in the mid 1st to mid 2nd century.

Cremation Burial [4245]

2.1 Joinery nails. Iron. Three head and shank fragments. (4244).

2.2 Upholstery nail. Iron. Large head, shape uncertain; short shank broken at tip. Head diameter (with corrosion products) 18mm. Found during osteological analysis. (4244) <56>.

2.3 Joinery nails. Iron. Six head and shank fragments including one retaining full length (75mm), eight shank fragments. Found during osteological analysis. (4244) <78>.

2.4 Hobnail. Iron. Complete pyramidal head and broken shank. Head diameter 10mm. Found during osteological analysis. (4244) <78>.

2.5 Hobnails? Iron. Corroded lump with features which suggest that it may consist of several hobnails corroded together. Found during osteological analysis. (4244) <78>.

2.6 Upholstery nail. Iron. Shallow domed head; short shank with tip bent sideways. Diameter of head *c* 16mm, length 21mm. Found during osteological analysis. (4244) <78>.

2.7 Upholstery nail. Iron. Upper part of shallow domed head. Head diameter *c* 18mm. Found during osteological analysis. (4244) <78>.

2.8 Hobnails or upholstery nail. Iron. Two corroded lumps representing either hobnails or studs corroded together. Found during osteological analysis. (4244) <78>.

2.9 Bead. Translucent (cloudy) mid-blue glass. Short square-sectioned; half extant split longitudinally. No obvious sign of burning. Length 4.5mm, width 3mm. Found during osteological analysis. (4244) <78>.

2.10 Bead? Corroded mass of iron retaining a small opaque piece of a mid-blue substance, material uncertain but might be glass. Found during osteological analysis. (4244) <78>.

Cremation Burial [4250]

2.11 Veneer (Fig 7.1). Bone, burnt white. Eight flat fragments decorated by single ring and dots. One fragment has two straight edges at 90^0, two others have one straight edges. One fragment has traces of brownish discolouration. Largest fragment 15mm x 15mm, total area *c* 8cm^2. (4249).

2.12 Veneer (Fig 7.1). Bone burnt white. Three small fragments with small ring and dot as 2.11 above. Found during osteological analysis. Area 1.5cm^2. (4248) <46A>.

2.13 Veneer (Fig 7.1). Bone burnt white. Three fragments from a disc of *c* 30mm diameter now distorted. Decorated by large double ring and dots, possibly a ring around the outside with one central one. (4249).

2.14 Veneer (Fig 7.1). Bone burnt white. Part of similar disc with large double ring and dot as 2.13 above. Found during osteological analysis. (4248) <46A>.

2.15 Veneer. Bone burnt white. Four small fragments with large double ring and dot motif as 2.13-14. Found during osteological analysis. (4248) <46A>.

2.16 Veneer. Bone burnt white. Irregular diamond shape, probably deliberately shaped. Dimensions 9 x 6mm. Found during osteological analysis. (4244) <78>.

2.17 Hobnails (5). Iron. One with pyramidal head. Diameter 9mm. Three with more shallow domed heads. One broken. Head diameters 9mm. Found during osteological analysis. (4244) <46>.

2.18 Ring (Fig 7.1). Iron. Square-sectioned. *c* 35% of circumference extant. Original diameter *c* 15mm, section 3mm. Found during osteological analysis. (4244) <46>.

2.19 Nails. Iron. Two head and shanks and twelve shank fragments. Burnt. (4248).

2.20 Nails. Iron. Head and shank fragment, also four shank fragments. Found during osteological analysis. (4248) <46>.

2.21 Hobnails (2). Iron. With pyramidal head. Diameter 10mm, 9mm. Found during osteological analysis. (4248) <46A>.

2.22 Nails. Iron. One head and 16 shank fragments. Found during osteological analysis. (4248) <46A>.

2.23 Nail. Iron. Corroded shank fragment. (4249).

2.24 Nail. Iron. One head and shank and two shank fragments. Found during osteological analysis. (4249) <47>.

2.25 Open work fitting (Fig 7.1). Iron. Triangular fragment of plate with projecting rectangular-sectioned, curved bar. Also one approximately circular piece of plate. Found during osteological analysis. (4249) <47A>.

2.26 Hobnail. Iron. Slightly domed head. Head diameter 9mm. Found during osteological analysis. (4249) <47A>.

2.27 Nails. Iron. Two head and shank and seven shank fragments.

Other Features

2.28 Melon bead (Fig 7.3). Frit. Surfaces slightly eroded but much turquoise glaze remaining. Length 10mm, diameter 14mm, perforation diameter 6mm. Fill (4132) of linear [4131].

2.29 Whetstone (Fig 7.5). Dark grey, very fine. Rectangular-sectioned with rounded angles; waisted through wear. Length 180mm, section 29 x 23mm. Fill (4027) of pit [4019].

Context	Complete	Incomplete	Total
4102	1	-	1
4140	-	1	1
4232	3	1	4
4235	-	1	1
4238	-	2	2
4240	-	1	1
Total	**4**	**6**	**10**

Table 7.1: Nails from the Cremation Zone other than those associated with cremation burials. Quantified by heads. The complete examples range from 40mm to 66mm in length.

Context	Runoff	Sheet	Total
4006	26	-	26
4161	13	-	13
4232	25	-	25
Total	**64**	**-**	**64**

Table 7.2: Lead runoff and sheet from the Cremation Zone. Quantified in g.

The Utilitarian Zone

Few of the items from the Utilitarian Zone are closely dated. The small brooch (3.1) (Fig 7.3) falls outside most of the standard classifications of Romano-British brooches. The type is discussed by Mackreth in connection with a larger example with a more pronounced headplate from Derby (Mackreth 1985, 293 no 33). It is not common and few have come from well-dated contexts. Such dating as there is suggests an origin in the 1st century with use also in the earlier 2nd century. Unfortunately this example was found unstratified and so can make no contribution to the dating problem.

There are two fragments (3.2-3) from blue/green glass bottles which were very common from the later 1st century into the 3rd (Price & Cottam 1998, 194-200). There is also a fragment from another blue/green glass vessel (3.4). The form cannot be identified but the colour indicates a 1st- to 3rd- century date.

The only other item of special note is the handle (3.5) (Fig 7.4), which might come from a latch lifter (Manning 1985, 88-9), and the iron shank (3.7) (Fig 7.4), which appears to be decorated with non-ferrous inlay. This is a decorative technique often used on styli but this shank appears to be too narrow for that.

Personal Ornaments

3.1 Unclassified bow brooch (Fig 7.3). Copper alloy. Surfaces obscured and details unclear. Head has projecting plate, probably part of a cast headloop. Back of the head has a perforated lug at either side to hold spring, some of which may be present. Oval-sectioned upper bow projecting forward and with sharp carination to lower bow with triangular cross-section; angle between the two marked by projecting boss; a rib across the lower bow and another wider beaded rib forming a footknob with a projecting spike below; small triangular catch plate. There may be two vertical ribs running down the upper bow. Present length 35mm, width of head 8mm. u/s, area B.

Household Items

3.2 Prismatic bottle; body fragment. Blue/green glass. Fill (2966) of ditch [2065].

3.3 Prismatic bottle; body fragment. Blue/green glass. Fill (3139) of ditch [3138] of U-shaped enclosure.

3.4 Body fragment. Blue/green glass. Fill (3213) of posthole/pit [3212] of Building 5.

Fasteners

3.5 Handle (Fig 7.4). Iron. Tang, one end broken, other bent round into loop. Length 88mm, width of handle 8mm. Pit [3029]. Identified from X-radiograph.

Buildings

3.6 T-clamp. Iron. Complete apart from broken end of one arm. Length 124mm, width of head 50mm. Pit [3084].

Miscellaneous

3.7 Pin (Fig 7.4). Iron. Tapering bar with transverse bands of non-ferrous inlay around widest part; both ends broken. Length 50mm, maximum width 4mm. Identified from X-radiograph. Posthole/pit [2039], Building 1.

3.8 Whorl (Fig 7.4). Lead alloy. Cylindrical. Diameter 25mm, thickness 4mm, perforation diameter 7mm, weight 16g. (3079) Pit [3078], Building 2?

3.9 Open socket? Iron. Identified only from X-radiograph. Length 60mm, maximum width 15mm. Fill (3162) of pit [3161], Building 2?

Context	Complete	Incomplete	Total
2039	-	2	2
2040	-	2	2
2048	1	-	1
2096	-	1	1
2134	-	1	1
2160	-	1	1
3002	-	1	1
3021	-	1	1
3070	1	-	1
3079	1	-	1
3137	-	1	1
3190	1	-	1
3192	-	1	1
3194	2	-	2
3230	-	1	1
3291	-	1	1
3332	-	1	1
3367	-	1	1
Total	**6**	**15**	**21**

Table 7.3: Nails from the Utilitarian Zone. Quantified by heads. The complete examples range from 45mm to 80mm in length.

Context	Runoff	Sheet	Total
2040	-	12	12
2063	9	-	9
2157	147	-	147
2159	22	-	22
u/s	752	-	752
3002	2118	-	2118
3079	9	-	9
3192	-	8	8
3196	-	637	637
3217	-	32	32
u/s	752	33	785
Total	**3809**	**722**	**4531**

Table 7.4: Lead runoff and sheet from the Utilitarian Zone. Quantified in g.

The Brine Industrial Zone

The bulk of the finds in this area were found in the two brine tanks and the recognisable material is summarised in Tables 7.5 and 7.6. Fragments from blue/green glass bottles (4.2-4 and 4.20-1) and other unidentified blue/green glass vessels (4.5 and 4.22-4) were found in both tanks, as were iron nails (Tables 7.7 and 7.9) and lead runoff and sheet (Tables 7.8 and 7.10).

The hobnails (4.1 and 4.16-17) would have derived from the same activity that deposited the leather-working debris in both tanks (see Chapter 9). One of the complete tools in tank [1182] also seems most likely to be associated with leather working (4.9) (Fig 7.5). The short blunt-ended blade could have been used as a creaser. These tools were used both to impress a line as a guide for sewing and to compress the edges of articles such as straps to harden them. They do not appear to have been noted on Romano-British sites before, although medieval examples are known (Goodall 1990, 249 no 335). Obviously the Roman leather worker would have been as much in need of such a tool as a medieval worker, and Mould has noted that the decoration on an upper found at the site would have been made with one (see Chapter 9). It is probably the exceptional preservational conditions within the tank that have enabled this tool to be identified for what it was. This makes judging how rare these tools actually were in Roman Britain difficult.

The commonest category of tool from the tanks were those connected with carpentry. Two complete adze-hammers came from tank [1207] (4.28-9) (Fig 7.6), together with a complete axe-head (4.30) (Fig 7.6). 4.28 retained traces of minerally preserved wood within the socket, so it is possible that all three were placed in the tank complete with their handles. 4.28 is extremely well preserved and retains lugs above the socket. Manning notes this as an early feature, suggesting that this example is of 1st- or 2nd-century date (Manning 1985, 18 no B16). Similar lugs are seen on the axe, which would again suggest a 1st- or 2nd-century date (Manning 1976, 24). The overall shape places it within Manning's type 2 category of axe-heads which could have been used either for felling or shaping timber (Manning 1985, 16). There is also a drill bit (4.7) (Fig 7.5), with a slightly expanded diamond-shaped point, from tank [1182]. Manning (1985, 26) notes that this is a form rarely identified in the archaeological record, probably because even small amounts of damage render them unrecognisable. Drills with spoon bits are far more likely to survive in a recognisable state. Again it is the exceptional preservation conditions of the brine tank that have preserved the piece in an easily recognisable form. 4.31 from the other tank might also have been a drill but it is in poor condition and this has to be a possible rather than a certain identification. The same is true of the possible paring chisel (4.8) from tank [1182].

The other activity represented by a tool is agriculture. 4.32 (Fig 7.6) is a reaping hook of Manning's (1985) type 2, a form that continues in use from the Iron Age into the

Roman period. The sockets are normally open but this appears to be closed, although it is covered with a thick corrosion crust. The wooden handle is still present and so the tool was clearly placed in the tank in a complete and functioning state.

All of these tools would have needed to be sharpened and a whetstone was recovered from each brine tank, neither of which showed any great indications of wear (4.10 and 4.33) (Fig 7.5).

A single item relating to transport came from each tank. 4.6 (Fig 7.5) from tank [1182] is half of a two-link snaffle bit, a common form in Roman Britain (Manning 1985, 66). By contrast, the shaped antler tine (4.27) (Fig 7.5) from tank [1207] is an uncommon find in the province, where antler was rarely exploited before the late Roman period. Such items are more common in the north-western and north central provinces on the continent, where they are generally identified as pack needles to help tie up the bundles that would have been transported by mule trains (Schenk 2008, 66-7). In his study of the ones conserved in the Landesmuseum Mainz, Mikler (1997, 56 n 83) drew attention to two from Britain, neither of which had been correctly identified before. One came from Newstead, possibly in an Antonine context (Curle 1911, 314, pl LXXXIII no 11), and the other from a 3rd-century context at Chalk (Johnston 1972, 139, fig 17 no 1). On the continent pack needles are found in contexts dating from the late Iron Age to the 3rd century. There is not as much evidence for the use of mule trains in Britain as there is on the continent (Johnstone 2008, 141), which might perhaps explain the rarity of such needles here. The presence of 4.27 might hint at a mule train at Nantwich. It also backs up the hints elsewhere in the assemblage that there was a military presence in the vicinity, as that would have been the organisation most likely to have had access to this type of transport. It is not known how the salt from the brine tanks was transported but, given the quantities they were capable of producing, a sophisticated transport network can be assumed.

The activities surrounding the extraction and dispatch of the salt would have involved record keeping and probably weighing, given that the most commodities in the ancient world were sold by weight. Record keeping is attested to by the stylus (4.26) (Fig 7.3) from tank [1207]. In the Manning typology this falls into to his type 4 category which includes all decorated styli (Manning 1985, 85). Major (2002) studied this category with regard to the examples from Heybridge and was able to define eight different types. 4.26 falls into her group 7 which have been found in 2nd- to 4th-century contexts. The Nantwich example has only a very thin corrosion crust but this would be sufficient to obscure any non-ferrous inlay decoration if it had been present. Unfortunately this item has not been the subject of X-radiography and so the presence of such decoration, which has been found on other examples, can neither be proved nor disproved. Weighing is possibly represented by 4.25 which may be a steelyard weight. A definite Roman steelyard weight was found at Kingsley Fields (5.4) (Fig 7.4) but from within which area of the excavation area is unknown.

As already noted in connection with 1.2 from the Road Zone, casseroles could have a variety of functions but the two found in brine tank [1207] (4.42-3) (Fig 7.7) were almost certainly connected with religious activity and would have been used to pour libations. Although no metal analysis has been carried out on them, it would appear that they are most probably made of lead alloy and not copper alloy as is far commoner. The burial conditions in the brine tank were such that the appearances of some metal objects are not typical of what is normally expected of particular alloys buried in archaeological contexts, but neither of these pieces has the slightest trace of green in any corrosion product. Their uniformly grey appearance would be typical of pewter.

The Romano-British pewter industry is generally associated with the late Roman period but it is clear that it was active earlier (Beagrie 1989, 175). The shape of these pans with waisted handles with perforated disc terminals, combined in the case of 4.42 with a steep side and strongly ribbed base, is the same as a very common form in copper alloy dated to the later 1st and 2nd centuries (Den Boesterd 1956, 7). The close copying of the imported copper alloy form is interesting. When casseroles in pewter have been found before this close adherence to the copper alloy original is not generally noted (see for example those from the Sacred Spring at Bath discussed below).

Though vessels of this form had a variety of purposes, ones made from pewter should almost certainly always be associated with sacrifice, as given their metal composition they would clearly be unsuitable for cooking. Lead, after all, has a low melting point and the walls of the vessels are thin, so the regular application of heat would not be a very good idea. In the Sacred Spring at Bath eight handled pans of a variety of shapes were found, most of which had inscriptions explicitly dedicating them to the goddess Sulis Minerva (Sunter & Brown 1988, 14-20). Of these five were pewter, two were silver and one was enamelled copper alloy. Pewter was thus explicitly thought of as being appropriate for the goddess. This might have been because when new it would have been shiny like silver but the element of sympathetic magic may also have played a part, as it did in using lead alloys for curse tablets (Tomlin 1988, 81-2). There has been much debate over whether pewter vessels in general should be seen as utilitarian alternatives to silver, or as something uniquely fitting for the deities that are found in watery places (see Earwood *et al* 2001, 280-1 for a summary of this). Certainly the recovery of these pans in this brine tank fits a common pattern of the deposition that can be associated with ritual activities, and they alone would be sufficient to suggest that the infilling of the tank included an act of structured deposition and not merely the tidying away of rubbish.

As items connected with what may be termed general household activity are rare in these tanks, the presence of the bucket escutcheon (4.19) (Fig 7.2) in tank [1207] is also of some interest. It is an ox-head escutcheon and would thus originally have been one of a pair for attaching the handle to the vessel (see for example Fox 1958, fig 46). Fox (1958, 73) made the attractive suggestion that they were fittings from milk pails, but what was in the vessels they decorated is unknown. The broken state of this piece and the fact that there is only one from the brine tank argues against it having been deposited as part of a complete vessel and so, like the vessel glass fragments, its deposition should probably be associated with rubbish disposal rather than having any ritual intent. Ox-head and more generally bovine escutcheons were a popular fashion that started to appear in the late Iron Age and were still being used in the late 4th century. They are widespread throughout the province and are a category of artefact where the numbers known have been greatly increased by finds recorded through the Portable Antiquities Scheme (Worrell 2006, 456-7). The depiction of the animal heads takes a large number of different forms and currently it is not possible to divide them into well-dated variants.

The brine tanks also included a number of iron fittings probably from boxes or items of furniture (4.12, 4.36-7) (Fig 7.3). Such an origin would also be appropriate for the copper alloy stud 4.34 (Fig 7.4) which retains wood and space for another element. Structural elements are indicated by such items as the bolt (4.38) (Fig 7.4) and nails. The number of the latter in brine tank [1207] was large (Table 7.9), especially when compared to the numbers from elsewhere on the site (see for example Table 7.3). Both in [1207] and [1182] (Table 7.7) the lengths of the nails are typical of general purpose joinery. Other items worthy of note include a possible padlock bolt (4.11) (Fig 7.4). This was only identified from the X-radiograph but has features that would be consistent with coming from a Manning type 1 padlock (Manning 1985, 95-6). Tank [1207] also contained one definite (4.39) (Fig 7.4) and two possible (4.40-1) lead pottery repairs and two lead whorls (4.45-6) (Fig 7.4). Pottery repairs were regularly encountered at this site. In addition to these, there was also one from the Road Zone (1.5) and a further ten were found unstratified and unlocated. Whorl 4.45 might possibly have been a spindle whorl. The other example could not have been, as the perforation diameter was too large for the normal Roman spindle, and so it is probably best to regard both of them as being of uncertain function. Again lead whorls of uncertain function were regularly found at this site, one coming from the Utilitarian Zone (3.8) (Fig 7.4) and ten being unstratified and unlocated. Given the presence of the definite 13th- to 14th-century lead spindle whorl noted in the introduction, not all of the latter group need have been Roman but many may have been.

An element of metalwork that was well represented in the brine tanks was lead alloy both in the form of sheet and runoff. As can be seen from Tables 7.8 and 7.10, over 1kg came from tank [1182] and nearly 4kg from tank [1207]. This type of lead was also well represented in the other contexts in the Brine Industrial Zone (Table 7.11). The Utilitarian Zone produced 4.5kg (Table 7.4) but further north the amounts drop off noticeably (Table 7.2).

	[1182]		[1207]	
	Complete	Incomplete	Complete	Incomplete
Hobnail	8	-	7	-
Bucket fitting	-	-	-	1
Weighing equipment	-	-	-	1
Writing equipment	-	-	1	-
Transport equipment	-	1	1	-
Carpentry tools	1	1	3	1
Leather tool	1	-	-	-
Whetstone	1	-	1	-
Fasteners	1	2	4	4
Agriculture	-	-	1	-
Casserole	-	-	1	1

Table 7.5: The small finds from brine tanks [1182] and [1207], excluding vessel glass, nails and unidentified miscellaneous items.

	[1182]			[1207]					
	(1279)	(1285)	u/s	(1208)	(1209)	(1239)	(1240)	(1337)	u/s
Hobnail	8	-	-	-	-	7	-	-	-
Bucket fitting	-	-	-	1	-	-	-	-	-
Weighing equipment	-	-	-		-	1	-	-	-
Writing equipment	-	-	-	-	-	-	-	-	1
Transport equipment	-	1	-	-	-	1	-	-	-
Carpentry tools	-	2	-	1	-	3	-	-	-
Leather tool	-	1	-	-	-	-	-	-	-
Whetstone	-	1	-	-	1	-	-	-	-
Fasteners	-	1	2	-	-	4	-	2	2
Agriculture	-	-	-	-	-	-	1	-	-
Casserole	-	-	-	-	-	1	1	-	-

Table 7.6: The small finds from brine tanks [1182] and [1207] by context, excluding vessel glass, nails and unidentified miscellaneous items.

All of this lead presumably derives from the pans in which the brine would have been evaporated and from the waste products from their manufacture. Lead brine evaporation tanks have been found in the vicinity of the excavations (Hassall & Tomlin 1984, 342 nos 18-19). A re-interpretation of the inscriptions on them has suggested that they might relate to clerical involvement in the industry in the 4th or 5th centuries (Shotter 2005, 44). If this is correct, they would not be directly relevant to the Kingsley Field finds. It is hard though to imagine any other reason for the very large quantities of lead found during the excavation other than the fact that lead pans were being used. This provides an interesting contrast to the contemporary industry at Droitwich where lead was very scarce (Woodiwiss 1992, 82) and the brine was clearly being evaporated in briquetage vessels. The ability to access large quantities of lead and to use that for evaporation might be another strand of evidence suggesting that there was military, or at least state, involvement with the industry at this site.

The contents of the brine tanks pose some interesting questions. Table 7.5 quantifies the items in the tanks according to whether they are whole or broken. If the hobnails and fasteners are set to one side, as they are often found whole in the archaeological record, it can be seen that both tanks contained a number of complete items of sorts that are not normally found in that state. Normally finding large complete items such as tools and casseroles would suggest that a ritual act of structured deposition was taking place, and certainly something of the sort would be likely to accompany the infilling of a large watery place such as the tanks. As already noted, the presence of a complete lead alloy casserole strongly suggests such an activity. That there is an element of what might be termed normal rubbish disposal is clear from the presence of the glass vessel fragments and items such as the bucket escutcheon and the possible padlock bolt. Given the scraps of leather which Mould has suggested to be workshop debris, it is possible that the complete creaser may have been an accidental inclusion bundled inside pieces of leather and out of sight. Possibly something similar might be envisioned for other small complete and still useable items such as the pack needle and the stylus. It is difficult to see, however, how items such as the adze-hammers, the axe, the complete drill bit and the reaping hook could have been casual inclusions. They were all still in a usable state and, even if no longer required, could have been used for scrap. On most sorts of site they would all suggest that structured deposition was taking place. On this site one note of caution does have to be sounded. Experimental work has shown that any tools used in the salt-making process rapidly become encrusted with salt as it crystallises and need to be constantly washed in fresh water (Fielding 2005, 63), so it is possible that some of the tools recovered from the tank were regarded as past their best, even if they were complete. The iron tools here, however, relate to carpentry and agriculture and are not obviously involved in salt making, though the adze-hammers would have been useful for breaking up large hard lumps of solidified salt. A utilitarian explanation for their disposal is hard to suggest and so they too, along with the complete casserole, are probably candidates for deliberate deposition. To them can probably be added the whetstones given that they showed little wear. The status of the other small but complete items (the creaser, the stylus and the pack needle) remains more ambiguous.

Relatively few finds came from the other contexts within the Brine Industrial Zone. There was one example (4.54) (Fig 7.3) of a small Polden Hill brooch (Bayley & Butcher 2004, 159-60). The date range for this type of the Colchester Derivative brooch which is typical of the western side of Britain runs from the Neronian period to the mid 2nd century. This example has no features which must be early and is most likely to date to the late 1st or 2nd century. There are also a few fragments of blue/green glass bottles (4.55-7, see above on 3.2-4) and two items of ironwork. One is a most unusual paddle-bladed tool (4.58) (Fig 7.6) that appears to have had an edge. The handle is a solid bar and was clearly not a tang for any other type of handle made of wood for example. It would be impractical to use as a knife or cleaver and it is tempting to suggest that it was used in the salt-manufacturing process. The experimental work showed that salt sets hard and breaking it up can be a problem. Perhaps this substantial tool was used to ensure that it was separated into the correct-sized blocks before it had set too hard. The other iron item (4.59) appears from the X-radiograph to have been a strap hinge (cf Manning 1985, 127).

Brine Tank [1182]

Personal Equipment

4.1 Hobnails (8). Iron. Best preserved has conical head. Length 9mm, head diameter 8mm. (1279). (The finds packaging indicated ten were present originally).

Household Items

4.2 Bottle; handle fragments (2). Blue/green glass. Edge of angular handle. (1188) and fragment of reeded handle (1277).

4.3 Prismatic bottle; body fragments (6). Blue/green glass. (1220).

4.4 Square bottle body fragment. Blue/green glass; strain-cracked to granular sugar texture. (1286).

4.5 Body fragment. Blue/green glass. (1220).

Transport Equipment

4.6 Snaffle bit (Fig 7.5). Iron. Rectangular-sectioned slightly curved bit, one end wrapped around circular-sectioned side ring; other end of bit curved and broken. Bit length 135mm, bit section 7mm x 5.5mm; side ring diameter 58mm, ring section 5.5mm. (1285).

Tools

4.7 Drill bit (Fig 7.5). Iron. Long pyramidal head; circular-sectioned stem; slightly expanded flat diamond-shaped bit. Length 169mm, section head 12mm x 11mm, section stem 7.5mm. (1285).

4.8 Paring chisel? Iron. Rectangular-sectioned bar expanding and tapering to broken blade end; other end broken. Length 48mm, section 12mm x 4mm, 17mm x 2.5mm. (1285).

4.9 Creaser (Fig 7.5). Iron. Narrow rectangular-sectioned

handle with short tapering, blunt-ended blade. Vivianite deposits. Length 130mm, length of blade 36mm, section of handle 6.5mm x 5mm. (1285).

4.10 Whetstone (Fig 7.5). Pale grey shelly limestone. Rectangular section, sharp angles, worn down at one end. Length 178mm, section 28mm x 19mm. (1285).

Fasteners and Fittings

4.11 Padlock bolt (Fig 7.4). Iron. Rectangular bar bending at 90 degrees with side view consistent with it being a barb spring. Length 75mm. Identified from X-radiograph. u/s in [1182].

4.12 Strap terminal. Iron. Diamond-shaped plate, one end broken, one with angular trefoil terminal. X-radiograph indicates the possibility of two shanks at the widest part, but there is no indication visible on the surface of this well-preserved piece. Length 116mm, maximum width 43mm, thickness 5mm. (1285).

4.13 Stud (Fig 7.4). Lead alloy. Circular cupped head; square-sectioned tapering shank. Length 43mm, head diameter 20mm. u/s in [1182].

Miscellaneous

4.14 Binding? Copper alloy. Two lengths, slightly curved, of cast asymmetrical 'V'-shape with flat base on one piece where it projects beyond inner edge. Lengths 150mm and 125mm. Also one corroded flat detached fragment. (1285).

4.15 Square-sectioned bar, both ends broken, part of rectangular element. Width 49mm. Bar section 19mm x 12.5mm. (1285). It is stated to have been found 'in context with large wood and metal object - possibly fragment of that item'. This has not been identified.

Context	Complete	Incomplete	Total
1283	2	1	3
1284	-	1	2
1285	1	1	2
1286	1	-	1
u/s	6	2	8
Total	**10**	**5**	**15**

Table 7.7: Nails from brine tank [1182]. Quantified by heads. The complete examples range from 39mm to 110 mm in length.

Context	Runoff	Sheet	Total
1183	146	206	206
1220	613	-	613
1277	-	1	1
1280	20	-	20
1281	92	-	92
1285	132	-	132
Total	**1003**	**207**	**1210**

Table 7.8: Lead runoff and sheet from brine tank [1182]. Quantified in g.

Brine Tank [1207]

Personal Equipment

4.16 Hobnail. Iron. Length 12mm, head section 7mm. (1239).

4.17 Hobnails (6). Iron. Length 11mm, head diameter 9mm. (1239).

4.18 Square buckle frame with traces of possible pin on one arm. Dimensions 35mm. Identified only from X-radiograph. (1239).

Household Items

4.19 Bucket escutcheon (Fig 7.2). Copper alloy; surfaces much obscured. Ox-head with broken attachment loop; tips of horns broken, possible beaded ridge between horns and more beading at nostrils; back hollow. Present length 49mm, width at horns 32mm. (1208).

4.20 Square bottle; body fragment. Blue/green glass. Strain cracked to granular sugar texture. (1236).

4.21 Prismatic bottle; one shoulder and two body fragments. Blue/green glass. (1225).

4.22 Body fragment. Blue/green glass. (1208).

4.23 Body fragment. Blue/green glass. (1224).

4.24 Body fragments (2). Blue/green. Strain cracked to granular sugar texture. (1225).

Weighing Equipment

4.25 Steelyard weight? Lead alloy. Small damaged sphere with area of iron? corrosion at one damaged point. Diameter 18mm. Weight 28mm. (1239).

Writing Equipment

4.26 Stylus (Fig 7.3). Iron. In two joining fragments. Waisted eraser with angular step junction with circular-sectioned shank; shank expands to swelling above the undifferentiated point. Three grooves below the eraser head; band of cross-hatched decoration between pairs of grooves on maximum part of swelling. Length 125mm, width of eraser 6mm, section at swelling 6mm. u/s in [1207].

Transport

4.27 Pack needle (Fig 7.5). Antler tine; rectangular notch on one face; vertically perforated above notch. High degree of polish on surfaces from long wear. Length 93mm. (1239).

Tools

4.28 Adze-hammer (Fig 7.6). Oval eye with long oval sleeve below and lugs on either side above; oval-sectioned hammer head with domed face; adze-blade rectangular-sectioned and is turned through sharp angle and has straight sides which splay out to give a wide, slightly convex edge. Interior retains minerally preserved wood, traces of vivianite. One large additional corroded lump on one side but otherwise pristine. Length 96mm, eye socket 28 mm x 26mm, width adze-blade 70mm, section of hammer 26mm x 21mm. Weight 0.988kg. (1239) SF 9.8.

4.29 Adze-hammer (Fig 7.6). Iron. Entirely enclosed in corrosion crust, X-radiograph reveals it to be complete; also shows a bright metal spot inside socket. It is probably not in such good condition as 4.28. Length 103mm, socket 29mm, section of hammer 19mm. (1239) SF 9.7.

4.30 Axe (Fig 7.6). Iron. Slightly concave front face, rear face sweeping back to provide wide cutting blade; oval socket with lugs above and below; back slightly rounded on one edge possibly from use as hammer. Blade broken at junction with rear face; otherwise complete. Length 155mm, socket diameter 30mm x 22mm, section of back 39mm, depth of blade 83mm. Weight 1.038kg. (1239) SF 9.9.

4.31 Drill bit head? Iron. Fragment with spalled surfaces; expanded triangular head square-sectioned shank with broken end. Length 140mm, maximum section 1mm x 7mm. (1208).

4.32 Reaping hook (Fig 7.6). Iron. Curved blade, recent breakage at tip; socketed handle with nail hole on one side retaining wood. Length 150mm, maximum width

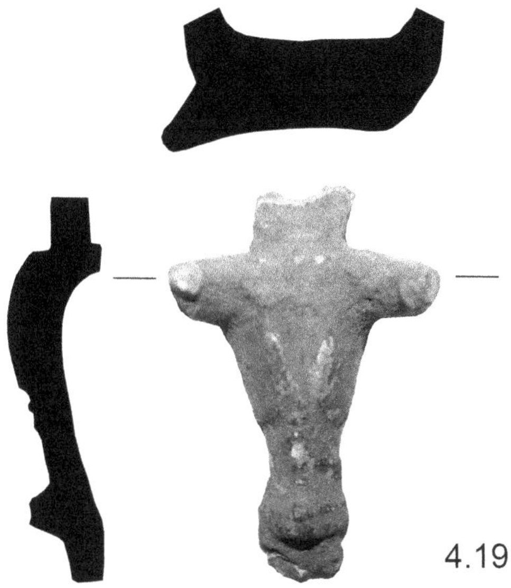

4.19

Figure 7.2: Small finds, ox-head escutcheon. Scale 1:1.

of blade 38mm, diameter of socket 25mm. (1240) SF9.5.

4.33 Whetstone (Fig 7.5). Grey very finely grained, micaceous stone. Rectangular section with one end rounded, no evidence of great wear. Length 59mm, section 16mm x 12mm. (1209).

Fasteners and Fittings

4.34 Stud (Fig 7.4). Copper alloy stud with integral washer retaining disc of wood and space for another layer between disc and washer. Length of stud 15mm, diameter of head 14mm. u/s in [1207].

4.35 Chain (Fig 7.3). Iron. Twelve figure-of-eight links joined to a larger terminal figure-of-eight link by narrow bar wrapped through two ends; larger link worn through. Total length 320mm; small link length 28mm, width 11mm; large link length 45mm, width 16mm. (1239) SF 9.4.

4.36 Box fitting. Iron. Strip with three perforations; projection at right angles at one end with cylinder at end; possibly part of hinge fitting. A patch of mineral replacement does not appear to be of wood. Length 254mm, depth of projection 40mm, strip section 25mm x 5mm. u/s in [1207].

4.37 Stud (Fig 7.3). Iron. Large circular head; short shank. Length 16mm, diameter 13mm. (1337).

4.38 Bolt (Fig 7.4). Iron. Cylindrical head; tapering square-sectioned shank. Length 170mm; head diameter 15mm. (1337).

4.39 Pottery repair cramp (Fig 7.4). Lead alloy. Rectangular bar with two small rounded lugs on underside. Length 47mm. Weight 20g (1239).

4.40 Pottery repair? Lead alloy. Subcircular, plano-convex. Diameter 37mm x 34mm. Weight 60g. (1239).

4.41 Pottery repair? Lead alloy. H-shaped plug; with additional loop on the underside. Diameter 27mm x 27mm. Weight 33g. (1239).

Religious Items

4.42 Casserole (Fig 7.7). Lead alloy. Complete apart from minor parts missing from lower part of body where wall is thinnest. Appearing grey and coated with thin gritty deposit. Waisted handle terminating in disc with central aperture; disc, aperture and sides of handle framed by ribs. Narrow out-turned rim, slightly convex-curved body; concave base. Interior and exterior have turned decoration. Below rim: groove on interior and narrow raised rib on exterior. Base: two raised ribs and central button internally, four deep concentric channels with central button and narrow grooves on ribs between channels externally. Turning marks visible in places internally. Height 91mm, rim diameter 160mm, base diameter 110mm, length of handle 130mm, section of handle at narrowest point 31mm x 6mm, weight 717g. (1239).

4.43 Casserole (Fig 7.7). Lead alloy. Complete handle retaining small part of wall. Encrusted with corrosion products. Waisted handle terminating in disc with (probably) small central aperture, which appears to be infilled with corrosion products, though this may be merely a small circular groove centrally rather than an aperture. Disc defined by pair of grooves on upper surface, groove parallel to either edge of handle and deeper groove marks the junction with the out-turned rim. Turning groove internally on the interior of the wall. Length of handle 100mm, section at narrowest point 25mm x 5mm, weight 115g. (1240).

Context	Complete	Incomplete	Total
1208	2	-	2
1239	21	34	55
1240	6	3	9
1337	4	1	5
1346	1	-	1
2517	-	1	1
u/s	2	-	2
Total	**36**	**39**	**75**

Table 7.9: Nails from brine tank [1207]. Quantified by heads. The complete examples range from 32mm to 130mm in length, with a median value of 71mm. The interquartile range was 58-88mm.

Context	Runoff	Sheet	Total
1208	-	103	103
1209	-	89	89
1210	7	-	7
1224	1	-	1
1225	7	45	52
1233	23	-	23
1239	2378	422	2800
1240	252	260	512
1336	-	36	36
1337	90	4	4
1346	57	59	116
Total	**2815**	**1018**	**3833**

Table 7.10: Lead runoff and sheet from the brine tank [1207]. Quantified in g.

Miscellaneous

4.44 Ring. Copper alloy. D-sectioned. Diameter 28mm. u/s in (1207).

4.45 Whorl (Fig 7.4). Lead alloy. Cylindrical. Diameter 22mm, thickness 5mm, perforation diameter 7mm, weight 16g. (1208).

4.46 Whorl (Fig 7.4). Lead alloy. Convex upper profile. Diameter 36mm; thickness 5.5mm; perforation diameter 10mm, weight 34g. u/s in (1207).

4.47 Cylinder. Lead alloy. Length 23mm, diameter 10mm. (1239).

4.48 Cone (Fig 7.4). Lead alloy. Elongated cone, rounded tip; partially perforated at base. Length 46mm, diameter

10mm x 9mm. (1239).

4.49 Cone of fired clay with a square-sectioned void (Fig 7.4). Length 27mm, maximum section 21mm x 18mm. Fill (1346) in chamber [1347].

4.50 Bar. Copper alloy. Circular-sectioned, both ends broken; slightly angled. Length 25mm, section 4mm. u/s in 1207.

4.51 Rod. Lead alloy. Circular-sectioned, bent. Length 191mm, section 6.5mm. (1208).

4.52 Bar. Lead alloy. Length 78mm. (1337).

4.53 Strip. Lead alloy. Edge of curved moulding. Length 50mm, weight 51g. (1337).

Other Contexts in the Brine Industrial Zone

Personal Ornaments

4.54 Polden Hill brooch (Fig 7.3). Copper alloy. Semi-cylindrical spring cover with closed perforated ends holding bar running through centre of spring of six turns, precise arrangement for holding chord of spring obscured, spring probably of different alloy; semicircular moulding on either side of head; bow tapering to foot which has two small transverse ribs forming a foot knob, front of bow has vertical groove centrally with diagonal grooves on either side; triangular catch plate; pin missing. Length 43mm. Fill (1286) of linear [1186].

Household Items

4.55 Prismatic bottle; body fragment. Blue/green glass. Heat affected. Fill (1110) of cut [1109].

4.56 Prismatic bottle; body fragment. Blue/green glass. Fill (1132) of linear [1131].

4.57 Prismatic bottle; body fragment. Blue/green glass. (1198) [1195].

Tools

4.58 Flat paddle-shaped blade (Fig 7.6). Iron; haematite deposits. Back flat, opposite edge appears to have had an edge; rectangular-sectioned broken handle. End of blade broken. Present length 256mm, maximum width of blade 80mm, handle section 30 x 13mm. Fill (1265) of wooden trough [1264].

Fasteners and Fittings

4.59 Strap hinge. Iron. Expanded arm tapering to rolled over loop hinge; 3 nails extant in arm. Length 100mm, maximum width 35mm. Identified from X-radiograph. Fill (1004) of [1003].

Miscellaneous

4.60 Ring. Iron. In two fragments, possibly not complete. Diameter 45mm, section 6.5mm. Fill (1161) of wicker-lined pit [1160].

Context	Complete	Incomplete	Total
1043	-	1	1
1080	-	1	1
1161	-	1	1
Total	**-**	**3**	**3**

Table 7.11: Nails from other contexts in the Brine Industrial Zone (quantified by heads).

Context	Runoff	Sheet	Total
1080	190	-	190
1173	10	-	10
u/s	2151	426	2577
Total	**2351**	**426**	**2777**

Table 7.12: Lead runoff and sheet from other contexts in the Brine Industrial Zone other than brine tanks [1182] and [1207]. Quantified in g.

Other Unstratified Finds of Note

Four items amongst the unstratified and unlocated material deserve special mention. The details of the central button that are preserved on 5.1 (Fig 7.3) place it amongst Bayley and Butcher's group A division of trumpet brooches (Bayley & Butcher 2004, 160) which were in use during the later part of the 1st century and into the mid 2nd century. The lower bow is definitely plain, the upper bow is obscured but unlikely to have been decorated. The casserole handle fragment 5.2 has already been discussed in connection with 1.2 from the Road Zone. Item 5.3 (Fig 7.3) is a small furniture mount with the type of head and shoulders bust that commonly decorates Roman items. The slot on the back of the shoulders would have allowed it to be attached as a terminal. Unfortunately the features are too obscured to show whether it was intended to be a male or female, but the hair detail at the back might be more appropriate for the latter. There is also a typical Roman steelyard weight (5.4) (Fig 7.4).

Other unstratified and unlocated Roman items, not catalogued in detail here, are lead alloy pottery repairs and whorls, ten examples of each of which were found. As noted in the discussion of 4.45-6, not all of the whorls need be of Roman date.

5.1 Trumpet brooch (Fig 7.3). Copper alloy. Corroded and many details obscured. Small trumpet head with cast lug behind retaining small part of spring, most of spring and pin missing. Upper bow corroded with blister removing upper part of central button which has a triple ribbed moulding centrally and double rib below; these may have traces of beading; moulding continues around the back of the bow; lower bow with sharply angled sides; double ribbed footknob; broken triangular catch plate. Length 53mm, trumpet head width 11mm.

5.2 Casserole. Copper alloy (Fig 7.7). Fragment of waisted handle retaining rim edge; two deep grooves parallel to each edge, one small punched ring by broken edge. Present length 45mm, handle section 21 x 4mm.

5.3 Furniture mount (Fig 7.3). Copper alloy. Head and shoulders bust of male or female. Facial features obscured, hair indicated as coming down onto forehead, and indicated on back as splaying out from a central parting. Square slot on back of shoulder. Length 30mm.

5.4 Steelyard weight (Fig 7.4). Lead alloy. Biconical; broken iron loop. Diameter 40mm. Weight 252g.

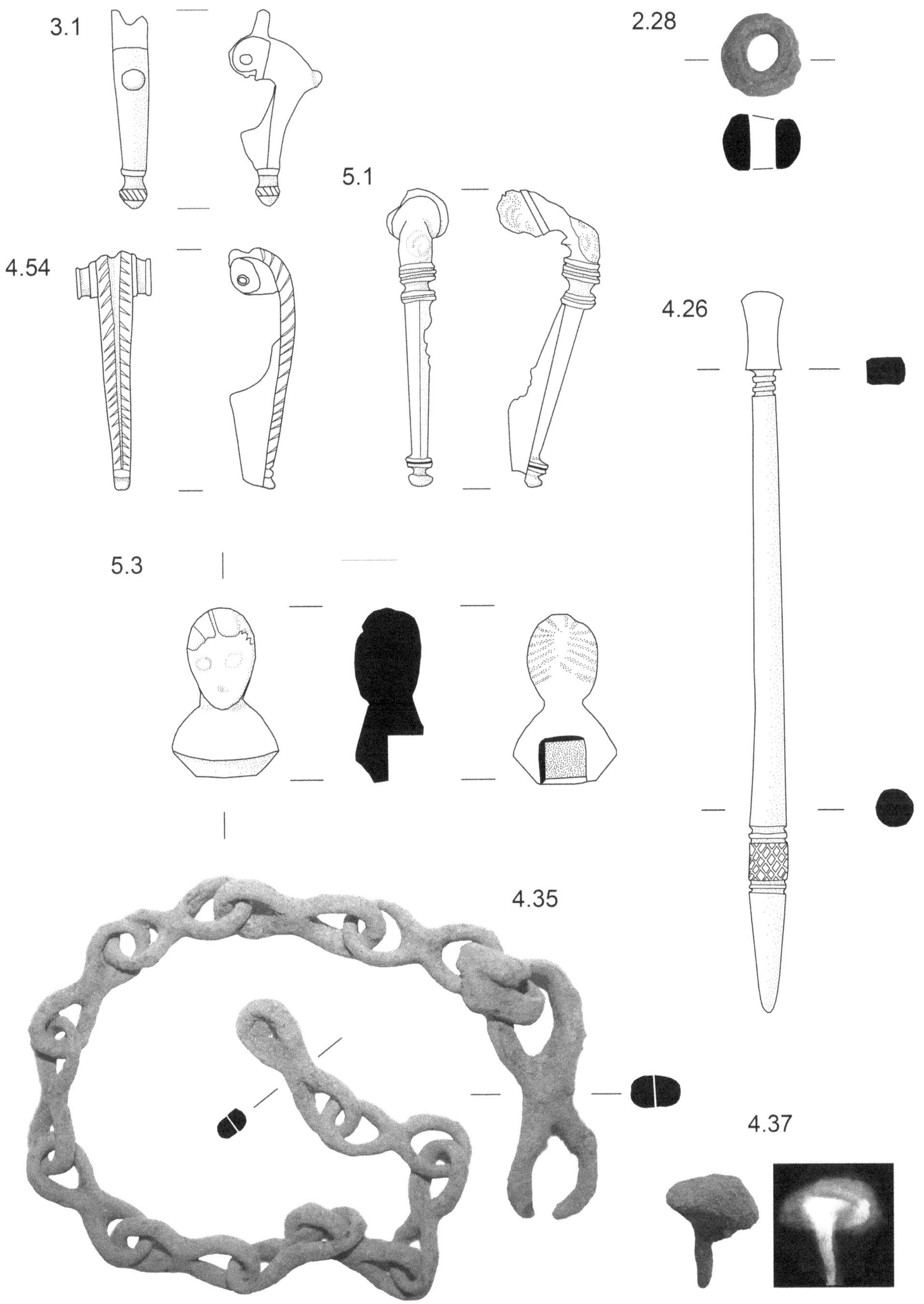

Figure 7.3: Small finds. Scale 1:1.

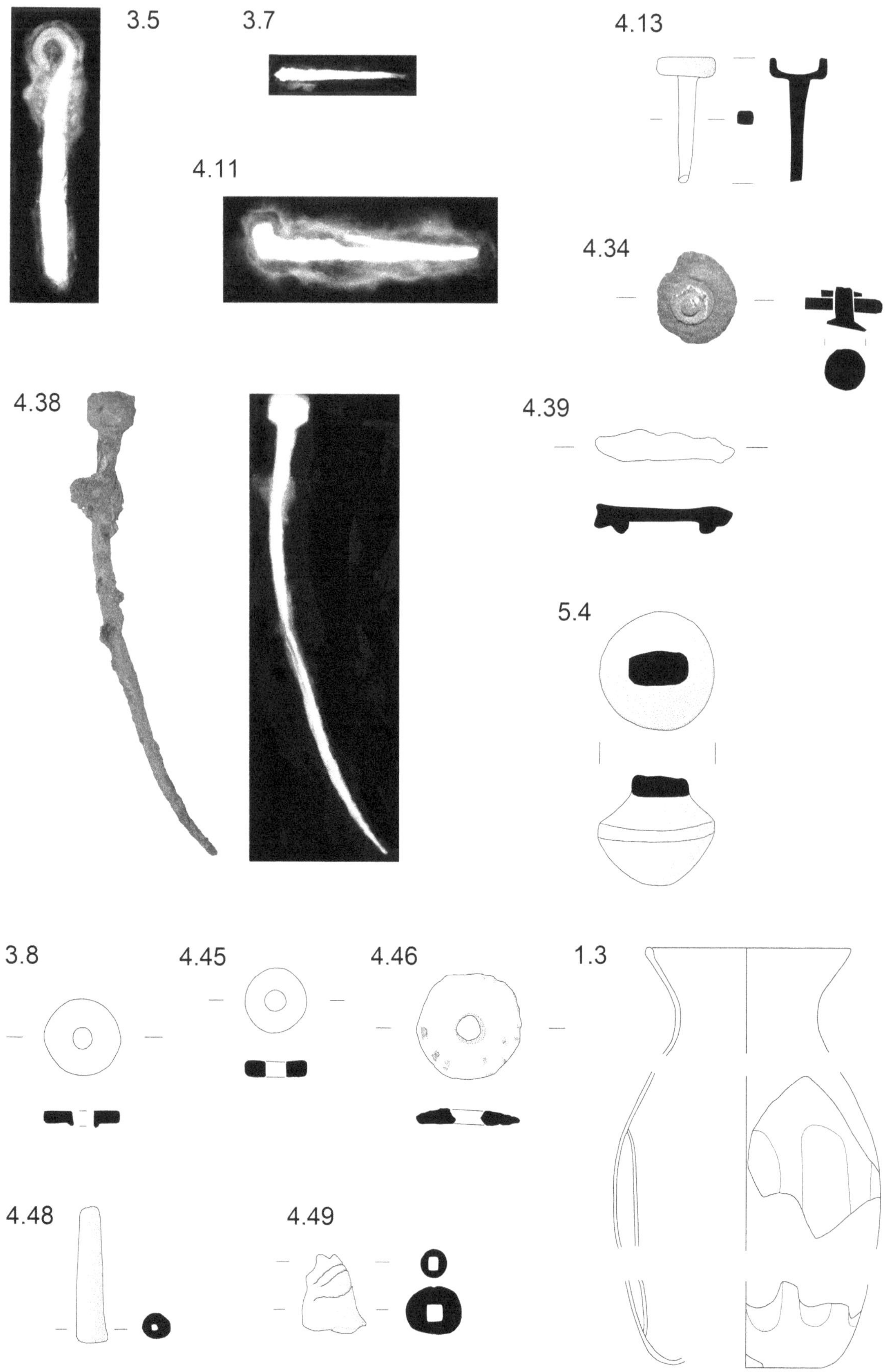

Figure 7.4: Small finds and vessel glass. Scale 1:2.

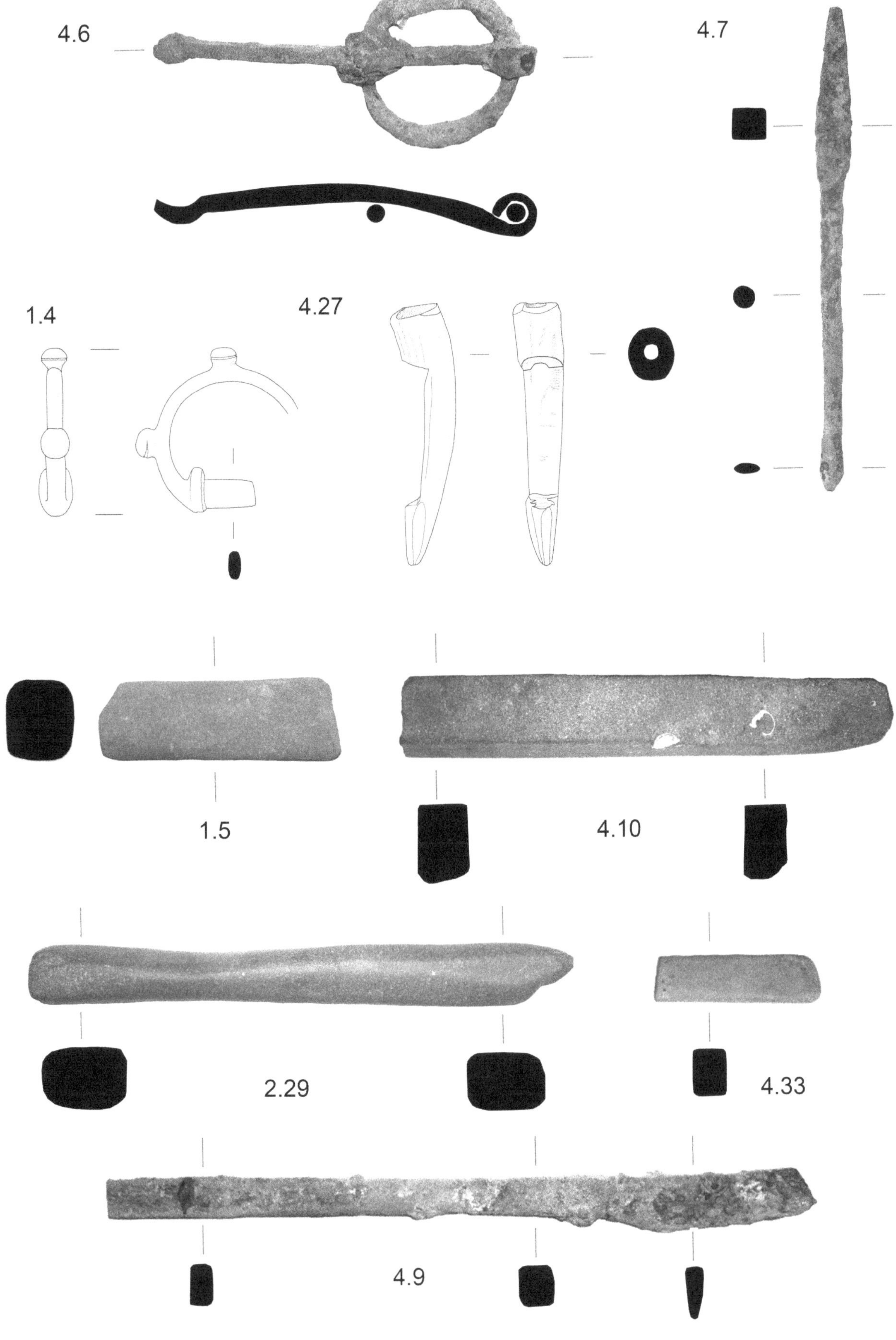

Figure 7.5: Small finds, transport and tools. Scale 1:2 except for 4.9 at 1:1.

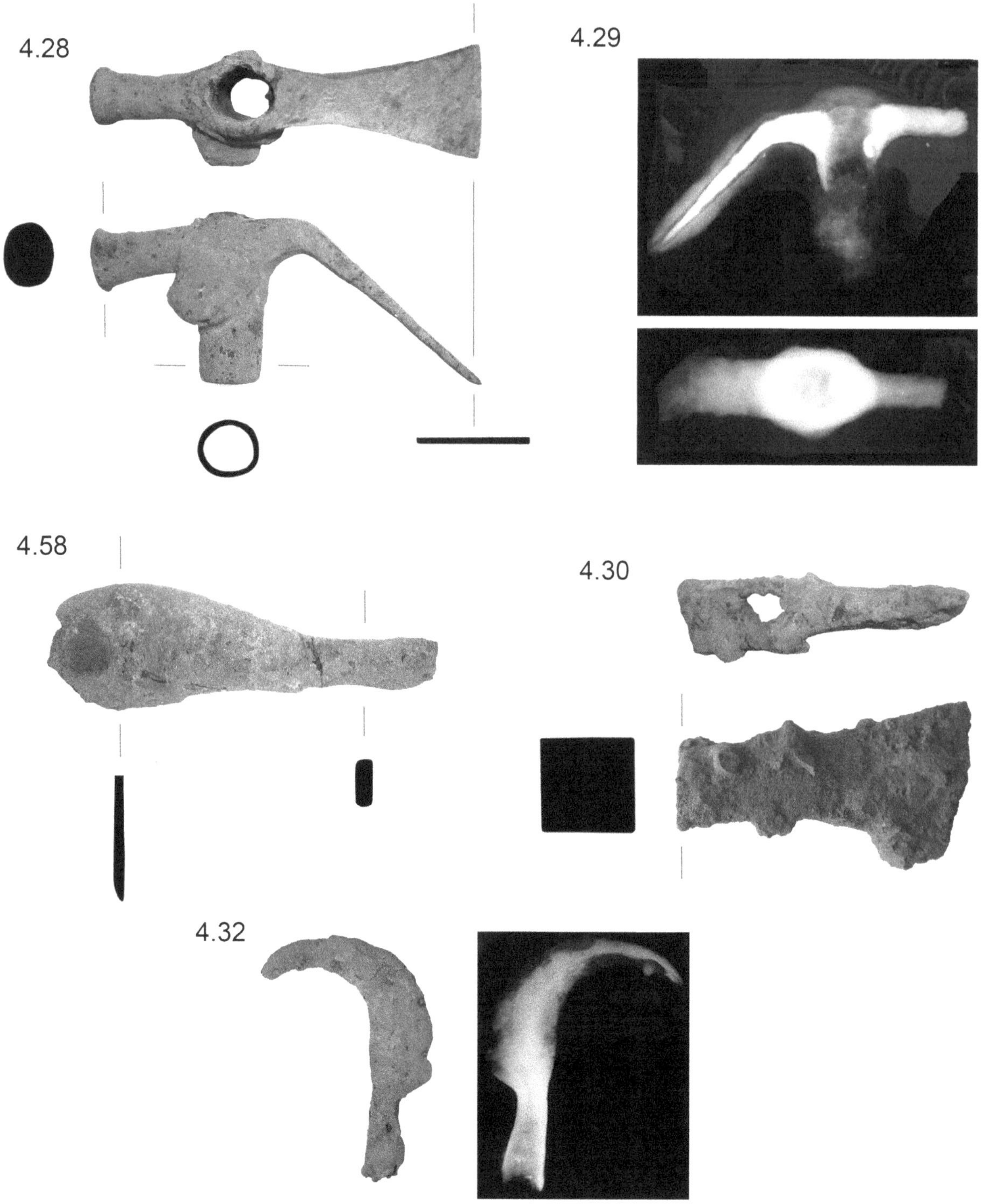

Figure 7.6: Small finds, tools. Scale 1:4.

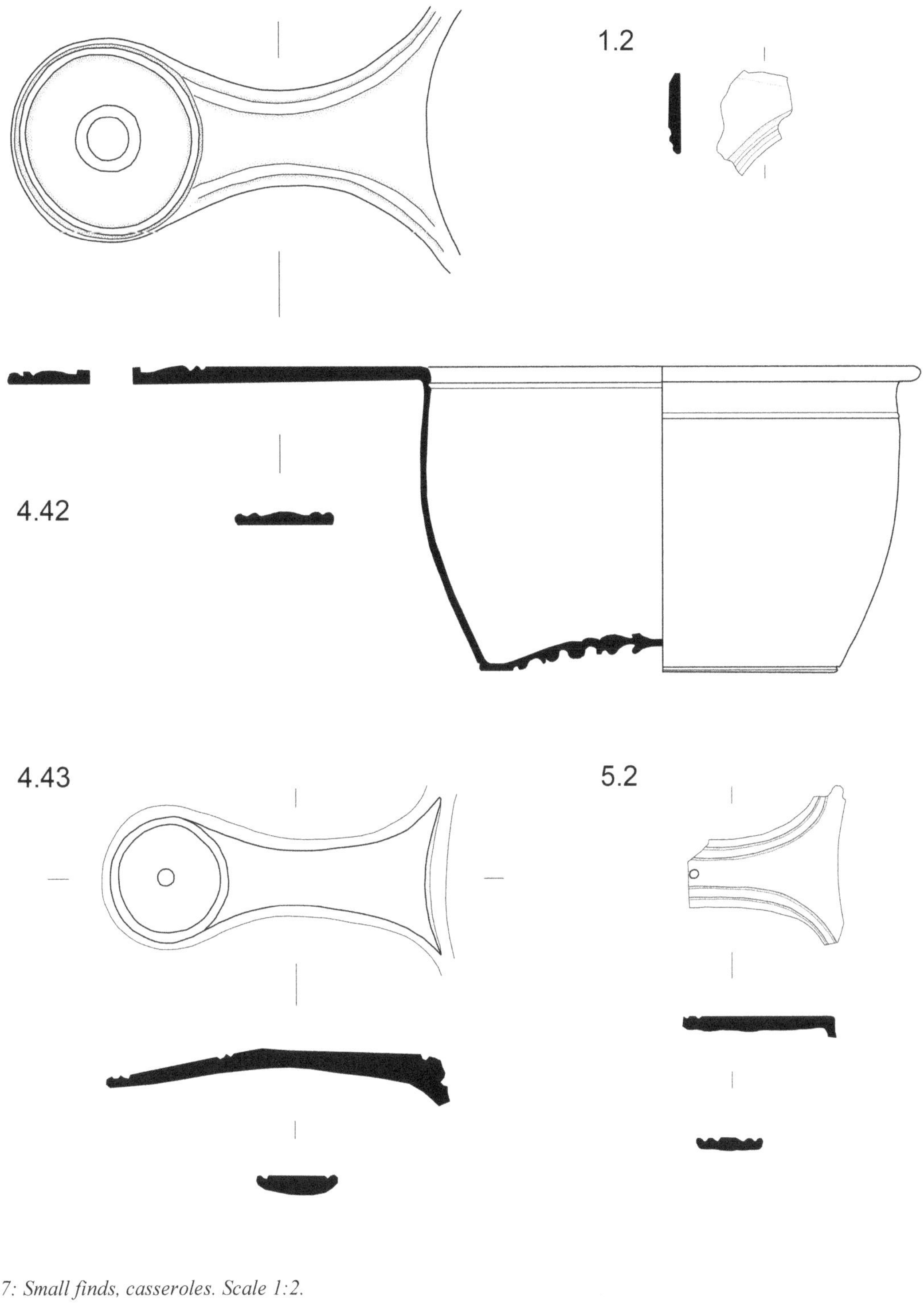

Figure 7.7: Small finds, casseroles. Scale 1:2.

8. Cremated Remains

Jacqueline I McKinley with contributions by Catherine Barnett and Ruth Pelling

Cremated bone from six contexts/subcontexts was subject to analysis. The deposits included the remains of one urned burial and two unurned burials with redeposited pyre debris. The nature of the remaining deposit is unclear.

Methods

The vessel from grave [4003] had survived largely intact (maximum height 150mm) and it, together with its contents, was lifted for micro-excavation by the writer. The fill was removed in a series of six spits, all of 20mm depth with the exception of the upper cleaning layer (Spit 0, 30mm depth not even across width of vessel). Spits 1-6 were subdivided into quadrants (a-d), the contents of each spitted quadrant being bagged separately for analysis. Annotated plans (1:2) were made at the top of each spit (see Appendix 8.2) and a photographic record was maintained throughout.

The whole-earth samples thus recovered were processed following the standard methodology of wet sieving to 1mm fraction-size and floatation using a 500 micron mesh for recovery of any charred plant remains and charcoal. The sieve residues >5mm mesh size were sorted and all non-osseous material removed; the <5mm sieve residues were retained and were subject to a rapid scan by the writer for the recovery of identifiable skeletal elements. A subjective note of the quantity of bone remaining amongst the unsorted residue was made and is presented in Appendix 8.1.

Osteological analysis followed the writer's standard procedure for the examination of cremated bone (McKinley 1994a, 5-21; 2000a; 2004a). Age was assessed from the stage of skeletal and tooth development (van Beek 1983; Scheuer & Black 2000), and the general degree of age-related changes to the bone (Buikstra & Ubelaker 1994). Sex was ascertained from the sexually dimorphic traits of the skeleton (*ibid*; Gejvall 1981), confidence levels (Table 8.1) '??' (most likely) and un-questioned.

The sub-contexts created during excavation of the urned burial remains were maintained in analysis to enable details of the burial formation process to be ascertained.

Results and Discussion

A summary of the results from the analysis is presented in Table 8.1. Full details are in Appendix 8.1.

Disturbance and Condition

Two of the features, [4245] and [4250], from which cremated bone was recovered had been cut by later features (Fig 3.6) but there is minimal likelihood of the resultant loss of much, if any, bone. The surviving depth of the features ranged from 0.05m to 0.27m, with only one at less than 0.1m. Cut [4239] had been subject to substantial horizontal truncation and it is possible that an unknown quantity of bone has been lost. It should be noted, however, that the bone within unurned burials has been found to be concentrated in the lower 0.08-0.1m of grave cuts (eg Dinwiddy & Schuster 2009, figs 2.34-8). Consequently the small amount of bone recovered from [4239] may be reflective of factors other than disturbance.

The bone is visually in good condition and trabecular bone, generally the first to be lost in an adverse burial environment (McKinley 1997, 245; Nielsen-Marsh *et al* 2000), is well represented in most deposits. Two bone fragments from grave [4245] (a fragment of radius and a fragment of fibula shaft), in contrast with all the other material, is worn and chalky in appearance. The described condition is indicative of burial within an acidic micro-environment. Since only these two bones are affected it suggests that they were originally deposited elsewhere and derived from a different cremation to the rest of the bone within this deposit. The recovery of 'alien' bone fragments from cremation burials has occasionally been recorded elsewhere. In some cases this could be reflective of contamination either from a neighbouring disturbed deposit or a reused and incompletely cleared pyre site. In this instance a different original burial environment is indicated and the inclusion could have been deliberate. A possible example of this latter potential form of 'token' symbolic or *memento mori* deposit was observed in one of the Romano-British burials from Hyde Street, Winchester (McKinley 2004b).

Demographic Data

A minimum of three individuals were identified, all adults comprising two young individuals of less than 30 years (one female and one male) and one older adult (>45 years) female. Graves [4245] and [4250] each contained two cremation-related deposits, the burial and redeposited pyre debris, but there is no evidence to suggest that they derived from other than the same cremation in each case; ie there was no duplication of skeletal elements or contradictory demographic evidence, and the condition of the bone and type of pyre goods are commensurate.

Context	Feature	Deposit type	Bone weight	Age/Sex	Pathology	Pyre goods
(4004)	[4003]	Urned burial	932.6g	Adult *c* 20-25 yr female	Infection - ?cuboid	45.6g animal bone - immature pig.
(4238)	[4239]	crd/?redep	9.8g	Subadult/adult >15 yr		
(4244) <56>	[4245]	rpd in ?grave fill	38.5g	= <78>		<0.1g animal bone u/id sp.
(4244) <78>	[4245]	Unurned burial + rpd	308.2g	Adult *c* 20-30 yr ??male		Blue glass ?bead frag; 5.8g animal bone inc immature ?pig.
(4248)	[4250]	Unurned burial + rpd	617.7g	Adult >40 yr ??female	*Ante mortem* tooth loss; degenerative disc disease - cervical; osteophytes - cervical articular process, lumbar body surface margins.	21.6g animal bone inc immature pig & ?woodcock; 1.5g/8 frags. worked bone; 2 Fe nails.
(4249)	[4250]	?rpd = (4248)	318.6g	Adult >45 yr ??female	Ankylosis - 2 cervical; osteophytes - thoracic articular process, rib; osteoarthritis - cervical; pitting - medial clavicles; enthesopathy - patella.	16.1g animal bone inc ?neonatal pig.

Table 8.1: Summary of results from analysis of cremated bone. rpd - redeposited pyre debris; crd - cremation-related deposit; u/id - unidentified.

The nature of [4239] is debatable. Since the fill of the 0.38m wide and 0.05m deep feature was lifted as a single sample rather than as quadrants (McKinley 2000d; forthcoming), the distribution of the small quantity of bone it contained is unknown and no details of the formation process can be deduced. The lack of fuel ash components within the fill indicates that it does not represent a deposit of pyre debris. The bone may represent the incidental inclusion of scattered/disturbed material derived from one of the cremations already represented within the assemblage (there is no duplication of skeletal elements between this deposit and any other from the site). Consequently it has not been included in the minimum number count.

The cremated remains were all recovered from features forming a small group situated within close proximity to each other, in an area measuring *c* 14m by 8m on the north-western margins of the site (Fig 3.5).

Pathology

Pathological lesions were observed in the remains of the two adult females, particularly those of the elderly individual from grave [4250] who had lost several maxillary teeth *ante mortem* (three of four molar sockets) and had numerous lesions associated with a variety of joint diseases (Table 8.1). Lesions indicative of osteoarthritis (Rogers & Waldron 1995, 32-46) were observed in at least one cervical vertebra; the ankylosis of two others (via smooth fusion of the facets and neural arch) and marginal osteophytes (new bone on joint surface margins) in another may be reflective of more extensive involvement. Degenerative disc disease (Rogers & Waldron 1995, 27), resulting from a breakdown in the intervertebral disc (generally reflective of age-related wear-and-tear), was recorded in one cervical vertebra. Lone osteophytes are largely seen as age-related but may, in synovial joints, be indicative of the early stages of osteoarthritis (particularly where there is already evidence for the condition). Enthesophytes (new bone at tendon insertions) may develop in response to a number of conditions and it is not always possible to ascertain the specific cause of individual lesions (Rogers & Waldron 1995, 245); the patella lesions seen in (4249) are most likely to be indicative of repetitive minor muscle stress.

A *c* 13mm diameter depression, juxta-articular to the distal end of the posterior surface of the ?right cuboid from (4404), has an uneven, porous surface with a marked rim and some disorganised new bone on the medial side; the lesions are suggestive of some form of infection.

Pyre Technology and Cremation Rituals

Oxidation

Although the majority of the bone is white in colour indicating a high level of oxidation, some slight blue/grey colour variations reflecting incomplete oxidation (Holden *et al* 1995a & b) were observed in a few bone fragments from all three graves. Variations were seen in one or more fragments of several skeletal elements but an entire element was never affected. The most extensive variation was seen in the bone from the urned burial where between one and five elements from each of the four skeletal areas were affected, including all parts of the axial skeleton; the ribs and fragments of femur were most frequently involved. Less extensive variation was seen in the remains from grave [4250]; a few bone fragments from all skeletal areas were affected, most frequently the hand bones. Only two skeletal elements (hands and femur) were affected amongst the remains from the male cremation, most commonly the finger bones. The comparative scarcity of variable oxidation to the skull (one fragment from each of two graves) is interesting given the involvement of this area of the skeleton commonly observed (McKinley 2008a, table 2).

Variable oxidation of the femur, seen in all three cases here, is frequently recorded (McKinley 2008a, table 2); this is largely related to the thickness of the overlying soft tissues which need to be removed before the bone itself can begin to burn. Elements of the hands and feet, affected in two cases here, also represent those bones commonly showing variable oxidation levels. This probably reflects their peripheral position on the pyre where the temperature would be lower and there is more scope for bones to fall or be flicked off. In the latter instance the pyres may not have been constructed wide/long enough, particularly not to accommodate the slight movement (flexion of legs and arms) which can often occur during cremation. In the former case a potential shortage of fuel is also suggested, the pyre not burning long enough to allow full oxidation of the thigh area. The more extensive variability seen in the remains of the gracile female from grave [4003] may also have been due to insufficient fuel/time for cremation but the frequent involvement of the axial skeleton suggests there may have been a reduced oxygen supply in this area. Were the body to have been placed on the pyre on a solid wood bier/couch or laid on a fur/leather wrapping, this would have been sufficient to muffle the corpse from the rising flame and oxygen, at least during the early stages of cremation.

Two fragments of skull vault from grave [4250] were largely unburnt and showed none of the classic dehydration fissures common to bone burnt fleshed or at least relatively green. They appear to have been burnt as dry bone fragments and are probably intrusive, although whether their burning and inclusion in the cremation graves were accidental or deliberate cannot be stated with confidence. They could represent some form of 'token' deposit of the sort described above.

Bone Weight

The weight of bone recovered from the graves has a range of 346.7-936.6g, ie between *c* 22% and *c* 58% of the expected average weight from an adult cremation (McKinley 1993), with an average of 738.6g. The limited disturbance and generally good condition of the bone render it probable that these weights are a close reflection of that included in the original deposits. It is interesting to note that the weight of bone from the two female graves (almost identical at 932.6g and 934.6g) is substantially greater than that from the male burial, although the sex of the individual has not been found to represent a consistent factor affecting the weight of bone included in the burial elsewhere (eg McKinley 2004c). The maximum and the average weight fall within the upper range for adult burials of this date (McKinley 2004c, 295-7). The type of burial may appear to show little difference regarding the quantity of bone recovered. However, the bone from grave [4250], although all from the same cremation, appears to have formed two separate deposit types, with the 'burial' containing *c* 66% of the bone and the redeposited pyre debris the rest. This suggests that, as is often observed, the urned burial included a greater quantity of bone that that made unurned.

Fragmentation

Fragmentation of cremated bone is influenced by a variety of intrinsic and extrinsic factors exclusive of human manipulation with the deliberate intent to fragment (McKinley 1994b; 2004c). The majority of the bone from all three burial deposits was recovered from the 10mm sieve fraction but the proportion varied widely from *c* 80% by weight of bone from the undisturbed urned burial to *c* 44-56% from the unurned burial deposits. The majority of bone from the non-burial deposits was recovered from the 5mm sieve fraction (45-46% by weight). The maximum fragment sizes show a similar pattern, that from the urned burial being 103mm, the unurned burials 51-55mm, and the non-burial deposits 20-39mm. As is often observed, the level of disturbance and deposit type appear to have influenced the degree of bone fragmentation. The bone within the urned burial will have been afforded additional protection from the burial environment by the presence of the vessel. The overall smaller size of bone fragments from the pyre debris deposits compared with that from the same cremation but recovered as part of the burial remains suggests some additional manipulation of the bone in the former. This was possibly associated with collection and storage methods (eg less careful handling or greater incidental trampling of the pyre site) rather than a deliberate intent to fragment.

Skeletal Elements

Most burials will contain a substantial proportion of 'unidentifiable' bone fragments (comprising long bone shaft or undistinguished trabecular bone), the proportion increasing where deposits were disturbed and the bone more fragmentary (McKinley 2004c, 298-9). This is demonstrated here by the relatively high proportion of identifiable bone from the largely undisturbed urned burial (*c* 68%) compared with the more usual 36-46% from the unurned burials. With the exception of deposits containing only small quantities of bone (<50g), a range of elements from all skeletal areas is generally recovered. Skull is often disproportionally highly represented due to ease of identification and elements of the axial skeleton are commonly under-represented because of the frequent preferential destruction of trabecular bone. This being so, the unusually low proportion of skull elements (9% of identifiable bone) from the urned burial (4004), particularly when compared with the relatively high proportion of axial skeleton (*c* 22%, though actually close to the 'normal' *c* 20% (McKinley 1994a, 6)), suggests deliberate selection/exclusion of the skull. The male burial contained the closest to 'normal' distribution but with more skull, apparently at the expense of upper limb. The two females both have a higher than normal proportion of lower limb elements apparently largely to the detriment of the upper limb, although the skull and axial skeleton proportions in (4244) are almost reversed from those seen in (4004) (ie similar to those often recorded).

An absence or paucity of skull elements has been observed on rare occasions in other Romano-British cemeteries including Puckeridge, Hertfordshire (Wells 1981, 291), East London (McKinley 2000b, 271), Brougham, Cumbria (McKinley 2004c, 301), and Wall, Staffordshire (McKinley 2008b, 136). The writer has previously suggested that, rather than being excluded from the burial, the symbolic significance of the skull may have resulted in its selection for some other ritual purpose in these instances, such as use as a *memento mori*.

Tooth roots and the small bones of the hands and feet are commonly recovered from cremation burials and it has been suggested that their frequency of occurrence may provide some indication of the mode of recovery of bone from the pyre site for burial (McKinley 2000a; 2004c, 303). These elements, particularly hand and foot bones, were recovered from all the deposits at Kingsley Fields, being especially common in all three burials (27-32 elements), though a large number (24) were also found within the pyre debris from grave [4250]. There was no noticeable difference in frequency between burial type or linked to the age/sex of the individual, although it should be noted that the substantially lower overall weight of bone from the male burial included a similar number of these small elements as the female burials. Few such elements were recovered from the burials at Brougham (McKinley 2004c, 298-301) and much smaller quantities were found at Wall (a maximum of 13 elements; McKinley 2008b, 136), suggesting some variation in the mode of collection of bone for burial between these different sites. The implication is that at Kingsley Fields collection was facilitated by raking-off and winnowing of the cremated remains rather than individual hand-recovery of fragments, thereby easing the recovery of the smaller skeletal elements as well as the larger ones.

Pyre Goods

Pyre goods were recovered amongst the remains of all three burials and from the redeposited pyre debris within the grave fills (Table 8.1). Artefactual remains were found in one male and one female grave in the form of a fragment of a blue glass bead, iron nails and fragments of worked bone derived from a funeral bier (see Chapter 7).

Fragments of cremated animal bone were recovered from all the deposits with the exception of (4238) (Table 8.1). Between 2% (in [4245]) and 5% (in [4003]) of the total weight of bone recovered from the graves was identified as animal. In each case immature pig, including at least one neonate, is represented (species identifications by Jessica Grimm; see Appendix 8.1). A few fragments of bird bone, possibly woodcock, were also included amongst the remains of the elderly female (4248).

The inclusion of animals, part or whole, on the pyre was a common facet of the Romano-British rite, although the frequency of occurrence varied widely, ranging from 3.5% of burials from Westhampnett, West Sussex (McKinley & Smith 1997) to 80% of urned burials from Wall (McKinley 2008b, 126-7). Although there are only three burials from Kingsley Fields, the consistent presence of animal bone suggests similarities with the latter site, which also represents one of its geographically closest contemporaries. Pig, particularly immature individuals, generally represents the most popular species in this period, possibly for ritualistic reasons linked to 'legalisation' of graves via pig sacrifice rather than simply as a food offering (Barber & Bowsher 2000, 72-3; McKinley 2004d; Toynbee 1996, 50). Birds also comprised a relatively common offering, with domestic fowl representing the usual species although others have been found (eg Barber & Bowsher 2000, table 25). Such offerings may represent 'food' for the deceased or have been symbolic in other ways (Toynbee 1996, 50). In this instance, the species identified suggests a food offering. Both pig and bird are regularly found with individuals of either sex. The inclusion of more than one species was not as common as it was to become in the Anglo-Saxon period but where this did occur it was generally limited to adult cremations (McKinley 2008b, 136-7). Elsewhere in Continental Europe at this time animal remains appear to have been consistently included on most pyres and subsequently within the burials (80%), with pig comprising the usual species (Wahl 2008, 150).

Formation Processes

Redeposited pyre debris was recovered from two grave fills. In the case of grave [4250], a heavily charcoal-rich deposit (0.06m deep) inclusive of 34% of the total weight of bone from the grave was placed in the base of the cut. A further charcoal-rich deposit (0.15m deep) apparently derived from the same cremation (see above) but inclusive of a greater proportion of the bone was subsequently deposited. It is not possible to deduce any further details of the formation process due to the recovery of both contexts as single entities. However, the surviving evidence suggests that the primary deposit represents pyre debris, followed by the burial and a subsequent undistinguished deposit of debris. The deposits in grave [4245] may have followed a similar pattern but in this case the single context was recovered as two separate samples and the writer has no record of how they were distinguished; again, further details of the formation process cannot be deduced.

The recovery of redeposited pyre debris from Romano-British grave fills is relatively common (McKinley 2000c; 2004c, 304-6) and its presence is taken as indicative of the relatively close proximity of the pyre site to the grave. The very large quantities of debris recovered from [4250] (*c* 7 litres), particularly from the lower of the two deposits (Tables 8.2 & 8.3), is, however, unusual other than from *bustum*-style pyre sites. Generally only a 'sample' of the pyre debris was included in the grave fill but in this case the quantity is close to commensurate with that recovered by the writer at the end of experimental pyre cremations, *c* 10 litres (McKinley forthcoming), and probably represents most if not all of the debris from the pyre. If the latter is the case, it is clear that some cremated bone was removed for distribution/ deposition elsewhere since even the combined amount from the burial and the pyre debris does not represent the total which would have been present at the end of cremation (see above).

The urned burial (4004) had survived intact to a maximum height of 150mm but it was only at the top of Spit 3 (80mm height) that the whole of the circumference was preserved. Distribution of the bone by weight was fairly even in both halves of the vessel; the greatest proportion of bone in one spit was the 22.5% by weight recovered from Spit 4 (vessel diameter 168mm, 40-60mm from the base), 40mm below the maximum diameter of 192mm in Spit 2. The distribution of skeletal elements was not uniform throughout the depth of the vessel. There was a mix of elements from all four skeletal areas in Spits 1-4 but skull was concentrated in Spits 0-3 and none was recovered from Spits 5-6. Fragments of axial skeleton were sparse in Spits 5-6 and most frequent in Spits 0-4 with a concentration in Spit 3. Elements of upper limb were also sparse in Spits 5-6, most being identified in Spits 3-4. Fragments of lower limb were concentrated in the lower two spits and were sparse in Spits 1-2. The animal bone was restricted to the upper half of the vessel (Spits 0-3). There was some evidence indicative of pairs of bones having been deposited together, eg the adjacent tali in Spit 4, but there were also joins between bones from different parts of the fill, eg fragments of clavicle (Spits 2c and 4b), femur (Spits 2d and 5c) and tibia (Spits 3c and 5b). The distribution suggests at least a partially ordered deposition, probably reflective of the order of collection and indicative of the bone being placed straight into the burial vessel rather than in an intermediate storage receptacle. Collection appears to have commenced at the foot end of the pyre and worked up towards the head. The location of the animal remains suggests that these were also placed towards the head-end of the pyre. Evidence for such an ordered deposition within Romano-British burials is currently sparse, though there is some evidence from at least two Kentish sites for such a practice and it is highly probable that there was a much broader geographic range (McKinley 2009).

Wood Charcoal Analysis

Catherine Barnett

Following assessment (Wessex Archaeology 2009) four samples of charred remains from the cremation graves and related features were chosen for wood charcoal analysis. Two moderately rich assemblages, (4244) in grave [4245] and (4240) in feature [4242] which cut [4245], and two particularly rich assemblages, (4248) and (4249) from grave [4250], were examined.

Methods

The smaller samples (44 and 78; Table 8.2) were examined following standard methods with 100 fragments >2mm identified from each. Due to the exceptional size of the two larger deposits from feature 4250 (at *c* 0.5 litre and >2 litres of charcoal respectively from 10-15 litres samples), and in an attempt to identify species other than the prevalent oak, rapid scanning using low-power stereo microscopy was first used to remove and quantify mature oak from a proportion of the >4mm fraction of the samples. Non-oak pieces were then

Sample	44	46	47	78
Context	(4240)	(4248)	(4249)	(4244)
Feature	[4242]	[4250]		[4245]
Deposit type	Large feature which cuts feature [4245]	Unurned burial with pyre debris	?Redeposited pyre debris	Unurned burial with pyre debris
Identifications				
Alnus glutinosa	47	-	-	-
Cf *Alnus glutinosa*	7	-	-	-
Betula pendula/ pubescens	9	33	25	-
Betula 3yr rwd	-	4	1	-
Cornus sp.	7	-	-	-
Corylus avellana	-	2	2	-
Cf *Corylus avellana*	2	-	-	-
Pomoideae	-	4	6	-
Prunus sp.	3	1	-	-
Quercus sp.	12	427	453	97
Quercus sp. rwd	1	28	12	-
Unidentified	12	1	1	3
Total no fragments (% of assemblage)	100 (*c* 15% >2mm)	500 (*c* 12.5% >4mm)	500 (*c* 12.5% >4mm)	100 (*c* 30% >2mm)
Bark	-	1	-	-
Comments	Smaller sample of small fresh fragments, friable, with extensive Fe dep in vessels.	Generally good condition, some Fe and other dep in vessels. Occ vitrified.	Large pieces, good condition, occ Fe dep, rarely vitrified. 1 nail head.	Fragmentary charcoal, many degraded, friable, Fe in vessels, conglomerated material.

Table 8.2: Wood charcoal identifications.

prepared for identification according to the standard methodology of Leney and Casteel (1975) (see also Gale and Cutler 2000). The fragments were fractured with a razor blade so that three planes could be seen: transverse section (TS), radial longitudinal section (RL) and tangential longitudinal section (TL). They were examined under bifocal epi-illuminated microscopy at magnifications of x50, x100 and x400 using a Kyowa ME-LUX2 microscope. Identification was undertaken according to the anatomical characteristics described by Schweingruber (1990) and Butterfield and Meylan (1980). Identification was to the lowest taxonomic level possible, usually that of genus and nomenclature according to Stace (1997), individual taxon (mature and twig) were separated and quantified, and the results tabulated.

Interpretation and Discussion

As shown in Table 8.2 the wood charcoal assemblages from the three features differ substantially in composition. The two contexts within grave [4250] proved to be very similar. Both are remarkably rich in charcoal and contain the same range of taxa and mix of mature and juvenile pieces. Large pieces of mature oak dominate the samples and probably represent substantial oak timbers. There is also a lesser component, comprising birch (*Betula pendula/pubescens*), including juvenile/roundwood pieces, hazel (*Corylus avellana*) and pomaceous fruit wood (Pomoideae, a group of anatomically similar taxa that include for instance hawthorn, apple and whitebeam). A single fragment of cherry type (*Prunus* sp.) was also found in the sample from context (4248).

These two contexts almost certainly relate to the same pyre. However, (4249) contained larger charcoal pieces and being the deeper of the two is likely to represent the initial removal of debris from the pyre. While mature oak timbers again formed the great bulk of the pyre, a minimum of four other species are represented in small numbers. Although it can be difficult to discern how each taxon was introduced to the pyre (for instance as offerings or as furniture such as biers or coffins), in this case all are common deciduous open woodland and hedgerow types and, given the presence of young roundwood, most probably represent the kindling used to start the fire under the oak timbers. The sheer volume of charcoal from feature [4250] is in itself exceptional at nearly 7 litres of charred flot from the two thin layers (total 0.21m thick). If it truly represents a single pyre, as it appears to (see McKinley, above), its removal and subsequent deposit must have been done with thoroughness and its preservation after burial is excellent.

The wood charcoal assemblage from grave [4245] comprised solely oak (*Quercus* sp.), all being mature pieces or pieces too small to gauge their maturity. The use of oak for the pyre is apparent but the lack of other species suggests that neither kindling nor any grave goods or pyre furniture are represented in the sample.

Context (4240) proved entirely different. It contained a minimum of six taxa, with oak present but not dominant. Alder (*Alnus glutinosa*) was most common, with dogwood (*Cornus* sp.), birch, hazel, cherry type and pomaceous fruit wood also present. The fragments were generally in a poor condition, soft and friable due to extensive iron redeposition along vessels. The range of types represented and the lesser importance of oak do not support the original interpretation of the feature as a pyre-related one. Instead it appears that a domestic fuel dump is represented, a hearth, although one perhaps related to feasting at the time of, or some time after, the interment of remains. Again common deciduous types are present but the substantial alder indicates that local wetland carr was exploited for fuel wood in addition to higher drier areas.

The use of oak for pyre building is commonly reported for all archaeological periods since it is a dense wood of high calorific value (Edlin 1949; Gale & Cutler 2000), and therefore capable of providing the high temperatures and sustained burn required to consume a body. Romano-British examples include material from the cremations at Springhead and Northfleet, Kent (Barnett forthcoming). What is particularly useful about this assemblage, however, is that the volume and preservation of charcoal from grave [4250] has allowed the lesser types used to act as kindling to be discerned. As such, it provides a clear insight into the local availability of woody species, their selection and use in pyre construction.

Charred Plant Remains

Ruth Pelling

Introduction

Six bulk samples were taken from four features containing the remains of cremation burials and/or related deposits. Two samples were taken from each of two graves, [4250] and [4245], while single samples were taken from features [4242] and [4239]. The samples were processed for the recovery and assessment of charred plant remains and wood charcoal. Assessment demonstrated that, in addition to the large quantity of charcoal in grave [4250] (see Barnett, above), a number of other plant remains (seeds and chaff) were present.

Methods

Bulk samples were processed by standard flotation methods; the flot retained on a 0.5mm mesh, residues fractionated into 5.6mm, 2mm and 1mm fractions and dried. Flots from feature [4250] were sorted under a x10-x40 stereo-binocular microscope. Charred seeds noted were extracted from the large (>1mm) fractions, while the presence and abundance of weed seeds noted in the 0.5mm fraction were recorded. Residues were scanned to 1mm for any charred seeds and chaff which did not float. Identifications and quantifications are given in Table 8.3. Nomenclature follows Stace (1997).

Results

The two samples from grave [4250] produced large flots dominated by oak charcoal (see Barnett, above). In addition a number of pulses, cereal grains, weed seeds, stems and rhizomes/roots were recovered (Table 8.3). The pulses, where whole, were generally spherical. Those which were sufficiently well-preserved to retain the hila necessary for positive identification were found to be of *Pisum sativum* (pea). A small number of individuals were more fragmentary and are therefore recorded as *Pisum sativum/Vicia faba* (pea/bean), although the absence of any positively identified *Vicia faba* does suggest *Pisum sativum* to be the sole species represented. The grain consisted of entirely of *Hordeum vulgare* (barley).

A number of weed seeds were present, most indicative of wet or damp grassland: small seeded grasses such as *Poa annua/Phleum* type (meadow grass/cat's tails), other grassland species such as *Hypochaeris glabra/radicata* (cat's ears) and *Vicia/Lathyrus* sp. (vetch/vetchling/tares), and wet ground or pond-side vegetation including *Persicaria maculosa* (red shank), *Carex* spp. (sedges) and *Isolepis setacea* (bristle club-rush). A number of grass stem and rhizome/root fragments were also present.

Discussion

The seeds and other non-charcoal plant remains from the deposit within grave [4250] are likely to consist of a mixture of pyre goods, plant material used as tinder or material including burnt grassland vegetation present in the immediate vicinity of the pyre. The presence of peas and barley grain is presumably related to some sort of food offering placed on the pyre. The presence of whole grain and pulses would suggest they were placed on the pyre in their raw state rather than as part of a prepared dish.

The provision of plant foods on Roman pyres and burials was a fairly common practice both in Britain and elsewhere within the Empire. The likelihood of material being represented archaeologically is, however, related to a number of preservation and taphonomic factors. The unusually substantial deposit of food plants from a high-status female burial in Watling Street, London (Giorgi 2000) is in part the result of the status and wealth of the deceased, but is also likely to be related to the use of a *bustum* (where the pyre is constructed directly over a pit which is then to form the grave), therefore resulting in the collection of a greater proportion of the pyre debris and incorporated offerings. The fact that such a large part of the pyre debris is represented in the samples from feature [4250] would indicate that the removal of the pyre debris and its subsequent deposition in the pit were carefully and completely executed (see McKinley, above). This thorough deposition of the pyre debris has clearly resulted in the survival of funeral offerings which might easily have been lost.

Pulses and cereal grains have been found at a number of sites including Ryknield Street (Site 12), Wall, Staffordshire (Gray 2008, 173-6; Stevens 2008, 459), the East London cemetery (Davis 2000, 369; Giorgi 2000) and from one cremation grave within a large cemetery at Pepper Hill, Kent (Davis 2006). Fruit and nut remains including exotic or imported species such as grape, fig, almond, mulberry, date and stone-pine have been recovered from a small number of sites, particularly within London (van der Veen *et al* 2008; Giorgi 2000; Davis 2000) but also from Pepper Hill (Davis 2006). Within much of continental Western Europe cereals and

Sample		46	47
Context		(4248)	(4249)
Feature		[4250]	[4250]
No of buckets		5	3
Food plants			
Hordeum vulgare sl.	Barley grain	-	24
Pisum sativum L.	Pea	2	7
cf *Pisum sativum* L.	cf Pea	18	17
Pisum sativum/Vicia faba	Pea/Bean	4	1
Weed Species			
Caltha palustris L.	Marsh-marigold	1	-
Rumex sp.	Docks	3	1
Persicaria maculosa	Red Shank	2	4
Plantago lanceolata/media	Plantain	8	1
Medicago/Trifolium/Lotus sp.	Medick, Trefoil, Clover etc	4	-
Vicia/Lathyrus sp.	Vetch, Vetchling, Tares etc	2	3
Hypochaeris glabra/radicata L.	Cat's Ear	-	1
Isolepis setacea	Bristle Club-rush	-	1
Carex sp. 3 sided	Sedges	4	7
Carex sp. 2 sided	Sedges	2	-
Poa annua/Phleum sp.	Meadow Grass/Cat's Tails	2	-
Poaceae	Grass, basal culms/rhizomes	8	3
Poaceae	Grass, culm segments	+	3
Unidentified	Seeds	4	4
Indeterminate	Seeds	3	1
Small weeds not extracted	Seeds	++	++

Table 8.3: Plant macrofossil identifications. Quantifications are based on seed, nutlet, grain and so on unless otherwise stated.

legumes appear to be the most frequently encountered funerary offering (Preiss *et al* 2005; Šoštaric *et al* 2006; Bouby & Marinval 2004; Marinval 1993). In contrast, in southern France and the Rhone Valley bread or pastry and Mediterranean fruits and nuts dominate while grain and pulses are rare, which may be related to more Romanised or Mediterranean traditions (Bouby & Marinval 2004).

It is difficult to assess the motivation behind the selection of particular food types for ritual or funerary offerings. Certain foods, particularly stone pine, grape, figs and dates, appear to have held strong ritual significance throughout the Roman world and are found as temple and high-status funeral offerings within Italy and the northern Empire including London and Mainz, Germany (Robinson 2002; Zach 2002; Giorgi 2000). It is likely that socio-economic factors were significant in the choice of foods offered. The recovery of lentils, fig and grape from Pepper Hill (Davis 2006) indicates that it is not solely within London that the exotic species are found (contra van der Veen *et al* 2008), although it is likely that such offerings are restricted to the elite. Peas, beans and cereal grains were available to the wider population and are more likely to represent more standard offerings. Certain traditions relating, for example, to the military may also be significant. It has been suggested that in France the prevalence of pulses and grains may be related to Iron Age customs (Bouby & Marinval 2004), and similar traditions may be seen in Britain.

The wild species present in the deposits are dominated by grassland plants and include the grass stems and rhizomes. Several of the plant species represented are indicative of wet or damp ground. Much or even all of this vegetation is likely to have derived from uprooted vegetation in the immediate vicinity of the pyre, possibly to create a fire break, particularly if constructed on areas with long grassland (Stevens 2008). The presence of such tubers is uncommon in Romano-British cemeteries, although they were present at Ryknield Street (Stevens 2008). It is likely that their presence is related to the nature of the site and an absence of formalised cremation areas (*ustrina*), and is in contrast to more urban sites such as the East London cemetery (Davies 2000) from which grass tubers were absent.

9. Leather

Quita Mould

Leather was recovered from six features, pits [1053], [1096] and [1157], brine tanks [1182] and [1207], and well [5052]. The overwhelming majority came from features in the southern end of the site (excavation area A), with most of the leather coming from the fill of the two brine tanks. A very small amount was recovered from well [5052] at the north of the site (area E) next to a Roman road which appears to represent the major approach to Roman Nantwich.

Methods

The leather was assessed and initially recorded when wet and washed, and was then examined after conservation by freeze drying and the records updated. The measurements quoted here are those made before conservation when the leather was wet. Any shoe sizing has been calculated according to the modern English Shoe-Size scale from the measurement of the insoles of shoes of multi-part constructions or the distance from the toe to the seat of shoes of one-piece construction. No allowance for shrinkage has been made. Leather species were identified by hair follicle pattern using low-powered magnification. Where the grain surface of the leather was heavily worn identification was not always possible. The grain patterns of sheep and goat skins are difficult to distinguish and have been grouped together as sheep/goat when the distinction could not be made. Similarly, the term bovine has been used when uncertainly arose between mature cattle hide and immature calfskin. Shoe bottom components and repairs are assumed to be of cattle hide unless stated otherwise.

The terminology used is that commonly employed for Roman shoe assemblages. Roman shoe constructions have been most recently summarised by van Driel-Murray (2001). Illustrated items are indicated by an asterisk (*) in the text and are catalogued below. A basic record of all the leather is provided in Appendix 9.1. In the illustrations, for the most part, the nailed shoe components and waste leather show the grain view and the one-piece shoes show the flesh side. The letter g denoting grain surface and f denoting flesh surface have been added for clarification where necessary.

Overview

The leather recovered comes principally from footwear. A small amount of waste leather, much of it characteristic of shoemaking, was also found along with a patched panel from an unidentified item. The shoes appear to be the result of general rubbish disposal. A small amount had been cut up apparently to recycle reusable leather before being thrown away; this suggests it to be the debris from a leatherworker engaged in salvaging leather probably for repair and refurbishment. Indeed one shoe appears to have been made using old shoe parts (8 SF423*). The waste leather is likely to derive from the occasional disposal of sweepings from the floor of a shoemaker's workshop. The leather finds all suggest that a leatherworker or leatherworkers, in particular shoemakers and cobblers, were active in the local area. The leather finds do not provide any evidence for the use of any of the excavated features in the tanning process.

Footwear

The fragmentary nature of many of the shoes makes reliable quantification difficult, as the number of shoe parts recovered is not necessarily representative of the number of shoes originally discarded. Consideration of the shoe parts found, however, indicates that at least 30 shoes were present and the figure is unlikely to have exceeded 37. The majority comprised shoes of nailed construction (approximately twenty examples), the most common Roman shoe type, and shoes of one-piece construction (at least ten examples) were well represented. A smaller number of stitched shoes and a single shoe of stitched and nailed construction were also found. As is commonly the case with Roman footwear of both nailed and stitched constructions, little of the shoe uppers had survived so that the styles of the shoes of these constructions are unknown.

Nailed Shoes

Approximately twenty shoes of nailed construction were found; components from at least sixteen were certainly present. They were represented for the most part by their shoe bottoms each comprising an insole, midsole or middle packing (lamina) and outer (tread) sole held together principally by hobnails. Carol van Driel-Murray's classification (2001, 351 fig 21) of nailing patterns has been used here. Lightly nailed shoes with widely spaced nailing (eg 1 SF404*) and heavily nailed examples (eg 2 SF421*, 3 SF431*) were both well represented. Only a single sole with a decorative tendril nailing pattern at the tread was noted (4 SF413* and Fig 9.2). Two thirds of the shoes (66%) were lightly nailed (pattern 1) and over a third (39%) had a nailing pattern that included a single nail at the waist (subtype C). In many cases the insole and midsole or middle packing were also joined by constructional thonging which had been used to hold the shoe upper and bottom parts together during manufacture before the shoe was nailed.

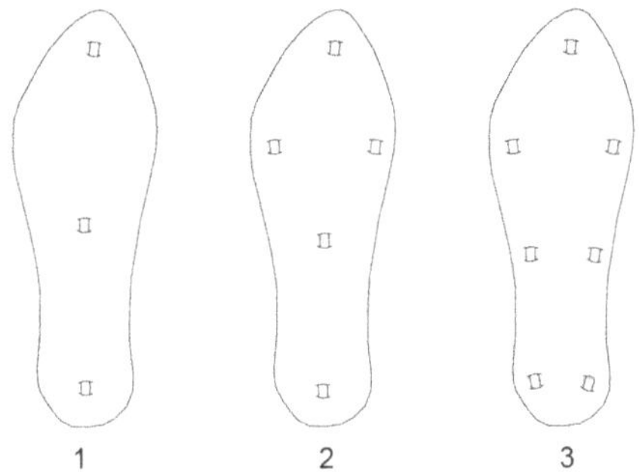

Figure 9.1: Constructional thronging patterns (type 1 thonging down the centre, type 2 around edge of tread and down centre of waist and seat, type 3 around the edge).

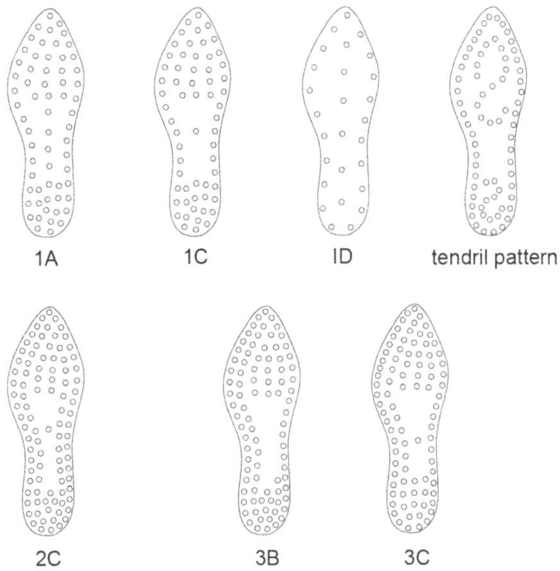

Figure 9.2: Nailing patterns.
Type 1A: widely spaced with infilling at tread and seat and line down centre.
Type 1C: widely spaced with infilling at tread and seat and single nail at the waist.
Type 1D: widely spaced with widely spaced line down the centre.
Tendril pattern at the tread and S shape at the seat.
Type 2C: double line around the edge infilling at tread and seat with single nail at the waist.
Type 3B: double line on outer edge, single line on the inner no nails at the waist.
Type 3C: double line on outer edge, single line on the inner and a single nail at the waist.

The constructional thonging patterns are shown in Figure 9.1 and the nailing patterns in Figure 9.2; all are commonly found types. A correlation between the type of constructional thonging and nailing pattern could be discerned in this assemblage (Table 9.1), although it has not been noted elsewhere, such as at the fort of Birdoswald, Cumbria (Mould 1997a, 331). Shoe bottoms with type 1D nailing had type 1 thonging, while no constructional thonging had been used on those with type 1A nailing. It was notable how many of the soles had no nails present nor iron staining marking their former position, suggesting that, in some cases, the nails may have been deliberately removed for recycling before the shoes were thrown away.

Feature/ Context		SF	Nailing	Thonging	Construction/ Comments
1053	1062	402	1D	1	Stitched and nailed construction
1096	1097	404	1A	None	Nailed
1157	1159	410	1A	None	Nailed
1182	1213	413	1C	2	Nailed, tendril pattern
1182	1279	415	1D	1	Nailed
1182	1279	416	1A	None	Nailed
1182	1283	421	3C	1	Nailed
1182	1283	422	1C	2	Nailed
1182	1283	423	None	3	Stitched construction
1182	1285	429	1A	Unknown	Nailed
1182	1285	430	1A	None	Nailed
1182	1285	431	3C	3	Nailed
1182	1285	432	1A	None	Nailed
1182	1285	433	1D	None	Nailed
1182	1285	434	3C	3	Nailed
1182	1285	435	2C	Unknown	Nailed
1182	1285	436	2C/3C	Unknown	Nailed
1207	1225	437	1A	N one	Nailed
5052	5108	447	3B	3	Nailed

Table 9.1: Shoe nailing patterns and constructional thonging.

Little survived that might tell us of the styles of shoe represented by the shoe bottoms found. The majority of the nailed footwear bottom units had distinctly pointed toes (eg 4 SF413*), only two were oval in shape (eg 1 SF404*), the treads were of 'natural' shape tapering slightly to a wide seat, and the waist area was not distinctly defined. Tunnel stitching aligned at right angles to the edge was present on the underside (flesh side) of many of the insoles showing where the lasting margin of the shoe upper had been attached. Fragments of shoe upper had been preserved sandwiched between the layers of the bottom components but few diagnostic features of the uppers survived (but see the stitched and nailed shoe below). Seven heel stiffeners were present, all of bovine leathers worn grain outward to the foot, and indicate that those shoes had closed uppers. One nailed shoe bottom comes from a sandal (5 SF429*) with decoratively notched toe and a second bottom component (SF430) may also come from a sandal, both of which were from fill (1285) of brine tank [1182]. They would have had lighter open uppers of strap-work.

Stitched and Nailed Shoe

The bottom of a small shoe of stitched and nailed construction (van Driel-Murray 2001, 348 fig 16.5) and fragments of decorated shoe upper which may come from it were found in wicker-lined pit [1053]. The components of the shoe bottom (6 SF402*) had thonging running around the edge as well as construction thonging running up the centre of the shoe (type 1) to hold the individual bottom parts together during manufacture. The sole was lightly nailed (type 1D) suggesting that it was worn outdoors but was not intended for heavy work. A seamed toe area broken from a shoe upper (7 SF403*) might belong to the shoe. The upper is decorated with an impressed oblique lattice design and a line of small dots. The linear tooled decoration would be made using a creaser, an example of which was found in brine tank [1182] (see Chapter 7).

Stitched Shoes

The remains of three shoes of stitched construction were recovered, two (8 SF423*, 9 SF443*) from brine tank [1182], while the toe of a third example sewn with running thonging was found unstratified in area A. All had pointed toes and are likely to be light indoor shoes. The insole of one example (8 SF423*) had paired grain/flesh stitching and constructional thonging (type 3) around the edge. Fragments of the lasting margin of the upper of sheep/goatskin were present, together with a low heel stiffener which suggests that the shoe had a closed upper originally. The shoe had pieces of upper from another shoe, again of sheep/goatskin, reused as middle packing between the insole and the sole (now missing). The midsole of another stitched shoe (9 SF443*) had tunnel stitching at a right angle to the edge to attach the shoe upper, but the insole had no stitching visible and appears to have been glued into position originally.

One-piece Shoes

The remains of at least ten shoes of one-piece construction were recovered, chiefly from brine tanks [1182] and [1207], with a single example coming from pit [1157]. The shoes were cut from a single piece of cattle hide with a back seam and a seam around the seat (heel area). The seams were for the most part whip stitched edge/flesh or grain/flesh seams (Rhodes 1980, fig 170 a & b). One example (10 SF420*) had a butted edge/flesh back seam also sewn with a running stitch of leather thong, another (11 SF439*) had both edge/flesh and grain/flesh stitching along the back seam, suggesting that these seams had been sewn twice to strengthen them. All had been very heavily worn and three (SF411, 10 420*,12 442*) had stitching from the attachment of repair patches to the sole areas. The fastening loops were plain with tooled edges to give them a rounded profile, a common feature on shoes on this type, and, where they survived, the toe was open. The majority had elliptical fastening loops. The front part of a one-piece shoe with plain elliptical fastening loops has been found along with a bone pin and Samian pottery dating no later than AD 160 at Snow Hill on the eastern side of the River Weaver in Nantwich (Mould 2004b). One-piece shoes with plain elliptical fastening loops are commonly found on both military and civilian sites across the country throughout the Roman period.

The one-piece shoes had little additional decoration, only one (SF441) having a decorative triangular cut out at the base of a fastening loop. Two shoes had slit straps close to their back seams (11 SF439*, 441). Slit straps also occurred on the more delicate strap-work of a one-piece shoe from the 1st- to 2nd-century footwear found at Billingsgate Buildings, London (Rhodes 1980, fig 70 no 644). One shoe (12 SF442*) had distinctive small semi-circular fastening loops, a feature also seen on a one-piece shoe found at the fort at Birdoswald (Mould 1997a, 339, fig 241 no 26) in a late 3rd- to mid 4th-century context, and others recovered recently (2009) from a well at Tollgate Farm, Uttoxeter, somewhat nearer to home (Dave Thomas, pers comm).

Patched Panel

A piece torn from a heavily worn panel of worn sheep/goatskin (13 SF428*) with a whip-stitched seam was found with footwear in fill (1285) of brine tank [1182]. The panel had remains of two patches sewn with a running stitch of leather thong present. A single whip stitch hole passes through the thong on one of the patches indicating that the patch had been in position before the seam was sewn and suggesting that the patch was not a secondary repair (Sue Winterbottom, pers comm).

Waste Leather

A small amount of waste leather was found associated with footwear in three features, [1096], [1182] and [1207], in area A with a single piece, also associated with shoe parts, coming from well [5052] in area E (Table 9.2).

The largest group of waste leather, 250 pieces, was thrown away along with at least one nailed shoe in pit [1096]. The group included primary waste from the initial trimming away of unusable parts of the tanned hides and secondary waste produced when cutting out pattern pieces and trimming them to size when manufacturing leather goods. The secondary waste included shapes which are characteristic of shoemaking resulting from the cutting out of shoe uppers; a selection is shown in Figure 9.6. The shoemaking off-cuts appear to have resulted from the cutting out of both uppers from nailed shoes and shoes of one-piece construction. The shapes are comparable with those found in large groups of shoemaking waste recovered elsewhere, including a large group from an early 2nd-century occupation layer at the fort at Carlisle (Mould 2009). Another group of waste leather (thirteen pieces) came from fills (1283) and (1285) of brine tank [1182]. These two groups of waste leather contained a range of bovine leathers, with two pieces of sheep/goatskin also present amongst that in fill (1285) in brine tank [1182]. Two pieces of primary waste, one a relatively large area of hide edge cut from a sheep/goatskin with a pegging-out hole present, were thrown away in fills (1336) and (1337) of brine tank [1207]. A single piece of heavily worn secondary waste leather from which a large disc (33mm in diameter) had been cut had been thrown into well [5052].

Waste leather has also been recovered from other comparable salt-making sites. A single piece of primary waste and a small amount of secondary waste, possibly from shoemaking, were recovered from the Roman levels at the multi-period salt production site at Upwich at Droitwich (Mould 1997b, 127). A small amount of shoe leather, a bound panel and a small group of secondary and primary waste was recovered from a Roman timber-lined well [328] at Middlewich (Mould 2007), although the secondary waste was not characteristic of shoemaking in this instance.

Feature/Context		Type	Details
[1096]	(1097)	Secondary waste	Shoemaking waste x50 Trimmings x11 Other shapes x42 Pieces with clamp marks x2 Piece with hide edge 1
		Primary waste	Hide edge with handling slit x1 (206x80x8.33mm)
		Other waste pieces	Cut and torn edges x60
		Scrap	All edges torn x3
[1096]	(1098)	Secondary waste	Shoemaking waste x8 Trimmings x5
		Primary waste	Hide edges x6 Unusable areas of hide x8
[1096]	(1142)	Secondary waste	Shoemaking x30 Other shapes x18
		Primary waste	Hide edges x8
[1182]	(1283)	Primary waste	Unusable area of hide x1
[1182]	(1285)	Secondary waste	Shoemaking waste x2 Trimming x1 Other shapes x5
		Primary waste	Hide edge (107x48x2.88mm) Hide edge (206x43x2.39mm) with teat Hide edge (201x114x2.20mm) Hide edge (90x114x2.20mm)
[1207]	(1336)	Primary waste	Hide edge with pegging out hole (280x120x1.6mm)
[1207]	(1337)	Primary waste	Hide edge (108x70x3.05mm)
[5052]	(5108)	Secondary waste	Rectangle with circle removed (117x80x1.66mm)

Table 9.2: Waste leather.

Catalogue of Illustrated Leather

1 Shoe bottom of nailed construction, left foot (Fig 9.4).
Insole, rectangular middle lamina and sole with oval toe, natural tread tapering slightly to a wide seat. Single line of nailing around the edge, single line down the centre with infilling at the tread and seat (van Driel-Murray 1A), no nails present. Tunnel stitching on the flesh side of the insole at right angle to the edge. Insole length 241mm, max width (tread) 88mm. Adult size 4(37). SF404. Fill (1097) of pit [1096].

2 Shoe bottom of nailed construction, right foot (Fig 9.4).
Insole, middle packing and sole worn away at the edge of the toe, natural tread tapering slightly to the wide seat. Double line of nailing along the outer edge, single at the inner, dense vertical lines of infilling at the tread and seat with group at the waist (van Driel-Murray 3C), some nails present. Insole with line of type 1 constructional thonging down the centre. Insole length 240mm, max width (tread) 93mm. Adult 4(37). SF421. Fill (1283) of brine tank [1182].

3. Shoe bottom of nailed construction, right foot (Fig 9.4).
Insole with broken toe and seat and fragments of midsole and middle lamina. Line of nailing along the inside edge and double line along the outer with infilling at the tread and seat with a nail at the waist (van Driel-Murrray 3C). Insole with type 3 constructional thonging around the edge. Remains of heel stiffener broken away before the lasting margin, surviving height 40mm, of bovine leather 2.08mm thick. Small fragment of upper fastening strap of worn bovine leather 1.39mm thick. Insole length 232+mm, max width (tread) 93mm. Large adult size. SF431. Fill (1285) of brine tank [1182].

4 Shoe bottom of nailed construction, right foot (Fig 9.4).
Insole and forepart of sole with pointed toe, natural tread, slight waist and wide seat. Single line of nailing around the edge with tendril pattern at the tread and 'S' shaped pattern at the seat (van Driel-Murray 1C), no nails present. Insole has type 2 constructional thonging. Fragile remains of upper lasting margin with a line of grain/flesh stitching torn along the nailing line, and other related fragments. Insole length 275mm, max width (tread) 84mm. Adult 8(42). SF413. Fill (1213) of brine tank [1182].

5 Shoe sandal sole, nailed construction, right foot (Fig 9.4).
Sole with oval toe with decorative notch at the great toe, tread tapering slightly to the seat. A single line of nailing around the edge with lozenge pattern infilling at the tread and a line running down the centre (van Driel-Murray 1A), four hobnails present. Length 250mm, max width (tread) 78mm. Adult male size. SF429. Fill (1285) of brine tank [1182].

6 Shoe bottom of stitched and nailed construction, right foot (Fig 9.4).
Shoe bottom with oval toe, natural tread tapering slightly to the medium seat. Insole with central line of constructional thonging type 1. Sole with widely spaced nailing around the edge and in central line (van Driel-Murray 1D), with a line of running stitching sewn with thong, 3mm wide, also running around the edge. Narrow middle lamina, seat area now detached with central thong holes and upper lasting margin present on the right side joined to the lamina with thonged stitching. Leather: upper bovine, insole (delaminating) cattle hide 2.38mm thick. Insole length 180mm, max width (tread) 67mm. Child size 10(28). SF402. Fill (1062) of pit [1053].

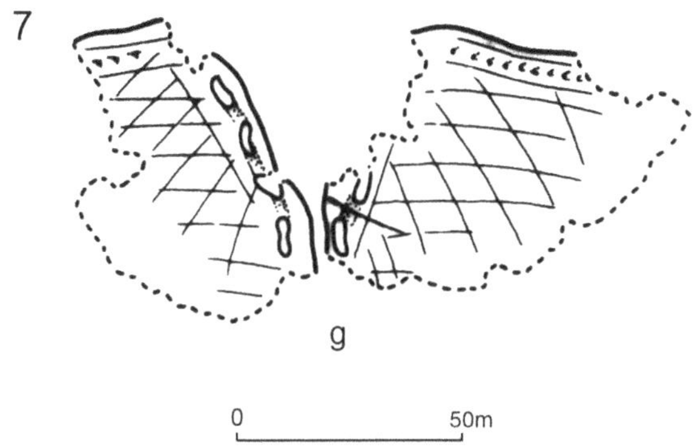

Figure 9.3: Leather 7.

7 Shoe upper fragments, decorated (Fig 9.3).
Two fragments each with remains of a grain/flesh, saddle stitched seam with large stitches, each 3-4mm long, likely to come from a central toe seam. A small area of original edge runs at a right angle to the seam with a line of impressed dots below, all other edges torn. Further decorated with an oblique lattice of impressed lines. Leather bovine *c* 3mm thick (now delaminated). 80x55mm; 68x50mm. SF403. Fill (1222) of pit [1053].

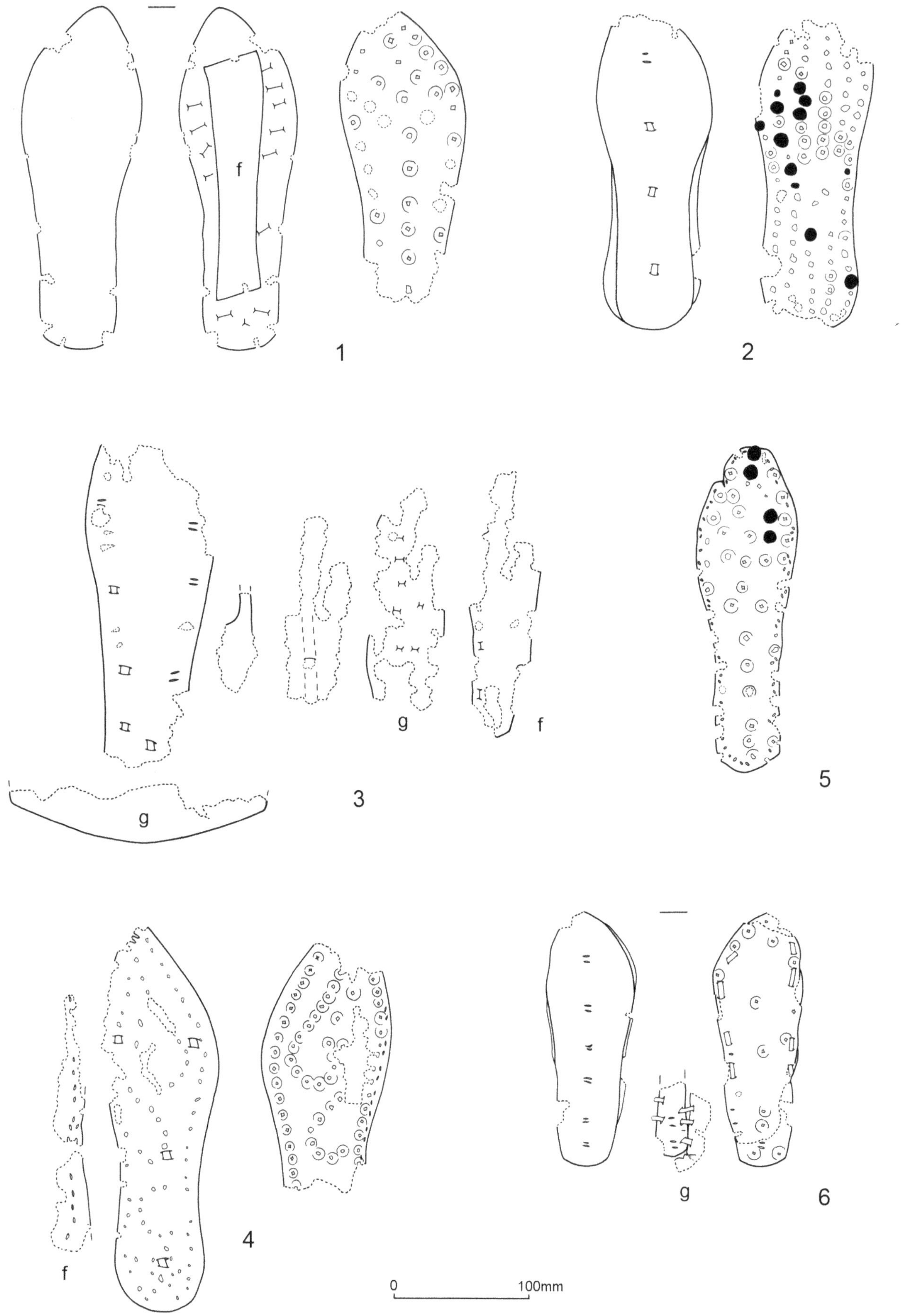

Figure 9.4: Leather 1-6.

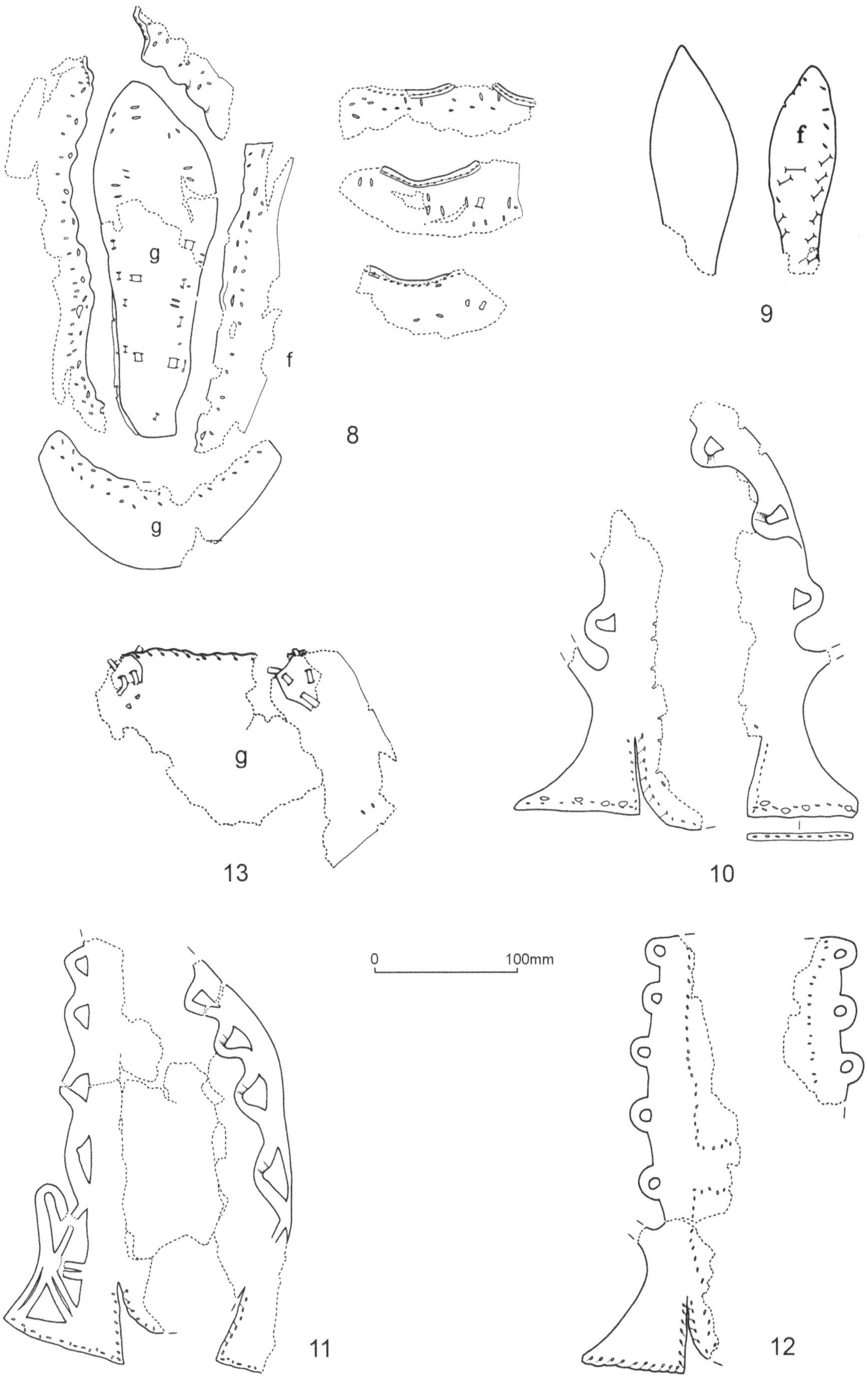

Figure 9.5: Leather 8-13.

8 Shoe, stitched construction, right foot (Fig 9.5).
Insole with pointed toe and paired grain/flesh stitching around the edge. Type 3 constructional thonging around the edge. Four pieces of upper lasting margin with holes from bracing thread and line of closely-spaced grain/flesh stitching, stitch length 4-8mm, running above, cut in areas and torn from the rest of the upper. Heel stiffener with braced and grain/flesh stitched lasting margin, grain side outward, surviving height centre back 30mm. Three pieces of shoe upper each with an area of hemmed top edge, grain/flesh seam with thread impression, stitch length 5mm. The upper pieces are middle packing, the constructional thonging slots matching those present in the insole indicating use in the shoe bottom. Upper leather sheep/goatskin 1.13mm thick. Insole length 247mm, max width (tread) 84mm, waist 58mm, seat 52mm. Adult 4(37). SF423. Fill (1283) of brine tank [1182].

9 Stitched shoe, forepart, right foot (Fig 9.5).
Insole and midsole forepart with pointed toe, broken away across the waist. The insole is unstitched, the smaller mid sole has tunnel stitching at an oblique right angle to the edge from the attachment of the upper and sole. Insole leather cattle hide 2.13mm thick. Insole surviving length 166+mm. Adult size. SF443. Fill (1290) of brine tank [1207].

10 Shoe, one-piece construction, foot uncertain (Fig 9.5).
Sole area worn and torn away. Peaked back seam with faint edge/flesh seam and running stitching sewn with thong 3mm wide, butted edge/flesh seam around the seat, stitch length 6-8mm. Plain fastening loops with tooled edges, two large stitches at the base of one loop from attachment of a sole repair. Leather cattle hide 5.38mm thick. Surviving length 273+mm, height at centre back 98mm. Adult 7(41). SF420. Fill (1283) of brine tank [1182].

11 Shoe, one-piece construction, right foot (Fig 9.5).
Sole area torn and worn, much of the forepart missing, worn through at the seat. Back seam with both widely spaced edge/flesh seam, stitch length 8mm, and a line of small grain/flesh stitching, stitch length 5-6mm present. Edge/flesh seat seam with whip stitching. Triangular cut-outs at the back seam with slit connecting straps. Plain, elliptical, fastening loops running down toward the toe. Leather worn cattle hide 3.90mm. Surviving length 280+mm, height centre back *c* 83mm. Adult 8(42). SF439. Fill (1239) of brine tank [1207].

12 Shoe, one-piece construction, foot uncertain (Fig 9.5).
Left side of shoe with sole area torn/worn away. Back seam with edge/flesh back seam with whip stitching, stitch length 8-10mm, with a similar seam around the seat. Top edge with plain, semi-circular fastening loops with a straight cut, open toe. Line of small stitching present from repair patches to the seat and forepart areas of the sole. Three matching loops from the toe end of the opposite side. Leather worn cattle hide 4.31mm thick. Length 296mm, height centre back 70mm. Adult 10(45). SF442. Fill (1240) of brine tank [1207].

13 Patched panel (Fig 9.5).
Subrectangular panel with a whip-stitched, grain/flesh seam, stitch length 12mm, and a cut edge along the opposite side, other edges cut and torn away. The remains of two small patches sewn with thong 3mm wide present. Leather much folded and distorted, heavily worn sheep/goatskin 1.40mm thick. Length 175mm, width 142mm. SF428. Fill (1285) of brine tank [1182].

14 Secondary waste, shoemaking (Fig 9.6).
Selection of secondary waste off-cuts characteristic of shoemaking. Upper two have marks produced by clamps used during the preparation of tanned hides, selection of shapes characteristic of cutting out fastening loops of shoe uppers in the centre, the lower two from cutting out the fastening loops of shoes of one-piece construction. Various bovine leathers. SF406. Fill (1097) of pit [1096].

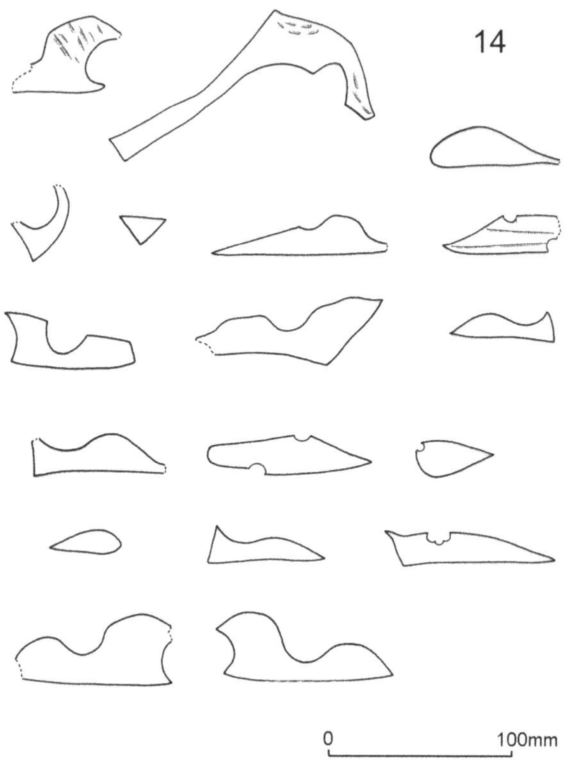

Figure 9.6: Leather 14, secondary waste.

10. Wooden Objects

Peter Arrowsmith & Steven Bellshaw

The remains of eighteen separate wooden objects were recovered from waterlogged contexts on the site (Table 10.1). These were conserved by the York Archaeological Trust, which provided species identification. Sixteen were found within the brine tanks, [1207] and [1182]. Of these, eight were recovered from [1207], with six objects from lower fill (1239), one from basal deposit (1336) and one unstratified. Seven of the eight objects from [1182] were derived from lower fill (1285), with the other being unstratified. Of the other objects, one was found within wooden trough [1264] and one in pit [1096] which also contained the remains of a wooden trough. The stratified finds from brine tanks [1207] and [1182] are from predominantly late 2nd-century fills.

Feature	Context	Wood no	Description
1207	1239	1	Brine paddle
		4	Slotted paddle
		7	Turned bowl
		9	Lid or seat?
		12	Board
		13	Lid or base?
	1336	16	Dagger-like object
	u/s	2	Tanged paddle blade
1182	1285	3	Single-piece shovel
		5	Blade of composite shovel
		6	Perforated stave
		8	Musical instrument?
		14 & 19	Staved vessel
		15	Scoop or platter?
		17	Turned box, or *pyxis*
	u/s	11	Vessel lid
1096	1098	18	Toggle?
1264	1268	10	Ratchet or cog?

Table 10.1: Wooden objects by feature and context.

The objects from the brine tanks include five 'shovels' or 'paddles' (Wood 1-5) (Fig 10.1), which invite comparison with wooden finds from other salt-producing sites. One of these (Wood 1) is possibly the earliest known example of a specialised implement which continued in use into the post-medieval period and is believed to have been used to stir the brine during the heating and evaporation process. A later, mid-Saxon to early medieval, implement of the same type is known from the multi-period salt-producing site at Upwich in Droitwich. Although the Droitwich example had a longer handle, the handle of the Nantwich paddle seems to have been broken or shortened in antiquity. This type of paddle is among the tools depicted in a 16th-century engraving of salt making in Georgius Agricola's *De Re Metallica* (Hurst 1997, 27, 107-10). It is also shown being carried by salt makers in a German illustration of 1616 (Emons 1984, 112 Bild 77).

A second object from Kingsley Fields (Wood 5) is the tanged blade of a composite, or separate-bladed, shovel. This form of tool is described by Morris (1980; 1981; 2000, 2313-15) and was made with a handle which was set at an angle into a bevelled hole within the blade and was also secured to the tang. Shovel blades of this type have a wide distribution in Britain and Ireland, and are also found in Sweden. The majority of dated examples fall within a time-frame beginning in the late 1st millennium AD and continuing into the medieval period. Manuscript evidence shows them still in use in the 14th and 15th centuries and Morris (2000, 2315) suggests that some Cornish examples may be post-medieval. Most examples have one or two peg holes set within the tang, or fixing end, and it is believed that cord was used as an additional fastening. The blades are generally made of oak. It is possible that the handles, which rarely survive, were of a different species such as ash (Morris 1981, 58).

Finds of separate-bladed shovels which can be dated to the Roman period are relatively rare. An example with a diamond-shaped fixing end (Morris's type 'b') was discovered in Chester in 1898 alongside Roman features (Newstead 1902, 84-5, fig 2) but its own dating to the Roman period was questioned by Morris (1980, 206 no 1) due to a lack of clear stratigraphical evidence. A more securely dated blade has been found at Exeter in a context of the late 2nd century AD (Earwood 1991, 275, fig 129). It has the bevelled hole characteristic of separate-bladed shovels but is broken at the position of a possible fixing end. In addition, an oval wooden blade with a bevelled hole but again lacking a fixing end was recovered from the fill of a late 1st- to early 2nd-century brine tank at the Old Bowling Green, Droitwich (Woodiwiss 1992, 17; Crone 1992, 107 fig 56, 109). The Nantwich blade is the most complete example known to have been recovered from a secure Roman context. It falls within the range of width and length of known blades of all dates, and its rectangular fixing end (Morris's type 'e') is found in several other examples including some, like the Nantwich blade, without peg holes.

Crone draws a possible comparison between the Droitwich blade and a 'loote', an implement described in Thomas Rastel's account of 1678 of the Droitwich salt industry as a board fixed slopeways on a haft *c* 1m long, which was used to ladle out crystals from the salt pans (Crone 1992, 107). 'Loots' are also documented in the accounts of 1671-2 of the Leftwich Eyes salt works near Northwich (Andrew Fielding, pers comm). However, a 1792 illustration of a German salt maker's '*Löit*' shows a tanged ladle with the handle fixed at a slight oblique angle through the tang (Häberlin 1934, 31).

Separate-bladed shovels seem to have been a general-purpose tool for lifting and working soft, loose materials. Although they were also utilised on non salt-producing sites, their use within the salt-making process is strongly suggested by their occurrence at both Nantwich and Droitwich. While possibly distinct from the 'loots' of the post-medieval period, they could easily have served the same purpose, and been used to lift and move salt crystals.

No parallels have been identified for a large tanged paddle blade (Wood 2) or a slotted paddle (Wood 4) which, like Wood 1, may be specialised implements. A fifth object, a single-piece spade or shovel (Wood 3), seems like Wood 5 to be a more general-purpose tool but again, like Wood 5, this may have been adopted within the salt industry.

A sixth object (Wood 6) (Fig 10.1) comprises a perforated stave and may also have been used in the salt industry. Wooden beams with multiple auger holes along their long axis have been found in Well 1 at the Roman villa at Dalton Parlours and in a well at Rothwell Haigh, West Yorkshire, leading to the suggestion that such timbers were perhaps associated with the use of wells or their superstructures (Morris 1990, 227 fig 139 no 87, 230). It is possible that the Nantwich stave had some use in the collection or storage of brine.

Comparison with the wooden objects from other salt-making sites also reveals some variations from the Kingsley Fields assemblage. In Droitwich forked staves were found within the fills of 1st- to 2nd-century brine tanks at both the Old Bowling Green and Friar Street, and objects of this type, which could have been used to suspend or rest brine or salt containers, are known to have been used in the post-medieval salt industry (Woodiwiss 1992, 17; Crone 1992, 109-10, fig 57). Another absent object is the rake. Excavations at Wood Street in Nantwich in 1979-80 produced thirteen semicircular perforated wooden objects, recovered from late 12th- to 13th-century contexts, which were interpreted as rake heads used for raking precipitating salt to the side of the salt pan (McNeil Sale 1980, 20-1; Hutchings 1983, 82-4). Other examples are known from mid-Saxon to 13th/14th-century contexts in Upwich (Hurst 1992, 107-8, fig 97). They are also found on non salt-making sites and seem to have had a wider range of uses, including as oven rakes for raking embers and ash (Morris 2000, 2319-20).

A second group of objects recovered from Kingsley Fields consists of various containers or parts of containers. It includes a turned box (Wood 17); a turned bowl (Wood 7); a stave-built vessel (Wood 14 and 19); a possible scoop or platter (Wood 15); a vessel lid (Wood 11); and a possible lid or base (Wood 13).

Wooden turned boxes, or *pyxides*, such as Wood 17 (Fig 10.3) are discussed in detail by Pugsley (2003, 66-84). Examples of such boxes, or their lids, are found widely on sites across the classical world including, in Roman Britain, London, Vindolanda, Corbridge, Carlisle, Bar Hill and Newstead (Pugsley 2003,153-5; Howard-Davis 2009b, 809, fig 483). As with the Nantwich example, they are commonly of boxwood but the use of other woods is known. Such boxes are believed to have been used principally for cosmetics or medicine.

Turned bowls again are known from a range of Roman sites. Earwood describes their typical form as shallow, flat-bottomed and with outward sloping sides (Earwood 1993, 91). Wood 7 (Fig 10.3) has a convex base and sides which curve outward from the base and slightly inward at the rim.

The stave-built vessel (Wood 14 and 19) (Fig 10.4) survives as three largely complete and two partial staves, which together would have formed the sides of a vertical-sided vessel with an internal diameter at the base of *c* 140mm. The base and the hoops, which were evident as two bands of rust discolouration on the exterior of the staves, were not recovered and the vessel seems to have been in a dismantled state when deposited in the tank. The interior face of the staves carries a single croze, or groove, for the base, while a small clasp, mainly of tin, is fixed on one stave just below the rim.

The size of the Nantwich vessel bears comparison with the stave-built tankard found in the Corbridge hoard, dated to the late 1st to early 2nd century, which had an internal diameter of 175mm and height of 159mm (Allason-Jones & Bishop 1988, 83, 86). A round oak base with a bevelled edge, with a diameter of 158mm, from the late 1st-century fort at Carlisle belonged to a stave-built vessel of a similar size (Howard-Davis 2009b, 808, fig 481). Staves from an early 2nd-century context at Ribchester may also be small enough to derive from tankards (Howard-Davis & Whitworth 2000, 330). The antecedents of such staved vessels lie in the Iron Age (Allason-Jones & Bishop 1988, 86) but their occurrence at Corbridge, Carlisle and Ribchester may be a hint of a military presence also at Kingsley Fields.

The vessel lid (Wood 11) (Fig 10.4) is a type of object found in Britain in contexts from the Roman period to the post-medieval, comprising small circular or subcircular wooden discs, either plain or with one or two holes. A similar lid with a single hole was found within deposits associated with the Phase 1 1st-century fort at Castleford, West Yorkshire (Morris 1998, 335, 336 fig 152 no 2), and another derives from the 1st-century fort at Carlisle (Howard-Davis 2009b, 815, fig 493 no 1). Others are known from Bar Hill (Keppie 1975, 52-4, fig 16 nos 9-12).

Wood 13 (Fig 10.4) is a circular board which is comparable in size with the bucket bases found within Well 1 of the Roman villa at Dalton Parlours, West Yorkshire (Morris 1990, 206-20). However, it lacks the bevelled edge typical of the base of these and other staved vessels and may perhaps be better identified as a lid.

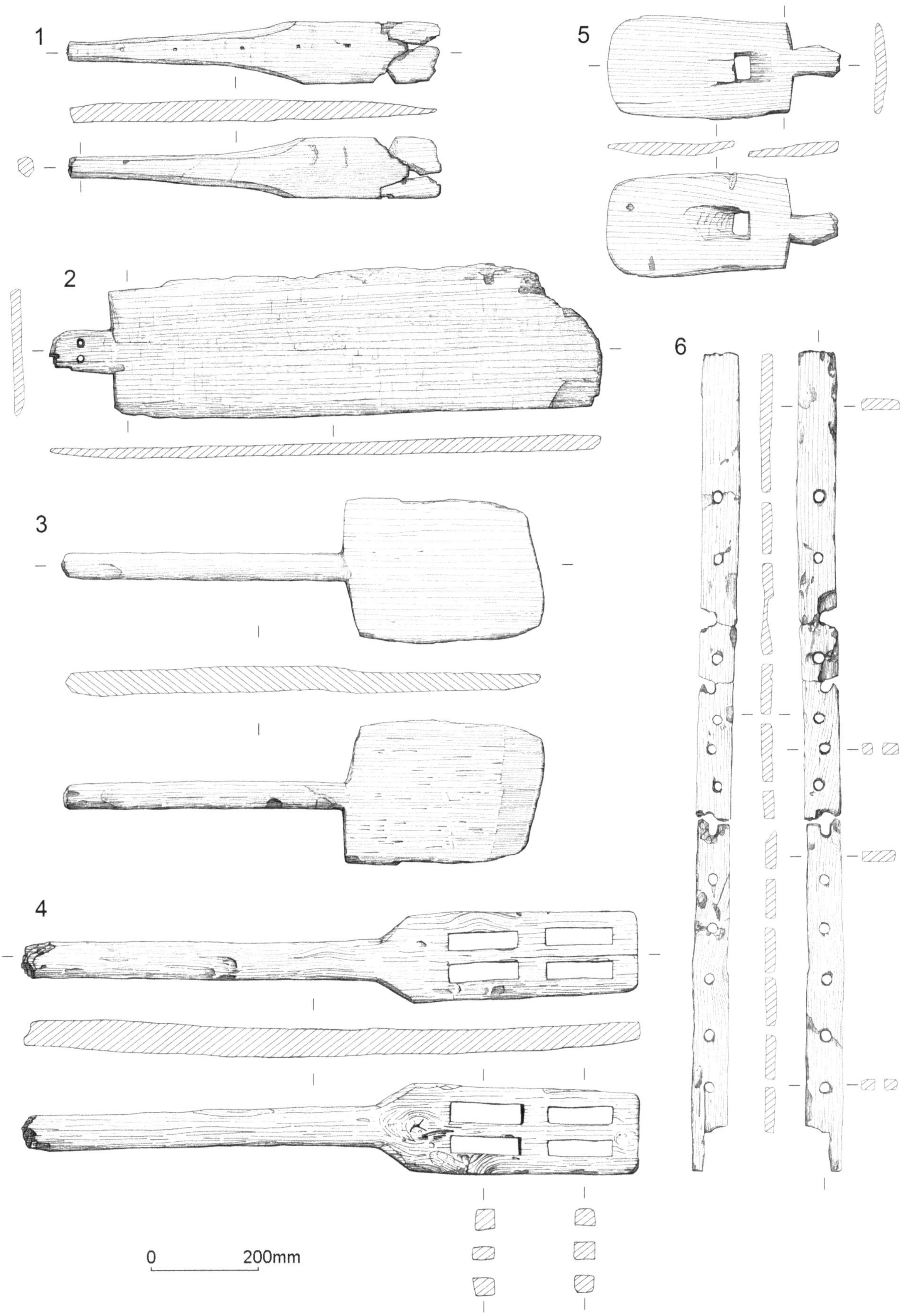

Figure 10.1: Wooden objects.

The remainder of the assemblage comprises several objects of uncertain function. Two are tentatively identified as a ratchet or cog (Wood 10) and a toggle (Wood 18) (Fig 10.4). A perforated pipe or rod, Wood 8 (Fig 10.3), may be a musical instrument but no parallels are known from Roman archaeology (Graeme Lawson, pers comm).

This group also includes a lid or seat-like object, with a central hole, and mortices and holes arranged around its semicircular edge (Wood 9), for which no Roman parallels are again known (Fig 10.2).

Circumstances of Deposition

The possible ratchet or cog (Wood 10) from pit [1096] and toggle (Wood 18) from trough [1264] may have been casually or accidentally deposited, perhaps after having a function in the operation of these features.

The objects from brine tanks [1207] and [1182] suggest a more varied picture. In some cases we may be dealing with no more than the disposal of rubbish. These include objects which are intact or virtually intact but originally were a part of a larger whole, such as the vessel lid (Wood 11), and perhaps the possible lid or base (Wood 13) and the possible lid or seat (Wood 9). They also include broken objects. The stave-built vessel, missing its base and hoops (Wood 14 and 19), can be included within this category, as can the possible scoop or platter (Wood 15), the fragment of board (Wood 12) and perhaps also the bowl (Wood 7) and box (Wood 17), although in the last two cases the relatively minor damage may have occurred after deposition.

The wooden tools, on the other hand, suggest a more structured deposition. Wood 3 and 4 still seem to have some working life, which may hint at a ritual element in their disposal. The composite shovel blade (Wood 5) has lost its handle but is in a better condition than many other examples of this type, which are commonly broken or split. Wood 1 and 2 both show damage to the blade but these objects may likewise represent more than casual disposal. It has been suggested that any wooden implement which suffered serious damage during salt manufacture would have been used as fuel for the hearth (Andrew Fielding, pers comm), making it likely that when deposited in the tanks these tools were still in a useful or repairable state. In both [1207] and [1182] the stratified wooden objects were recovered from lower fills which appear predominantly to represent the primary infilling after the tanks had ceased to be used for the managed storage of brine and which also contained a number of other finds seemingly indicative of a ritual deposition. It seems probable that the deposition of salt-making tools was part of the ritual of closure. The presence in the brine tanks at Droitwich of a composite shovel blade and forked staves, both of which are likely to have been used in salt making, may be due to similar circumstances.

Catalogue of Wooden Objects

1. Brine paddle (Fig 10.1). Oak (*Quercus* sp.). A long cylindrical handle which broadens out into a flat blade. The end of the blade survives as two fragments which taper to a worn edge with a maximum thickness of 3mm. There is evidence of tool marks on the faces of the blade. The handle appears to have been broken or shortened in antiquity. Length 670mm, width 112mm, thickness 28mm. Fill (1239) in brine tank [1207].
2. Tanged paddle blade (Fig 10.1). Oak (*Quercus* sp.). A long flat blade tapering in thickness towards one end where there is a tang. The tang itself tapers to a thickness of 10mm and is pierced by two rivet holes, one 12mm and the other 8mm in diameter, both slightly countersunk. The edge of the blade has some damage. Total length 1030mm, width 245mm, thickness *c* 28mm. Tang: length 120mm, width 75mm, maximum thickness 20mm. u/s in brine tank [1207].
3. Single-piece shovel (Fig 10.1). Oak (*Quercus* sp.). A cylindrical handle continuing as a blade with a slightly convex profile, with the blade end chamfered on the underside and tapering to *c* 10mm thick. There is some evidence of tool marks on the blade. Total length 900mm, width 250mm, blade length 370mm, handle diameter 50mm. Fill (1285) in brine tank [1182].
4. Slotted paddle (Fig 10.1). Oak (*Quercus* sp.). A long cylindrical handle which flattens out to form a rectangular paddle, pierced with four rectangular slots. The paddle is slightly convex, with a maximum thickness of 40mm narrowing to 35mm at the sides. Length 1140mm, width 150mm, handle diameter 54mm. Rectangular slots: length 130mm, width 35mm. Fill (1239) in brine tank [1207].
5. Tanged blade of a composite shovel (Fig 10.1). Oak (*Quercus* sp.). The blade is slightly concave in shape with a tapering end. It is pierced by a hole, *c* 50mm long and 30mm wide, the edges of which have been bevelled by 60-70mm towards the tang on one side and by 60-70mm towards the end of the blade on the other. There are tool marks around the hole and bevelled areas. Total length 430mm, width 180mm, thickness 20mm. Tang: length 52mm, width 42mm. Fill (1285) in brine tank [1182].
6. Stave perforated with fourteen auger holes (Fig 10.1). Oak (*Quercus* sp.). The holes are *c* 18-20mm in diameter, with five towards the centre of the stave being set in a zig-zag arrangement *c* 40mm apart, and the others located centrally along the long axis of the stave, *c* 70-90mm apart. One end of the stave has a flat edge, the other has been broken. Total length *c* 1260mm, width 70mm, thickness 18-19mm. Fill (1285) in brine tank [1182].
7. Bowl (Fig 10.3). Oak (*Quercus* sp.). Turned, with a convex base. The sides curve outward from the base and slightly inward at the rim. There are two external grooves 1-2mm and 5-6mm below the rim, a groove around the interior of the bowl *c* 2mm below the rim, and scoring is evident around the exterior and interior of the vessel. There is a split from the base to the rim and also damage to the base. Height 65mm, maximum wall thickness 9mm. Diameter: maximum 183mm, base 91mm, internal rim 166mm. Fill (1239) in [1207].
8. Musical instrument? (Fig 10.3). (*Pomoideae* sp.). A pipe with turned bevelled ends and with three holes, *c* 6mm in diameter, offset to one end. It is oval in section, probably a result of deposition. Length 138mm, maximum diameter 20mm. Fill (1285) in brine tank [1182].

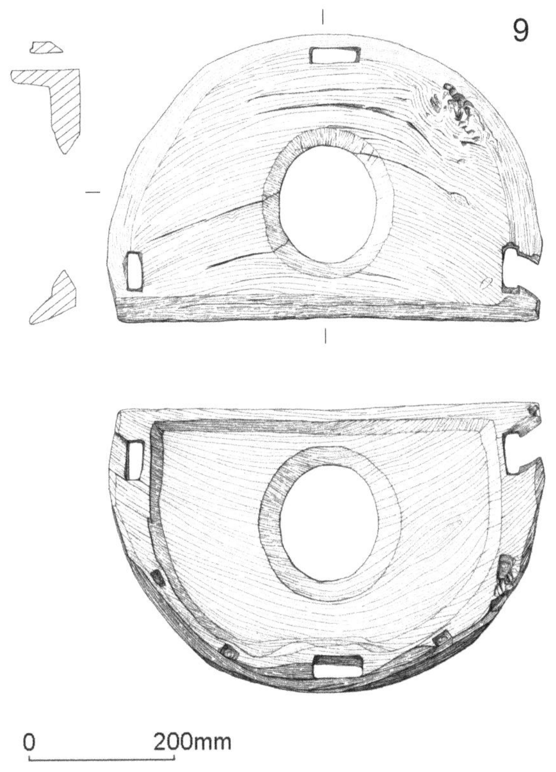

Figure 10.2: Wooden object 9.

9. Lid or seat? (Fig 10.2). Oak (*Quercus* sp.). Semicircular object, with a central hole 140mm in diameter with chamfered edges. The upper? surface of the object also has a chamfered outer edge. On the underside? is a raised rim with chamfered sides. The curved edge of the object has three evenly spaced rectangular holes, *c* 50-60mm long and 20mm wide, which on the underside? pass through the raised rim. That rim also contains four smaller mortices, two measuring 20mm x 20mm and two 20mm x 15mm. These mortices are 15-20mm deep and are positioned in a symmetrical arrangement between the outer holes. Length 585mm, width 345mm. Thickness: rim *c* 85mm, centre *c* 40mm. Fill (1239) in brine tank [1207].
10. Ratchet or cog? (Fig 10.4). Oak (*Quercus* sp.). Cruciform object which is a composite of two layers of wood joined using three evenly spaced dowels, visible on each side, and is pierced by an irregularly cut central square hole, 27mm x 30mm. The edges of the object are heavily abraded. Length 136mm, width 127mm, thickness *c* 34m. Fill (1268) in wooden trough [1264].
11. Vessel lid (Fig 10.4). Oak (*Quercus* sp.). A disc perforated by an oval off-centre hole, 22mm x 27mm. The object has been damaged and abraded around the edges. Diameter 85mm, thickness 6mm. u/s in brine tank [1182].
12. Fragment of tangentially faced board (Fig 10.4). Elm (*Ulmus* sp.). Flat on one side and heavily abraded on the other. Possible tool marks, which consist of diagonal scoring, are evident across the flat surface. Height 80mm, width 80mm, thickness 27mm. Fill (1239) in brine tank [1207].
13. Lid or base? (Fig 10.4). Oak (*Quercus* sp.). A circular board, with linear tool marks evident on both sides and uneven heavy abrasion around the edge. Diameter *c* 196mm, thickness 10mm. Fill (1239) in brine tank [1207].
14. Three staves from a coopered vessel, one with a metal clasp (Fig 10.4). Oak (*Quercus* sp.). A total of five staves were recovered from this vessel, three complete (Wood 14) and two fragmented (Wood 19). The staves average 100mm wide and are 130mm high and *c* 13mm thick, with bevelled sides. The interior walls taper towards the rim, which has a maximum thickness of 2mm. At a height of *c* 10mm from the bottom edge is a croze (groove), *c* 4mm wide and *c* 3mm deep, which would have provided the seating for the base. The exterior of the staves has evidence of rust discolouration in two bands, indicative of the position of iron hoops; one located *c* 18mm above the base, 15-16mm wide, and the other 90mm above the base, again *c* 16mm wide. A metal clasp is fastened to the interior of one stave fragment near the rim. The clasp measures *c* 24mm x 11mm and is held in place using a small circular rivet through a circular section of the clasp. The clasp is partially curved over the rim into a small groove on the exterior of the stave. Scanning electron microscope analysis by the Institute of Archaeology, University College London, found that the clasp consists mainly of tin, giving it a grey/white colour. Fill (1285) in brine tank [1182].
15. Scoop or platter? (Fig 10.4). Oak (*Quercus* sp.). This object has been broken leaving a flat area with a raised lip to one side. Length 122mm, width 100mm, thickness of base *c* 10mm. Fill (1285) in brine tank [1182].
16. Unidentified object (Fig 10.4). Oak (*Quercus* sp.). A flat dagger-like object with a broad 'blade', a 'handle' and 'pommel', but the handle is lobed on one side and originally also seems to have had a projection on the other, now broken off. Length 216mm, width 40mm, thickness 9mm. Fill (1336) in brine tank [1207].
17. Box, or *pyxis* (Fig 10.3). Boxwood (*Buxus sempervivens* L.). Turned, with a circular base, the interior of which is slightly conical with a central conical depression. The rim is oval, probably a result of distortion following deposition. The base has a slight projecting band above which the sides slope outward to the top, where there is a stepped rim with a chamfered outer and inner edge, designed to receive a lid flush with the sides. The lid was missing. There is evidence of scoring around the interior *c* 22mm below the top of the rim. Around the edge of the base of the interior is a fine black gritty residue, which has not been analysed. There is damage to the base and wall. Height 47mm, wall thickness 6-7mm. Diameter: top 47mm, base 43mm. Fill (1285) in brine tank [1182].
18. Toggle? (Fig 10.4). Willow (*Salix* sp.). This object is irregular in shape and pierced by a *c* 20mm diameter hole near the centre. A *c* 2-3mm deep notch has been worn or cut on one face and runs centrally across the hole. Length *c* 98mm, width 36mm, thickness 16mm. Fill (1098) in pit [1096].
19. Four fragments of two staves from a coopered vessel (see Wood 14). Fill (1285) in brine tank [1182].

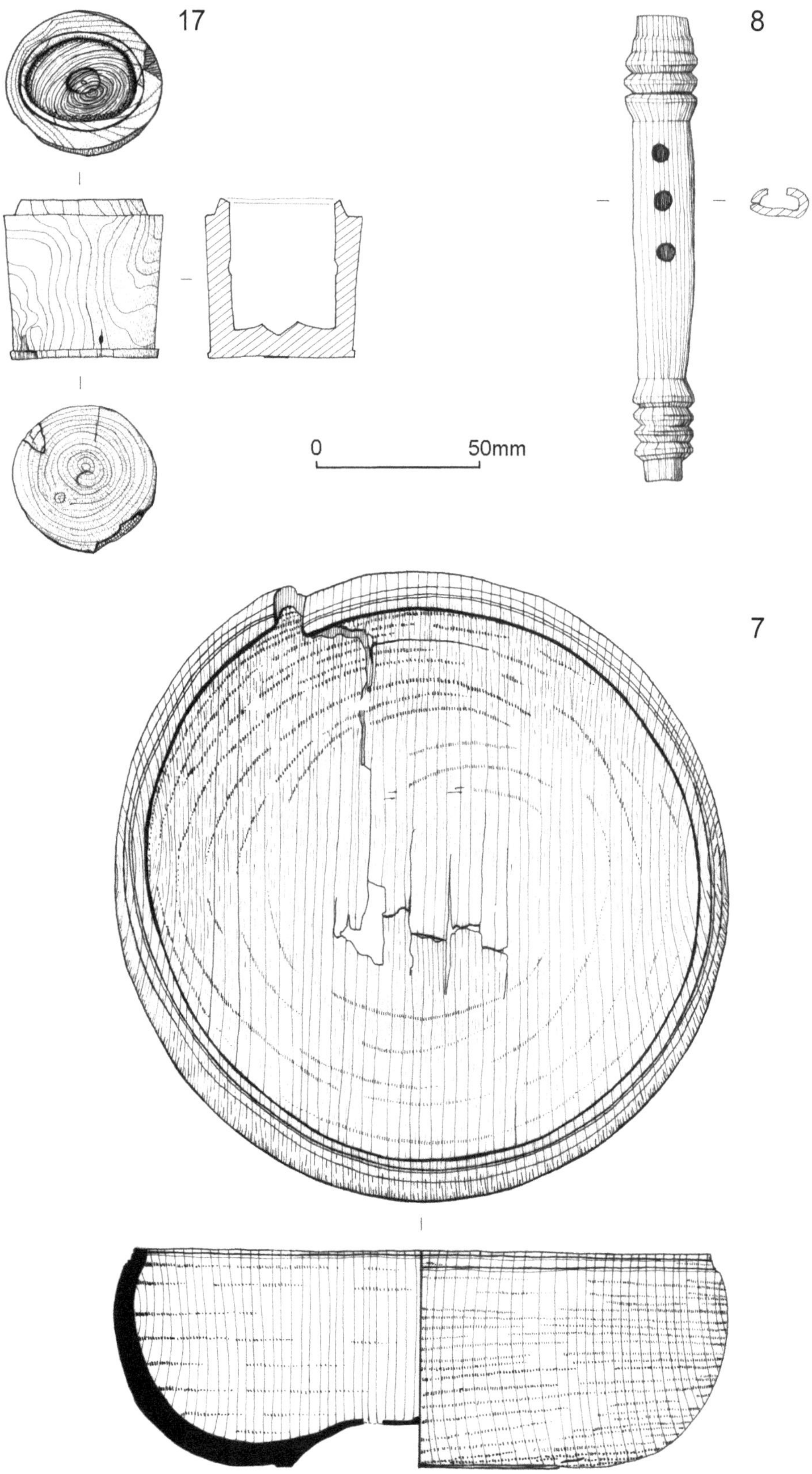

Figure 10.3: Wooden objects.

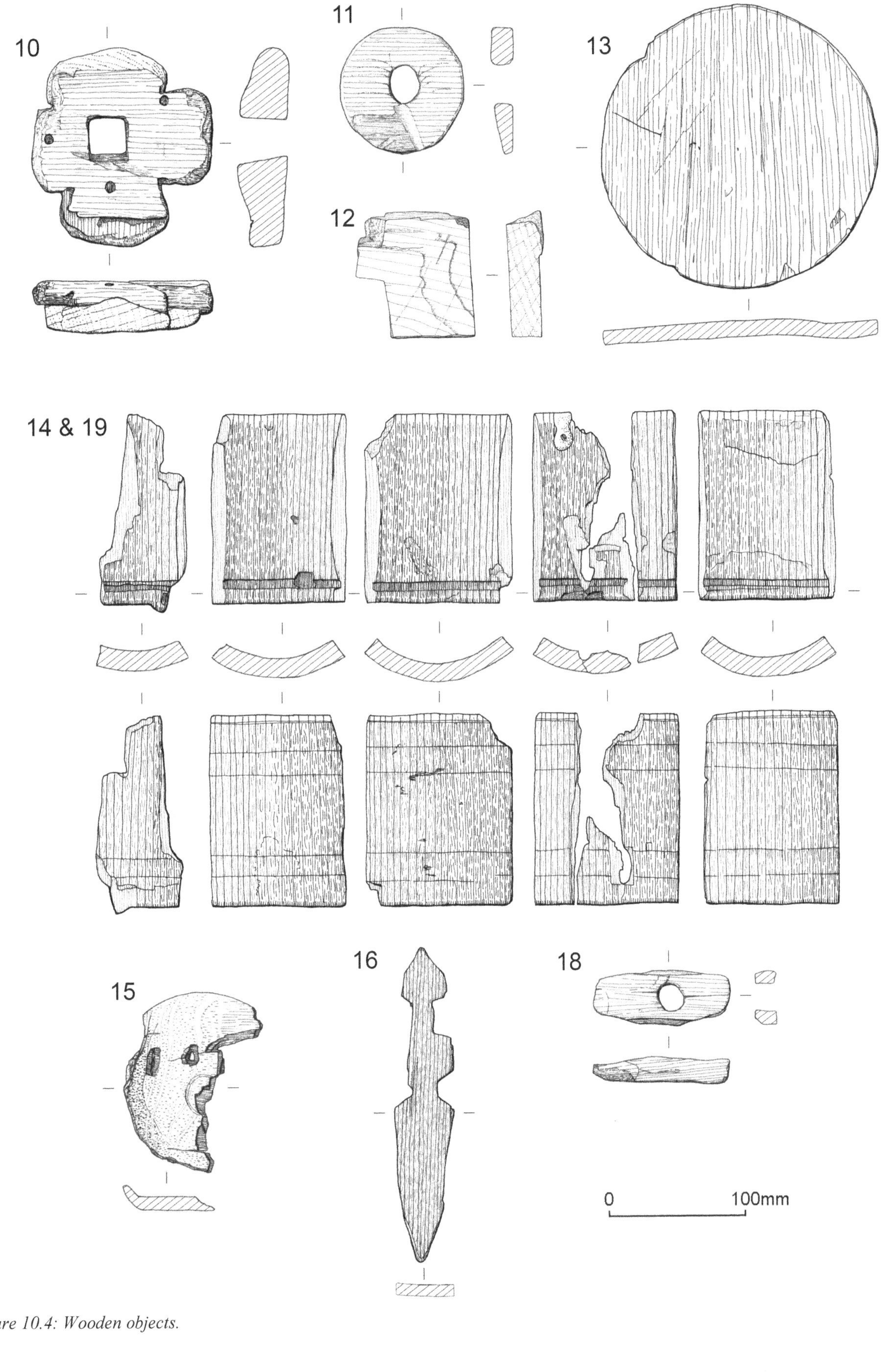

Figure 10.4: Wooden objects.

11. Quernstones

John Cruse

The excavation produced a quernstone collection which includes four intact or near intact stones (Q1-Q4), several diagnostic fragments (Q5-Q12) and a miscellaneous group of small and less diagnostic pieces (Q13-Q25). The intact or near intact stones were recovered from brine tanks [1207] and [1182] and form two likely matching pairs of upper and lower stones, one pair from each tank.

In terms of typology the collection represents a surprisingly uniform collection of disc querns. In the typical size range for normal hand querns (370-430mm diameter) are the two sets of matching pairs (Q1 and Q2 and Q3 and Q4) and one upper stone (Q5). Three stones are in the large hand quern category (440-490mm), one upper (Q6/Q7) and two lowers (Q9 and Q10/Q11). An apparently unused stone (Q12) may be either a large hand-mill or, with its significant thickness, a small millstone. The collection is rounded off by evidence of lava quern use (Q24 and Q25), but no saddle querns or beehive quern derivatives (normally common on early Romano-British sites) are present. In the absence of specialised geological analysis, it would seem that the main raw materials were locally available. The total weight of the artefacts from the site is 146kg, representing an original estimated total weight of 260-275kg, a significant resource.

Four upper stones have radial dressing on their upper surfaces (Q1, Q3, Q5 and Q6/Q7) and both sets of quern pairs also have vertical grooves on their skirt. Interestingly such decoration is absent from the larger diameter stones, which may be chronologically significant, indicating a somewhat later date. Linear decoration of the upper surfaces is common on lava stones and, as Parkhouse (2000, 454) has noted, 'vertical striae engraved upon the circumference of the upper [lava] stones are imitated [by local copyists] even though they serve no useful purpose'. Such decorative traits mimicked those on contemporary imported lava querns, which were favoured on early Roman military and high-status sites but are rarely found on later Roman querns.

Variable skirt height (ie Q5 and Q10/Q11) is indicative of either uneven wear or of poor manufacturing practices. In general, the querns are 60-85mm thick and seem to have a lot of 'life' remaining in them.

The absence of any beehive derivatives or saddle querns argues against this being a domestic assemblage and the scarcity of imported lava querns does not indicate the presence of military or high-status users. The comparatively restricted range of hand quern types, made from local stone, some with lava quern decorative antecedents, could perhaps be consistent with use by an organised grouping of workers, issued with standardised grinding equipment by their owners/employers.

Elizabeth Wright has endorsed Curwen's observation that 'through time, querns became gradually both larger and thinner' and that 'hand querns of 500mm diameter or more are generally absent from early Roman contexts' (Wright 2002, 271). On this basis the Kingsley Fields quern types, having absorbed several early Roman decorative features, sit comfortably in a mid Roman context since they are beginning to be larger but show few signs of later Roman slimness.

The end of life of Roman disc querns has to date been little studied. David Heslop's investigation into the usages of North Yorkshire beehive querns (2008, 68-72) has revealed that during the later Iron Age/early Roman period two types of quern breakage took place prior to final deposition. In a first phase several different varieties of 'detachment' are commonly found (including deliberate removal of the grinding surface edge) whose purpose seems to be to ensure that the stone could not be further used. This was often followed by 'division', where the unusable quern was 'divided vertically into fractions'. As only single fragments frequently survive on site, Heslop assumes that the missing elements have been moved off-site for a range of social reasons.

The collection of rather later, mid Roman, disc querns from Kingsley Fields includes several examples of deliberate 'detachment' of the grinding surface edges, some of which (Q6/Q7, Q8 and Q10/Q11) can be seen to precede an act of 'division'. This may indicate that some of these earlier practices continued or, alternatively, that these actions were merely a coincidental outcome of straightforward artefact disposal. With the exception of the two sets of quern pairs, the Kingsley Fields stones were typically broken into 20-30% elements before disposal. The survival of two probable matching pairs from the same or adjacent contexts is thus doubly significant, as such behaviour has been rarely observed on Roman sites and is clearly contrary to the prevailing site practice. Like much other material from the brine tanks they each appear to form part of a ritual deposition. If it is accepted that most of the fragmented querns were not worn out, then a functional (ie non-ritual) explanation for their destruction might be that it was the consequence of the abandonment of the site and an unwillingness to transfer the heavy querns to a new location.

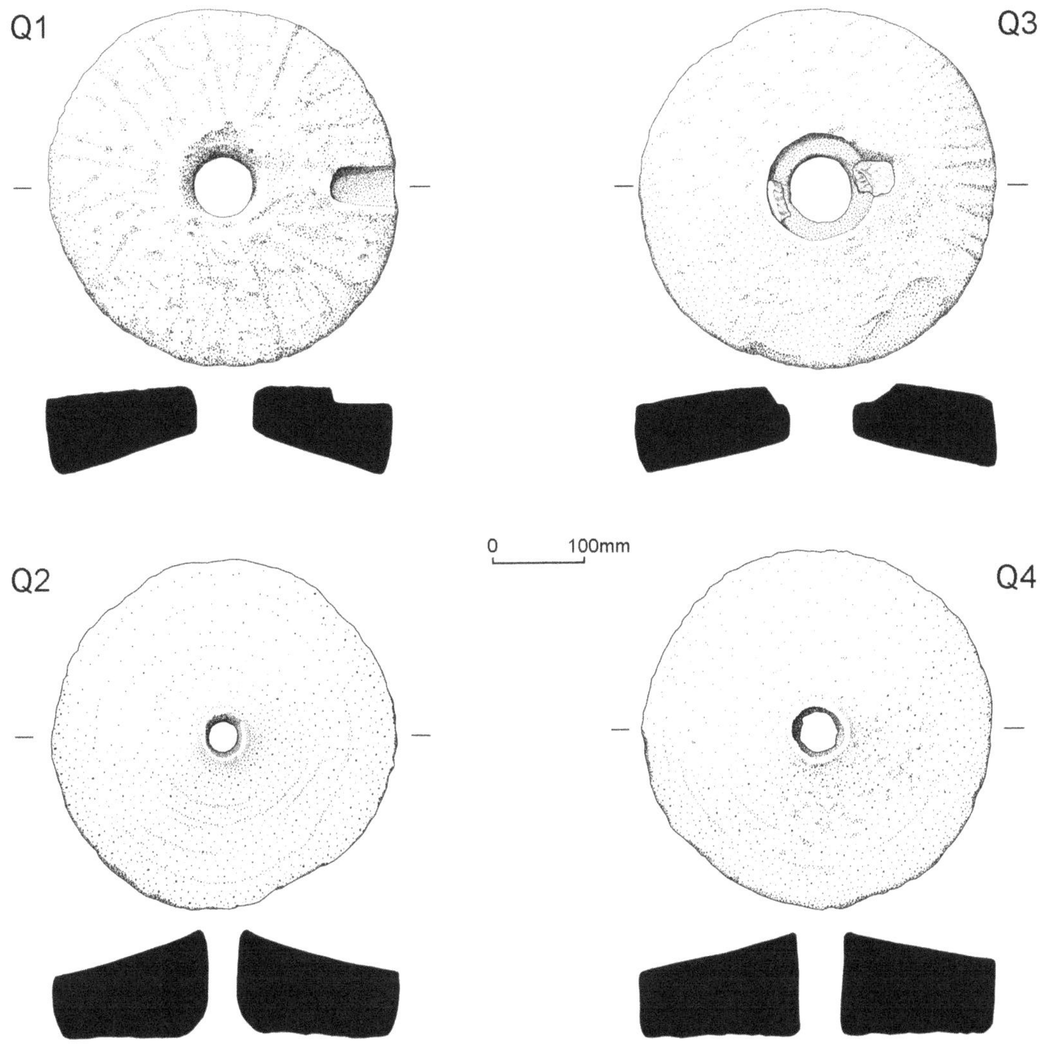

Figure 11.1: Quernstones, matching pairs.

Catalogue

Q1-12 are listed in order of ascending diameter.

Q1. 100% complete, drum-shaped upper stone (Fig 11.1). Fine/medium grained gritstone. Upper surface: peck-dressed, decorated with shallow grooves in radial lines. V-shaped horizontal slot handle hole set into rim. Skirt: vertical sides, dressed with vertical grooving. Grinding surface: worn flat, concave (15°). Diameter 360mm. Height: centre 35mm, rim 85-95mm. Concave hopper: diameter 110mm, depth 15mm. Feed-pipe diameter 60mm. Handle-hole: 73mm long; 18-15mm deep; width outer 45mm, inner 35mm. Weight 18.2kg. Fill (1239) of brine tank [1207]. Likely to be a pair with Q2.

Q2. 98% intact, drum shaped lower stone (Fig 11.1). Fine/medium grained gritstone. 10% of grinding surface edge removed by a single impact. Grinding surface: worn smooth, convex (13°), with a central, 10mm wide 'lip'. Skirt: vertical sides, dressed with vertical grooving. Base: roughly hammer-dressed, flat. Diameter 360mm. Height: centre 110mm, rim 65mm. Diameter of central perforation: top 40mm, base 70mm. Weight 16.9kg. Fill (1239) of brine tank [1207]. Likely to be a pair with Q1.

Q3. 97% intact, drum-shaped upper stone (Fig 11.1). Medium/coarse grained gritstone. 15% of grinding surface edge removed by a single impact. Upper surface: gently curving convex, random peck-dressed, with nine shallow radial grooves evident in one quadrant only. Skirt: vertical, dressed with shallow vertical grooves at 15mm intervals. Grinding surface: originally peck-dressed, worn flat, especially at edge, convex (12°). Diameter 390mm. Height: centre 60mm, rim 65mm. Vertical sided hopper: diameter 115mm, depth 25mm. Feed pipe diameter 70mm. No handle holes. Weight 16kg (estimated original weight 16.5kg). Unusually, this stone preserves evidence of its rynd attachment. At the base of the vertical sided hopper, at each side, a fragment of the iron rynd is visible (*c* 40mm wide) sealed by lead, which conceals the setting into the stone. On one side a 45mm wide, 50mm long, rynd chase has been carved but on the other no chase exists, so perhaps here the hopper is undercut, to retain the rynd. Fill (1277) of brine tank [1182]. Likely to be a pair with Q4.

Q4. 100% intact, drum-shaped lower stone (Fig 11.1). Medium/coarse grained gritstone. Grinding surface: flat, worn, convex (12°); no central 'lip'. Skirt: dressed into vertical grooves. Base: roughly hammer-dressed, flat. Central perforation retains marks of nine vertical grooves; these are unlikely to be decorative and thus may be evidence that it was initially chiselled out or was required to increase the 'grip' of an inserted wooden bush, as found at Nendrum (McErlean & Crothers 2007, 189). Diameter 390mm. Height: centre 105mm, rim 65-70mm. Central perforation diameter 55mm. Weight 20.1kg. Fill (1280) of brine tank [1182]. Likely to be a pair with Q3.

Q5. 50% fragment of an upper disc quern, broken across diameter, through the slot handle. Very coarse gritstone. Upper surface: slightly convex, flat, dressed with nineteen shallow radial grooves. Horizontal slot handle cut into rim of outer surface. Skirt: vertical sided. Variable height indicates uneven wear. Conical hopper is cut by opposing rynd chases, 30mm wide, 10mm deep. Grinding surface: outer 45mm band is flat and well worn, with the inner area concave (10°). Diameter 390mm. Height: centre 35mm, rim 40-55mm. Hopper: width 120mm, depth 15mm. Feed pipe diameter 50mm. Slot handle-hole: 65mm long, 25mm deep, >8mm wide. Weight 5.8kg (estimated original weight 12kg). Fill (3017) of linear [3016], Building 2.

Q6/Q7. *c* 40% survival of upper disc quern in two non-joining fragments. Medium grained gritstone. 100% of the grinding surface edge has been removed from Q7, presumably before the quern was broken radially into four+ pieces. Upper surface: peck dressed flat, with shallow, radial dressing. Skirt: vertical sided but curving on top. Grinding surface: originally peck-dressed but well worn, very slightly concave. Q7 has a 45mm long, 25mm wide, 25mm deep, rectangular rynd chase cut into the grinding surface, but none is visible on Q6. Diameter 430mm. Height: centre 85-90mm, rim 90-95mm. Hopper: width 120mm, depth 30mm. Feed pipe diameter 60mm. Total weight 13.7kg (estimated original weight 35kg). Subsoil (3002).

Q8. 10-15% fragment of lower disc quern/millstone. Medium grained gritstone. Broken radially (after?) complete removal of grinding surface edges. Grinding surface: flat surface, convex (12°). Base: hammer-dressed; outer 100mm is flat, then the inner area is concave to the central perforation and has apparently been perforated from above. Diameter >430mm. Height: centre 80mm, towards rim 90mm. Central perforation diameter: top 30mm, mid 20mm, base *c* 60mm. Weight 4.8kg, (estimated original weight 32-50kg?). Subsoil (3002).

Q9. 30% fragment of the lower stone of a disc quern, broken radially (Fig 11.2). Coarse gritstone. Grinding surface: slightly convex, grooved with one complete and two partial 'harps' (restorable as a six 'harp' pattern). Unusually two 'harps' are directly opposed, with the third indicating anti-clockwise rotation. Base: flat. Diameter 450mm. Height: centre 95mm, rim 70mm. Central perforation: diameter top 40mm, base 70mm. Weight 9.5kg (estimated original weight 32kg). Fill (3085) of pit [3084].

Q10/Q11. 40% survival, in two joining fragments, of a lower disc quern, which had 50% of the grinding surface removed, before being broken radially (Fig 11.2). Gritstone. Grinding surface: slightly convex, grooved with two complete and two partial 'harps' (restorable as a six 'harp' pattern); all indicate anti-clockwise rotation. Skirt of variable height, curving down towards a flat base. Diameter 480mm. Height: centre 100mm, rim 45-80mm. Central perforation diameter: top 100mm, base 120mm. Weight 10.5kg (estimated original weight 26kg). Fill (3085) of pit [3084].

Q12. *c* 5% rim fragment of a large disc quern or small millstone, possibly result of deliberate grinding surface removal. Medium/coarse grained gritstone. Both flat

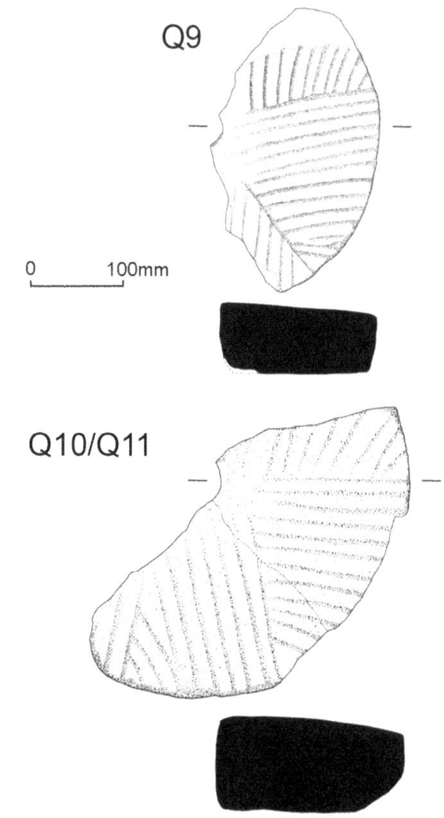

Figure 11.2: Quernstones with 'harps'.

surfaces and the gently curved skirt are dressed smooth and a clear grinding surface cannot be identified. As the base of lower stones are usually less carefully finished, this may be derived from an upper stone. Diameter *c* 500mm. Thickness 110mm. Weight 2.2kg (estimated original weight *c* 45kg). Fill (3213) of posthole/pit [3212], Building 5.

Miscellaneous Small Fragments

Q13. Fragment. Coarse gritstone. Weight 0.1kg. Fill (1065) of pit [1064].

Q14. Fragment. Fine/medium grained gritstone. Weight 0.3kg. Fill (1280) of brine tank [1182].

Q15. Fragment. Coarse gritstone. Weight 0.07kg. Fill (1209) of brine tank [1182].

Q16. Fragment. Coarse gritstone. Weight 0.15kg. Fill (1239) of brine tank [1207].

Q17/Q18. Two fragments of lower stone. Fine/medium grained gritstone. Total weight 0.6kg. Fill (1208) of brine tank [1207].

Q19. Fragment of upper stone. Fine/medium grained gritstone. Upper surface has radial grooves and the skirt vertical grooves. Weight 0.5 kg. Fill (1346) of brine tank chamber [1347].

Q20. Fragment. Coarse gritstone. Weight 0.7kg. Fill (3085) of pit [3084].

Q21. Two fragments. Coarse gritstone. Weight 0.2 kg. Fill (4232) of pit [4237].

Q22. Fragment. Coarse gritstone. Weight 0.05kg. Fill (5021) of posthole [5020], Building 3.

Q23. Fragment. Fine/medium grained gritstone. Weight 0.4kg. Area G, subsoil (7002).

Q24. 27 fragments of lava stone. Total weight 0.8kg. Fill (3402) of pit [3401].

Q25. 43 fragments of lava stone. Total weight 0.6kg. Fill (2006) of posthole [2205], Building 1.

12. Architectural and Altar Fragments

Peter Arrowsmith

The excavation produced two large fragments of dressed red sandstone (Fig 12.1). One was an unfinished 'architectural' piece from fill (4232) of pit [4237]. The other was an unstratified altar base, which was found in the spoil following machine stripping of excavation areas A and C.

Altar Fragment

The altar fragment has been carved and dressed from a fine-grained red sandstone. The piece has been damaged by ploughing and comprises the lower *c* 420mm of a small altar. It measures *c* 250mm by *c* 245mm at the base and tapers slightly towards decorated bands, above which it is 215mm by 200mm in cross-section. The decoration comprises a double moulded band and, below this, a series of chip-carved bands, consisting of zig-zags, horizontal indents and inverted triangles (Plate 12.1). There is no evidence of an inscription, which could have been lost from one of the damaged and broken faces.

The diagnostic feature of this piece is the use of chip-carving. This technique, defined by Blagg as 'the execution of leaves, rosettes and simple geometrical patterns with angular facets [which] has frequently been likened to the chip-carving of wood', is found on two types of stonework in Roman Britain. One use is on the edge of limestone slabs, probably table tops, in south-west Britain in the 3rd and 4th centuries. The other is essentially as a decorative feature of altars as here, with a distribution which relates primarily to Hadrian's Wall and the northern frontier zone. It is probable that the technique originated in Germany and arrived with army units transferred to Britain in the reign of Antoninus Pius. The technique is found on an altar from Chester dated AD 154 and may have been introduced not long before this. Its known date range continues until the AD 230s or later (Blagg 2002, 74-5).

The zig-zag motif used on the Nantwich altar is found on a number of other examples. These include two altars from Chester, one of the late 2nd or 3rd century, the other that of AD 154 (Henig 2004, 2 no 5 (*RIB* 448), 8-9 no 20 (*RIB* 452), pls 2 & 10). Other instances are found on altars from: Maryport (*RIB* 814, 816 & 827); Vindolanda (*RIB* 1685, 1686 & 1692; Tomlin & Hassall 2003, 366-7); Corbridge (*RIB* 1122); Carvoran, dated *c* AD 163-6 (*RIB* 1792); Lanchester, dated AD 238-44 (*RIB* 1074); and Birrens (*RIB* 2092). The most common application of the zig-zag motif is on the altar capital but in some instances it is also used for a band on the base, as at Corbridge and Vindolanda (*RIB* 1685 & 1686). In the case of the Nantwich altar the motif may likewise have been repeated on the capital, now lost.

Prior to the discovery of the Nantwich stone, the only altars with chip-carved decoration known from outside the frontier zone are believed to have been the two examples from Chester (Blagg 2002, 74). The existence of this altar in Nantwich corresponds with other evidence from the site to suggest a military presence.

Apart from the major assemblage from Chester, only a small number of altars have been found in Cheshire. Two of these derive from close proximity to Chester. One is from the satellite settlement at Heronbridge (Mason 2003, 56-7 (*RIB* 571)). The other, dedicated to the 'Nymphs and the Fountains', is said to have been found at a site *c* 2km to the east of the city (Henig 2004, 6). Two others have been found at Helsby and Wilderspool, both of which are small and uninscribed (Henig 2004, 36 nos 166 & 177).

Plate 12.1: The altar, detail of the carved decoration. Scale 1:2.

Architectural Fragment

The architectural fragment from (4232) in [4237] is also carved from a fine-grained red sandstone. Information on the stone-working techniques employed on this piece was kindly provided by Susan Sportun, The Manchester Museum.

The piece measures *c* 470mm high and *c* 300mm wide, with an irregular hexagonal cross-section. The faces show varying degrees of completion, two being worked, two partly worked and the others unworked. The wider of the two worked faces has chisel marks. The other has point marks and at one end is decorated with moulding which is unpolished and unfinished. Of the two ends of the piece, one is either unworked or broken; the other is more irregular and has been partly worked above the moulding. It is possible that the piece was intended for completion but was abandoned, one reason perhaps being that the bedding of the unworked part of the stone seems vulnerable to fracture. Another possibility is that this was simply a practice piece.

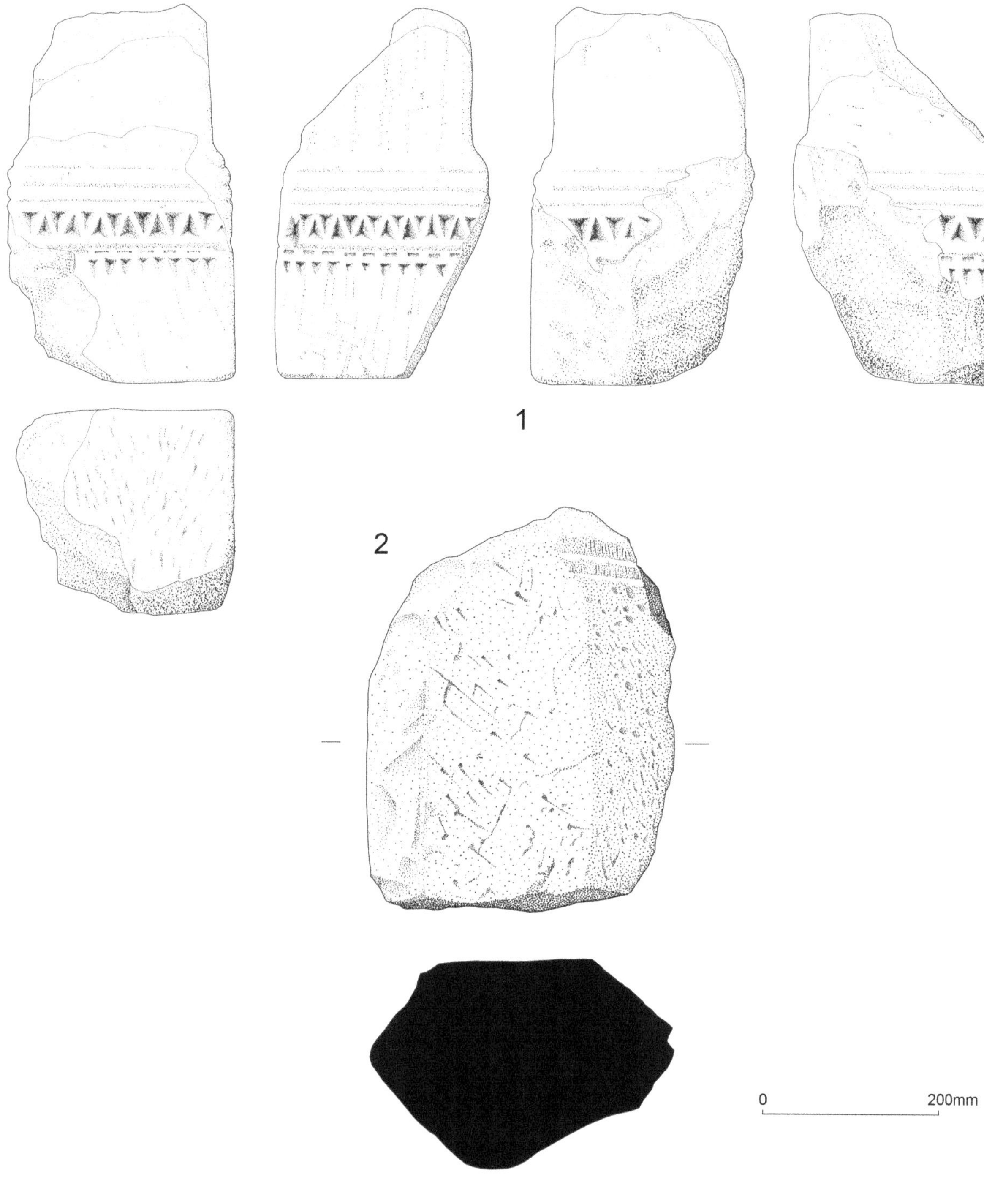

Figure 12.1: Altar fragment (1) and architectural fragment (2).

13. Fired Clay and Briquetage

Cynthia Poole

Fired clay and briquetage were recovered from all areas of the excavation amounting to a total of 1091 fragments weighing 28.797kg (Table 13.1 & Appendix 13.1) generating 237 records. Apart from a small quantity either from medieval/post-medieval deposits or unphased, the majority was recovered from a range of Roman phase contexts. Although fired clay occurred in all excavation areas the major concentrations were found in areas A and C (Table 13.2).

The assemblage is well preserved with a mean fragment weight of 26g and little or no abrasion on a high proportion of fragments. It has been subdivided into broad class categories of structural (walls, lining and industrial), oven furniture (plates and supports), containers and indeterminate. All the material appears to be fired and a high proportion has a cream or buff surface that is commonly found on briquetage.

Class	Type	Count	Wt (g)
Structural	Wall with wattles STR1	127	2540
	Wall: Flat and curving STR2, STR3	36	2675
	Wall: Corner / edges STR6	13	3258
	Oven / hearth floor STR4	17	484
	Misc STR9	13	406
	Lining: LN1, LN2	13	218
	Furnace wall / lining	8	145
Furniture	Plates / floors	134	6767
	Perforated plates	52	1108
Supports	Pedestals	27	1420
	Small plates PL1, PL2, PL3	95	2460
	Wedges W0, W1, W2, W5	74	1104
	Stablilisers	4	58
Containers	Briquetage	1	5
	Mould / crucible	1	4
Indeterminate	Moulded surface FC01-05	244	3858
	Amorphous FC98-99	227	1661
Total		**1091**	**28797**

Table 13.1: Quantification of fired clay by class and form types.

The Fabrics

Fabrics were characterised on macroscopic characteristics as well as with a microscope at magnification from x15 to x25. Four main fabrics (A, B, D, V) were identified and two minor fabrics (C, L) accounting for only a few pieces. The fabrics are quantified in Table 13.3 and tabulated according to the main functional classes. All the clay has been well fired in oxidising conditions with the majority of the material in the red-orange colour spectrum, with cream, buff or pale grey surfaces. There appears to be no underfired or non-fired material retained during excavation, although a light brownish red plastic clay is recorded as lining brine tanks [1182] and [1207].

Fabric A

Colour: red, orange, pinkish/reddish brown, light brown; some surfaces cream, buff, light brown.

At a macroscopic level this appears to be a fine-grained clay with no or very few obvious coarse inclusions. At magnification rare to low densities of poorly sorted sand, mostly fine to medium size and rare coarse grains are visible. The majority of the sand is quartz plus rare dark or black grains, possibly an iron rich mineral. The sand is predominantly rounded/sub-rounded but some angular grains are present. Clay pellets or mudstone grits are occasionally present.

Fabric B

Colour: red, orange, yellowish red, reddish yellow, weak red, pink, brownish red; brown light brown, yellowish brown, reddish brown. Surfaces are commonly cream, buff or light brown.

It contains a moderate to high density of medium-coarse sand, though in some samples the coarse sand component is low or infrequent. The sand content varies from well to poorly sorted and the sand grains are generally rounded/sub-rounded although some of the coarse sand (1-2mm) and small grits up to 4mm are angular/sub-angular. The sand is predominantly quartz, with occasional calcite and dark or black grains, possibly an iron-rich mineral. Grits of 1-7mm size are rare and usually less than 4mm. In general where inclusions larger than sand-size are present in any quantity the fabric has been designated D.

Fabric C

Colour: dark grey (surface), weak red (margin), bright red (core).

The clay contains a high density of coarse sand. This is predominantly angular quartz, rare mica and a low density of coarser grits 2-4mm of sandstone or quartzite.

Fabric D

Colour: grey/light yellowish brown (heavily fired); brown, pale pinkish brown, orange, bright red; yellowish red, reddish yellow, brownish red, reddish brown, grey (average firing); buff, yellowish buff, light brown, grey (surfaces).

The clay matrix is essentially the same as fabric B. The sand is frequently poorly sorted. The component of coarse sand and angular grits 2-5mm occurs in greater density generally compared to fabric B. Coarse inclusions comprise a low density *c* 5-10% of rounded stones and waterworn pebbles *c* 3-20mm with stone voids suggesting that some were up to 32mm in size;

most pebbles appear to be a fine-grained rock, probably a fine sandstone. Also present are fragments of mudstone (sub-angular/sub-rounded) 3-12mm, small angular/sub-angular grits of sandstone <10mm and rare angular calcite grit *c* 2mm.

Fabric L

Buff-grey surface; grey body. Small limestone inclusions or voids 1-3mm.

Fabric V

The basic clay matrix of this fabric is essentially the same as fabric B with a similar range of colours. This is differentiated by the addition of vegetal temper in the form of chaff or chopped straw or hay resulting in a porous fabric.

Discussion of the Fabrics

Apart from fabric L, which must originate from outside the immediate area, all the fabrics are likely to be of local origin. Fabrics B, C and D are characterised by the presence of frequent sand, whilst A is more clayey with a finer and less dense sand component, suggesting that two different clay sources may have been targeted. However some examples, intermediate between A and B and A and D, suggest that the two either occurred in close proximity or were being exploited at the same time resulting in some mixing. Fabric V utilises the same clays with the addition of a vegetal component.

The local geology supports the hypothesis of two possible clay sources. The Kingsley Fields site lies on river terrace sands overlying boulder clay, while to the east there are alluvial deposits along the River Weaver. The mixed character of boulder clay is compatible with it being the source of the sandier clay fabrics (B, C, D), whilst A is probably derived from the alluvium. There is evidence that the clay lining of the brine tanks [1182] and [1207] was quarried from the area of the tanks and this suggests that clay was utilised where features cut through suitable deposits, although there may also have been areas of dedicated quarrying for extracting clay either from a single quarry area or from several exposures.

Class		Excavation areas								Total	Total %
		A	B	C	D	E	F	G	Other		
Structural: Wall	Count	102	26	42	39	3	-	-	-	212	19.4%
	Wt (g)	5245	258	3688	650	168	-	-	-	9989	34.7%
Structural: Lining	Count	12	1	-	-	-	-	-	-	13	1.2%
	Wt (g)	208	10	-	-	-	-	-	-	218	0.76%
Structural: Furnace lining	Count	2	-	6	-	-	-	-	-	8	0.7%
	Wt (g)	10	-	135	-	-	-	-	-	145	0.5%
Furniture: Plates	Count	62	-	105	-	1	-	17	-	185	17.0%
	Wt (g)	3606	-	4132	-	29	-	108	-	7875	27.3%
Furniture: Supports	Count	88	10	82	1	9	1	7	2	200	18.3%
	Wt (g)	2945	217	1411	40	283	17	101	28	5042	17.5%
Containers	Count	1	-	-	1	-	-	-	-	2	0.2%
	Wt (g)	4	-	-	5	-	-	-	-	9	0.03%
Indeterminate	Count	89	61	228	35	38	6	-	14	471	43.0%
	Wt (g)	683	539	3455	268	327	46	-	211	5519	19.2%
Total	Count	356	98	463	76	51	7	24	16	1091	
	Wt (g)	12701	1024	12781	963	807	73	209	239	28797	
Total %	Count	32.6%	9.0%	42.4%	7.0%	4.7%	0.7%	2.1%	1.5%		
	Wt (g)	44.0%	3.6%	44.4%	3.3%	2.8%	0.3%	0.7%	0.8%		

Table 13.2: Quantification of major classes of fired clay in relation to excavation areas.

Fabric		Structural: Wall	Structural: Lining	Structural: Industrial	Furniture: Plates	Furniture: Supports	Containers	Indeterminate	Total	Total %
A	Count	2	1	-	-	2	-	43	48	4.4%
	Wt (g)	88	10	-	-	25	-	481	634	2.2%
A/B	Count	15	-	-	1	7	-	3	26	2.4%
	Wt (g)	420	-	-	125	114	-	28	687	2.4%
B	Count	142	10	7	75	142	1	266	646	59.2%
	Wt (g)	3105	126	130	1511	3548	4	1924	10368	36.0%
C	Count	-	-	1	-	-	-	2	3	0.3%
	Wt (g)	-	-	15	-	-	-	18	33	0.11%
D	Count	49	2	-	76	32	1	149	309	28.3%
	Wt (g)	5829	52	-	4158	1097	5	3003	14174	49.22%
V	Count	4	-	-	33	16	-	4	57	5.2%
	Wt (g)	547	-	-	2081	147	-	14	2789	9.69%
L	Count	-	-	-	-	1	-	-	1	0.09%
	Wt (g)	-	-	-	-	81	-	-	81	0.28%
-	Count	-	-	-	-	-	-	1	1	0.09%
	Wt (g)	-	-	-	-	-	-	31	31	0.11%
Total	Count	212	13	8	176	200	2	480	1091	
	Wt (g)	9989	218	143	7296	5042	9	6098	28797	

Table 13.3: Quantification of fired clay by fabrics and class.

Function and Forms

The majority of the fired clay has been interpreted as oven or hearth structure and furniture, although diagnostic characteristics and forms are not well preserved. This may reflect the general character of activities with items made to serve an immediate need rather than prepared to a predetermined design. The assemblage has been categorised according to broad classes with further divisions into subtypes and form categories.

Structural

Oven Wall and Superstructure

This has been subdivided into two broad categories: pieces with evidence of wattle/timber impressions (STR1), and pieces with two parallel surfaces, either flat (STR2) and with evidence of squared edges or corners (STR6) or curving (STR3).

The first category comprises fragments with flat exterior surfaces varying from smooth, even or fairly regular to rough, irregular or undulating. On the opposed surface are impressions of wattles (STR1.1), and in only a small number of examples is the evidence sufficient to suggest that they formed part of an interwoven framework (STR1.2) or had been laid flat side by side or criss-crossed (STR1.3). The only example of shaped timbers suggests the presence of two planks or squared timbers set at right angles (STR1.4). The majority of fragments with wattle impressions (STR1.1) range in thickness from 10mm to 35mm, whilst the pieces with interwoven wattles are 15-42mm thick, those with parallel wattles 30-60mm and the piece with plank impressions 36mm thick. A total of 37 impressions could be measured and the overall range of sizes is 6-33mm in diameter. This range is very similar to the sizes found in a study of the roundwood withies from the wicker-lined pits (see Chapter 18).

The second category of structural material with two parallel surfaces (STR2, STR3) usually has fairly roughly moulded surfaces, frequently irregular or undulating, though sometimes having one smoother surface. Thickness varies from 40mm to 90mm. One example measuring 30-55mm thick has a diameter of *c* 200mm at the base decreasing to *c* 160mm higher up the wall. The wall base splays out and thins at the edges where it had been plastered over some other surface (Fig 13.1 no 1). All the examples came from brine tank [1182].

Large flat blocks with evidence of edges or corners (STR6) are 40-90mm thick. It is unclear whether these formed part of the wall or base of free-standing structures or some other element of the superstructure. The largest blocks came from pit [3191] in area C (Fig 13.1 no 4).

Floor/Base

Pieces assigned to this category (STR4) have a single smooth flat or slightly concave surface, in one case burnt or sooted black and in another fired to a cream colour. The opposed surface is irregular, sometimes with deep depressions possibly from a stone base or convex from being plastered into a concave hollow. This type could have formed the basal floor of an oven or a simple hearth surface.

Lining

A small quantity of material was identified as oven lining, although some of the material categorised as wedges or small plates could include lining. This is characterised by one moulded surface flat or concave with finger marks, either fingertip depressions or linear grooves resulting from shaping and smoothing the surface (LN1). The opposite surface preserves evidence of being plastered against some other material. On one piece this may be the grain of split or cut timber, whilst on most the back is rough and irregular, possibly from being plastered over the ground surface or a primary clay structure. These pieces range from 15mm to 32mm thick. A thinner variety (LN2) thins to an acute angular edge and probably represents limited areas of repairs or relining where the clay was smoothed out and thinned to the edges as it merged with the original wall. It is possible that material categorised as wedge type W2 is in fact this type of lining. Most of the lining came from brine tank [1207].

Industrial

A few small pieces had been very intensely fired resulting in a dark grey vitrified or vesicular surface layer grading through a purplish margin into a red-orange core, typical of furnace lining. This was found in pits [1055] and [1160] in area A and in Building 5 posthole [3212] and large enclosure ditch [3206] in area C. Such material is usually associated with industrial processes such as metal working. A small fragment of a thin walled vessel (7mm thick) in a greyish brown sandy fabric was also found in pit [1160] and may be part of a crucible or mould.

Oven Furniture: Plates and Floors

The plates in this category are between *c* 20mm and 60mm thick and have flat or slightly curving surfaces, usually fairly smooth or even on one side but generally rougher or more irregular on what is probably the underside. Some have shallow recessed depressions 2-7mm deep and one measures 35mm long. Two pieces from pit [3191], probably part of the same plate, have a straight bevelled edge and one has some incised or impressed lines marking the surface, possibly forming a chevron pattern. Some pieces from brine tank [1207] have wattle impressions on the underside which appear to be laid parallel, not interwoven, possibly to support a floor or plate. The wattle sizes fall within the same range as observed on the wall structure.

The overall size of the plates is uncertain though the thickness of some suggests they could be fairly large.

Two, however, appear to be relatively small oval plates; one from brine tank [1182] measures 25mm thick by 110mm wide by >90mm long. Another from brine tank [1207], measuring 50-70mm thick by 125mm wide, is slightly domed and roughly moulded with finger marks across the surface which has polygonal hollows, possibly from the attachment of pedestals or other structural elements.

A small number of fragments ranging from *c* 20mm to 75mm thick have evidence of a perforation 30-60mm wide, circular or oval in form. One from large enclosure ditch [3195] in area C is biconical, 25mm diameter in the centre widening to *c* 40mm at the surface (Fig 13.1 no 3). In some cases the clay forms a thickened ridge around the base of the perforation. The plates are interpreted as forming a suspended floor above the main firing chamber of the oven. Some with edges appear to be portable items but it is possible that some of the thicker examples were built as an integral part of the oven structure.

Oven Furniture: Supports

A range of forms are included in this class including pedestals, small plates or discs, wedges and stabilisers. Firebars or rods appear to be absent; some poorly preserved fragments hint at their presence but there is nothing that can be allocated to such a category with any certainty.

Pedestals

Six examples of pedestals of varied form were recorded. None are complete or well preserved and the shape is poorly defined in most cases. One of uncertain form (Ped0) (1182), but possibly rectangular 60-70mm thick, may have been pierced by a perforation *c* 14mm in diameter. The circular end of a pedestal 50mm in diameter may have been part of a pedestal of cylindrical (Ped4) or tapering conical form (Ped5) (3102). Two roughly moulded objects with plano-convex surfaces have been tentatively identified as 'hand-squeezed' pedestals (Ped6), a type associated with briquetage production in the Fenland (Morris 2001, fig 114 PD1). The fragments from [1207] measured 45mm x 40mm wide by >30mm high and 35mm x 50mm by >30mm high. Another from cut feature [5088] measured 45-70mm wide by 10-43mm thick and over 75mm high. Two further pieces distinguished by a lip at one edge (Ped7) were identified as pedestals on the basis of comparison to examples with a similar feature from the Fenland briquetage sites (Morris 2001, PD3). One of them was unstratified but the better preserved example was found in deposit (1285) in brine tank [1182] (Fig 13.1 no 2). This piece is a large block measuring 85mm x 90mm with two flat surfaces at right angles joined by a curving edge with a lip or ridge curving up at the opposite edge to a height of 30mm. The back/underside is undulating, possibly having been moulded over some other structural element. The clay fabric of this piece appears to be built up in a series of layers, suggesting that it had been resurfaced on a number of occasions. Morris (2001) has interpreted this type of lipped pedestal as being used for supporting large evaporating pans during salt production.

Small Plates and Discs

Most pieces assigned to this category are incomplete but the general character of the pieces suggests that they formed small rounded, circular or oval plates or discs, though a few were possibly subrectangular. Most have roughly parallel surfaces (PL1, PL2), though some narrow towards the edges and a few have a distinctly lenticular cross-section (PL3). The first flat grouping was subdivided on the basis of thickness: PL1 <0.15mm and PL2 15-27mm. No other dimensions survive complete for type PL1 and just a small number for types PL2 and 3; the dimensions are tabulated in Table 13.4. The surfaces are roughly moulded and undulating, some with finger marks and depressions, and with one side usually being smoother than the other. Edges tend to be roughly rounded or fairly angular where the surfaces converge to the edge. A few have chaff, straw or grass impressions covering the surfaces.

Type	Thick-ness (mm)	Width/ diam (mm)	Length (mm)	Feature	Comments
PL1	7	>40	>49	5017	
PL1	8-16	>45	>65	1182	
PL1	9-15	65	75	1207	
PL15	7-10	55		1207	Shallow depression on surface
PL2	16	45	>50	2158	Fig 13.2 no 5. Chaff and straw impressions
PL2	20-25	53	77	1207	Straw impressions
PL2	23-25	*c* 100		3125	Fig 13.2 no 6. Cut grooves, probably a small plate or wedge
PL2	26	48	>62	5001	
PL3	25	75	>95	1182	
PL3	30	85	>105	1182	Fig 13.2 no 7. Finger marks and rare organic impressions
PL3	15	55	80	1207	Fig 13.2 no 8. Shell-shaped; covered in dense grass/ straw impressions
PL3	10-24	30		2039	Fig 13.2 no 9. Sub-oval, convex upper surface with finger tip depressions

Table 13.4: Dimensions of small plates.

The exact function of these objects is uncertain, but similar objects have been observed in assemblages of Iron Age fired clay (Cunliffe & Poole 1991, 149) and they are similar to kiln setters or spacers found in association with Late Iron Age and Roman pottery kilns (Swan 1984, 40-1). They probably functioned as props or stabilisers. It is possible that some of the pieces assigned to this category and also to the wedges are in fact lining material.

Wedges

These objects have smooth flat or concave surfaces diverging from a narrow edge, either rounded (W2) or angular (W1). Most range from 5mm to 35mm in thickness, or rarely up to 42mm maximum thickness. A few have a slight lip suggesting they had been attached to another clay structure. A couple have straw or organic impressions on the surface. Size measurements for the more complete are shown in Table 13.5. In many respects these are similar to the small plates and probably had a similar function.

Type	Thick-ness (mm)	Width/ diam (mm)	Length (mm)	Feature	Comments
W1	30	31	>65	3020	Fig 13.2 no 10
W1	7-20	57	>65	3273	
W2	3-35	*c* 50	>55	1182	Straw impressions
W5	15-20	>50	>65	3092	

Table 13.5: Dimensions of wedges.

Stabilisers

Three objects have been classified as stabilisers though this identification is not certain. One small piece from brine tank [1207] has a flat top, rough flat base and concave edge and measures 20mm thick. This is similar to the spacer clips used to stabilise adjacent brine evaporating troughs in the Fenland areas (Morris 2001, 370, fig 113).

Two objects of quite different form have been included here, though their function is uncertain. Both are very similar being shaped like a hook. One from posthole/pit [2039] in area B measures 12mm thick, 22mm long, *c* 26mm high with the semi-circular hook end having a diameter of 21mm (Fig 13.2 no 12). The example from pit [3191] in area C is slightly larger, 11mm thick, 40mm long and 35mm high, with a hook end forming a curve of 20mm diameter (Fig 13.2 no 11). They bear some similarity to the end of a stilt or tongue of a 'thimble' used to separate plates during firing in pottery kilns.

Containers

Two small fragments have been identified as containers. One may be a fragment of briquetage container from area D, whilst the other could be a crucible or mould from area A. The virtual absence of briquetage vessel fragments, both of the troughs for evaporating brine and the containers for transporting cakes of salt, is significant and indicates that other materials were used for these functions.

Indeterminate

Material in this category forms a significant proportion (40% by count, 21% by weight) of the overall assemblage. It has been broadly subdivided into material largely amorphous with no (FC99) or very little (FC98) evidence of shaping and those pieces with some evidence of shape (FC01-05). The latter group was subdivided on the basis of a single flat surface (FC01), a curving convex (FC02) or concave (FC03) surface, two adjacent surfaces at right angles (FC04) and pieces with a flat surface with a shallow hollow or depression (FC05). The thickness measurement was incomplete, ranging from <10mm to over 35mm. Much of the material is likely to derive from structural categories and to some extent from the larger forms of furniture.

Discussion

The fired clay assemblage consists almost entirely of structural material and furniture from ovens or hearths. The clear evidence on site for salt processing suggests that much of the fired clay is likely to derive from structures for the evaporation and drying of salt, but the possibility that other activities may be represented by the material should at least be considered. The main concentrations of fired clay focused on the brine tanks in area A and contexts in area C in roughly equal quantity, accounting for almost 90% (wt) of all the fired clay. A third much smaller group occurred in area B.

The largest groups were found in the fills of the two brine tanks and in area A. Brine tank [1182] contained 91 fragments (6903g) comprising wall structure, including fragments with wattle impressions, oven plates, pedestals, small plates and wedges. Tank [1207] produced a similar quantity (127 fragments, 4231g) with a slightly varied range of material composed of wall structure predominantly with wattle impressions, wall lining and relining, oven plate and perforated oven plate, a pedestal, small plates, wedge and a spacer clip.

In area C the single largest deposit (140 fragments, 8182g), comprising wall structure, oven plates and perforated plate, one hook and miscellaneous pieces, was found in pit [3191]. The largest single group from area B occurred in posthole/pit [2039] (40 fragments, 405g) and was composed of wall structure, small plates, a hook and undiagnostic pieces.

No significant variation can be observed in the combinations of structural elements and furniture within the main groups which could suggest different designs of structures. This consistency in the assemblage can be interpreted as evidence of one activity and the large quantity of fired clay dumped in the brine tanks suggests a real association with salt production. It could be argued that the absence of briquetage containers and troughs for evaporating the brine does not support such a conclusion, but the characteristics of the assemblage are consistent with a Roman industry using lead evaporating pans.

Later Roman lead pans have been discovered in the Nantwich area and the excavation itself produced a significant quantity of lead runoff and sheet (see Chapters 1 and 7). The emphasis on structural elements is a typical feature of this type of production. At Middleton, Norfolk, a late 3rd- to 4th-century Roman saltern site dominated by structural material, including walling, lining, pedestals and rods, and with a similar absence of ceramic evaporating containers has been interpreted as using lead pans for evaporation (Percival 2001). The apparent use of lead pans in Cheshire earlier than elsewhere in the country is possibly due to a military association.

The only hearths or oven bases encountered in the excavations occurred in area C and only three produced any fired clay, which comprised pieces with a single plain surface (FC01) in hearth [3218], oven plate (OP) with hollows in the surface from attached clay structure or furniture in hearth [3216], and wedge fragments from hearth [3242]. These hearths had long narrow linear flues, somewhat longer in size than those found at Middleton in Norfolk (Crowson 2001), but similar in other respects to the saltern structures. The fired clay debris found in area C, especially that in pit [3191], may have been demolished superstructure and broken furniture related to these hearth or oven bases. Although no analysis of carbonised plant remains was undertaken for the hearths, samples from Building 1 suggest that this building may have been used as a granary (see Chapter 18) and it could be argued that the hearths to the north were in fact associated with crop processing rather than brine evaporation. However the fired clay found in area C suggests more complex superstructures than is normally associated with crop drying and the fired clay associated with Building 1 is no different from the rest of the assemblage. It is likely that this area was at least partly used in the salt production process.

The clear separation of the fired clay in area C from that in area A suggests that the latter relates to brine evaporation in a location closer to the brine tanks, possibly in the unexcavated area between the tanks and the road.

The overall design of the evaporating hearths or ovens can be postulated on the basis of the fired clay forms. These structures probably combined subsurface flues lined with clay with substantial above-ground walls at least partly reinforced with wattles. The walls probably supported a suspended floor, possibly perforated, formed by the oven plates resting on or built into the walls above the flue. The shallow hollows on both structural elements and the small plates and wedges are probably scars left by pedestals set in place and moulded to the structure while the clay was wet. The pedestals probably supported the evaporating pans, possibly in conjunction with the small plates and stabilised by the wedges. This suggests that the evaporating pans were not in direct contact with the fire but used an indirect heat source and that the process at this site used saltern ovens, as defined by Morris (2001).

Figure 13.1: Fired clay. Scale 1:2.

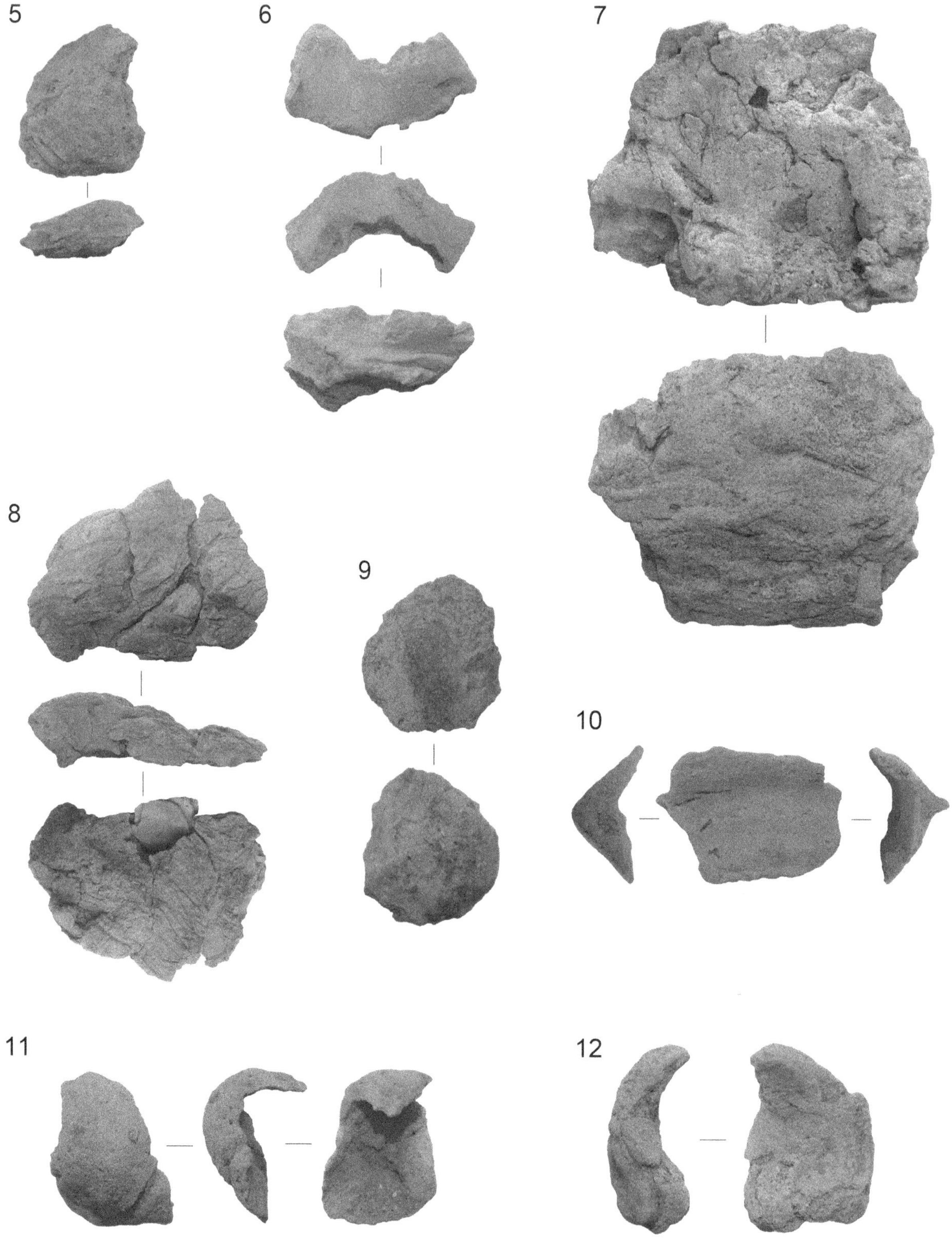

Figure 13.2: Fired clay. Scale 1:2, except no 12 at 1:1.

14. Industrial Debris

Mark Adams

The assemblage presented for examination consists of a total of 74 pieces weighing 3.665kg. This material was assessed by a visual examination and classification of the material supplemented where required by examination under a hand lens.

Of this material, twelve pieces weighing a total of 0.344kg consist of metallic iron or fragments of iron concretion. They include at least one fragment of iron strip, from (1240) in brine tank [1207], and iron plate, from upper layer (1237) also in [1207]. There is also one fragment of fuel ash weighing 0.006kg from trough [1264], which need not be derived from a metallurgical process.

Three crucible fragments weighing a total of 0.024kg from context (1183) in brine tank [1182] are probably from the same vessel, though there are no joining breaks. The largest of these fragments is a conical base and is probably from a crucible similar in form to that illustrated in Tylecote (1986, 96 fig 50 nos 5 & 9). Similar crucibles have been excavated at contemporary settlements at Irby, Wirral (Philpott & Adams 2010), and Court Farm, Halewood (Adams, in prep). The interior of the vessel is in a pale purplish grey fine-grained fabric with a vesicular texture. On the exterior the surface has been heavily vitrified to a glassy green fabric. There are no visible metal deposits surviving on any of the fragments.

The assemblage includes eight smithing hearth bases weighing a total of 0.893kg and measuring up to 110mm across. Four were found in brine tank [1207], although this is likely to be a result of the relatively large volume of this deposit rather than any direct association with metallurgy.

The bulk of the remainder of the material (41 fragments, weighing 2.023kg) consists of non-diagnostic fragments of slag up to *c* 100mm across. Most of these fragments are grey, probably fayalitic, with a slightly vesicular texture and are weakly to moderately magnetic. However, some of the material is heavily corroded to a dense 'sandy' structure with extensive orange iron corrosion products. Some of this material is virtually unrecognisable as industrial waste and has degraded to a very pale brown, almost white, concretion of powdery corrosion products. This corroded material occurred in brine tank [1207], pit [1160] and posthole/pit [2039]. Tank [1207] also contained slags coated with a surface deposit of blue corrosion product. These effects were probably a result of the waterlogging of the deposits in [1207].

The remainder of the assemblage consists of fragments of partly vitrified hearth wall; the unvitrified portions are in a coarse grained dark orange fabric.

The assemblage of industrial waste largely results from ferrous metallurgy, exclusively the smithing of small objects. There is nothing within the assemblage to suggest the smelting of any metals. The crucible fragments suggest the casting of small items, although the metal being worked was not determined. The assemblage is small and was widely spread across a number of contexts and therefore need not suggest that any of this activity was taking place within the excavated area.

15. Brick and Tile

Jeff Speakman

The excavation produced a total of 164 Roman brick and tile fragments, weighing 16487.9g (Table 15.1), with a further nine pieces (weight 102.5g) which may be medieval in date. Amongst the 164 there are seven fragments of incised flue tile (weight 1231.4g), six of which were recovered from the subsoil [7002] in excavation area G, and the seventh fragment (weight 254.1g) from pit [6018] in area F. There are no complete tiles and no visible stamps, although there are a number of possible signature marks, albeit incomplete and heavily abraded.

A minimum of two fabrics seem to have been used within the brick and tile, one being soft, underfired, oxidised and orange, and the other a hard, high-fired pink/purple, often partially reduced, fabric which often sheers in very sharp lines. A possible third fabric may have been used for the flue tile.

The majority of the brick and tile is quite fragmentary with many of the lower-fired oxidised fabrics being very soft and abraded. The size and thickness of many of the fragments within this category are too large to represent roofing tiles and the presence of one fragment with a flat upper surface and two straight edges confirms the presence of brick within the assemblage (SF143, area G subsoil (7002); dimensions: no length, 125mm width and *c* 45mm depth).

Whilst only small in number, the brick and tile points to a possible substantial building close to or on the site, although it could also be an indication of robbing of material from elsewhere.

Feature/ Context		**Type**	**No of fragments**	**Weight (g)**
	u/s	Rooftile	6	547.5
Topsoil	1001	Rooftile	10	609.2
Topsoil	1001	Rooftile/brick	1	69.8
Subsoil	1002	Rooftile	2	84.9
1005	1006	Rooftile	1	54.4
1068	1069	Rooftile	1	57.0
1072	1073	Rooftile	1	114.0
1079	1080	Rooftile	4	228.0
1079	1080	Rooftile?	1	40.3
1079	1080	Tile	2	37.7
1099	1100	Brick	1	431.6
1157	1158	Brick	1	99.0
1157	1158	Rooftile	1	65.4
1182	u/s	Rooftile	1	29.4
1182	1188	Rooftile	2	43.7
1207	1208	Rooftile	1	104.6
1207	1209	Rooftile	1	42.1
1207	1225	Rooftile	3	337.9
1207	1225	Rooftile/brick	1	107.4
1207	1239	Rooftile/brick	2	283.6
1207	1240	Rooftile	1	60.8
1182	1285	Rooftile	1	153.7
Topsoil	2001	Rooftile	6	263.2
Subsoil	2002	Rooftile	3	212.2
Topsoil	3001	Rooftile	3	33.9
Subsoil	3002	Rooftile	10	394.6
Subsoil	3002	Rooftile/brick	2	442.1
3100	3101	Rooftile	2	60.4
3168	3169	Rooftile	1	88.8
3212	3213	Rooftile/brick	2	51.3
3302	3284	Rooftile/brick	1	104.8
Topsoil	4001	Rooftile	14	602.5
Subsoil	4002	Rooftile	13	1076.3
4237	4232	Rooftile	2	88.4
Topsoil	5001	Rooftile?	5	209.6
Subsoil	6002	Rooftile	6	171.2
Subsoil	6002	Rooftile	2	176.8
Subsoil	6002	Rooftile	1	5.9
6017	6018	Flue tile	1	254.1
Topsoil	7001	Tile	1	88.8
Subsoil	7002	Flue tile	6	977.3
Subsoil	7002	Roof?tile	5	108.5
Subsoil	7002	Brick	16	5232.0
Subsoil	7002	Brick/tile	8	372.6
Subsoil	7002	Rooftile tegula	3	392.0
Subsoil	7002	Rooftile tegula?	1	66.7
Subsoil	7002	Rooftile/brick tegula	1	1225.0
Subsoil	7002	Rooftile?	1	68.9
Topsoil	8001	Rooftile	1	97.5
8103	8102	Rooftile	1	10.2
u/s		Rooftile	1	10.3
Total			**164**	**16487.9**

Table 15.1: Brick and tile by type, feature and context.

16. Structural Timbers

Peter Arrowsmith, Michael Nevell & David Power

A total of 355 structural timbers were recovered from the site and recorded. They derived from the main structural features, namely brine tank [1207], its north-eastern chamber [1347] and failed wall [1348], brine tank [1182] and wooden trough [1264]. The vast majority of the timbers (334) were excavated as in-situ structural components of these features. A small number (21) of unstratified structural timbers were also recovered from the same features, together with other fragmentary pieces. In addition, a few timbers were excavated within the remains of the other wooden troughs, the wicker-lined pits and the well [5109].

Methods

Adverse weather conditions made the exposure and recording of the wooden structures across the site a challenging task for the team of excavators. Lateral pressure generated by heavy rainfall and the surrounding soils resulted in localised failure of the structures during excavation. This was most evident in brine tank [1207] where an arrangement of props and sandbags was employed as both a control measure and a method of minimising damage during the excavation process.

Where it was possible, the timbers of the main structural features were numbered, photographed and drawn in situ. After excavation, both the in-situ and unstratified timbers found within these features were removed from the site for further analysis. Record sheets were completed for individual timbers and included their function within each feature, evidence of carpentry, and any evidence of burning and possible reuse. Timbers were individually drawn and their faces photographed.

An Overview of the Timbers

A total of 236 in-situ and 18 unstratified structural timbers were recovered from brine tank [1207], chamber [1347] and failed wall [1348]; 68 in-situ timbers from brine tank [1182]; and 30 in-situ and 3 unstratified timbers from wooden trough [1264]. Summary details of all of these timbers are given in Appendices 16.1 and 16.2. The unstratified timbers were found both during the initial phases of excavation of the brine tanks by machine (identified by the prefix W in Appendix 16.2) and within the fills of the main structural features. None of the unstratified timbers could be identified with any certainty as having been originally part of these features. The following discussion relates only to the in-situ timbers unless stated otherwise.

All the timbers were generally well preserved, although the upper level of planking and the top of the posts showed significant damage caused by post-Roman agricultural activity. All the timbers were worked in some way, usually showing evidence for the use of saws (124+4? timbers) and axes/adzes (148+4?), often both (63+1?), but more rarely for the use of chisels (5+2?). The most common carpentry and joinery techniques were tenons (82+1?), mortices (36+1?), lap joints, including scarf joints (25), and bevelling/chamfering (45+1). Some timbers (the majority of the posts and a few beams) although shaped were left virtually whole, some with their bark still on.

Almost all of the timber used was oak (*Quercus* sp.). The use of birch (*Betula* sp.) was noted in trough [1264] and chamber [1347]. There was also a single Scots Pine (*Pinus sylvestris* L.) timber from [1207]. Its occurrence is highly unusual within a Romano-British context.

	[1207]	[1347]	[1348]	[1182]	[1264]	Total
No of timbers	133	71	32	68	30	334
Species						
Quercus sp.	132	68+1?	32	68	27	327+1?
Pinus sp.	1	-	-	-	-	1
Betula sp.	-	2	-	-	3	5
Function in feature						
Beam	23	5	4	13	-	45
Plank	52	44	20	25	19	160
Squared post	4	7	1	22	7	41
Roundwood post	36	12	5	7	3	63
Brace	6	2	1	-	1	10
Wedge	12	1	1	1	-	15
Carpentry evidence						
Saw	51+2?	31+1?	18+1?	19	5	124+4?
Axe/adze	52+1?	29+1?	18	40+1?	9+?	148+4?
Chisel	3+2?	-	-	2	-	5+2?
Mortice	16+1?	10	3	7	-	36+1?
Tenon	36+1	10	6	30	-	82+1?
Lap joint	10	5	3	6	1	25
Bevel/ chamfer	21+1?	11	6	5	2	45+1?
Rebate	9	3	-	2	-	14
Groove	5	1	1	2	-	9
Notch	6+3?	3	2	3	-	14+3?
Nail hole	3	3	-	1	1	8
Nail	1+1?	3	1	-	-	5+1?
Scribe marked	1	1+1?	-	-	-	2+1?
Other						
Burnt	7	2	-	-	3	12
Reused	14+4?	13+2?	2+3?	4+1?	1+3?	34+13?

Table 16.1: Summary of details of in-situ structural timbers.

Table 16.2: Summary of details of in-situ structural timbers by feature and timber function within the feature.

			Beam	Plank	Squared post	Roundwood post	Brace	Wedge	Total
Brine tank [1207]	**Species**	*Quercus* sp.	22	52	4	36	6	12	132
		Pinus sp.	1	-	-	-	-	-	1
		Betula sp.	-	-	-	-	-	-	-
	Carpentry evidence	Saw	7+2?	30	-	11	2	1	51+2?
		Axe/adze	10	20	2	14+1?	3	3	52+1?
		Chisel	2	1	-	-	2?	-	3+2?
		Mortice	10	1+1?	-	1	3	1	16+1?
		Tenon	1?	1	3	32	-	-	36+1?
		Lap joint	6	1	1	1	1	-	10
		Bevel/chamfer	4	13+1?	-	4	-	-	21+1?
		Rebate	3	6	-	-	-	-	9
		Groove	2	2	1	-	-	-	5
		Notch	2+3?	3	-	1	-	-	6+3?
		Nail hole	1	1	-	1	-	-	3
		Nail	-	1+1?	-	-	-	-	1+1?
		Scribe marked	-	-	-	-	1	-	1
	Other	Burnt	1	3	-	1	2	-	7
		Reused	3+2?	5+2?	-	2	4	-	14+4?
Chamber [1347]	**Species**	*Quercus* sp.	5	43	6+1?	11	2	1	68+1?
		Pinus sp.	-	-	-	-	-	-	-
		Betula sp.	-	1	-	1	-	-	2
	Carpentry evidence	Saw	2+1?	19	4	5	1	-	31+1?
		Axe/adze	3	15+1?	4	6	1	-	29+1?
		Chisel	-	-	-	-	-	-	-
		Mortice	3	4	2	1	-	-	10
		Tenon	-	1	1	8	-	-	10
		Lap joint	3	2	-	-	-	-	5
		Bevel/chamfer	1	10	-	-	-	-	11
		Rebate	-	3	-	-	-	-	3
		Groove	1	-	-	-	-	-	1
		Notch	-	2	-	-	1	-	2
		Nail hole	-	2	1	-	-	-	3
		Nail	-	2	1	-	-	-	3
		Scribe marked	1	-	1?	-	-	-	1+1?
	Other	Burnt	1	1	-	-	-	-	2
		Reused	1	8+2?	3	1	-	-	13+2?
Failed wall [1348]	**Species**	*Quercus* sp.	4	20	1	5	1	1	32
		Pinus sp.	-	-	-	-	-	-	-
		Betula sp.	-	-	-	-	-	-	-
	Carpentry evidence	Saw	3	11+1?	1	2	-	1	18+1?
		Axe/adze	3	9	1	4	-	1	18
		Chisel	-	-	-	-	-	-	-
		Mortice	3	-	-	-	-	-	3
		Tenon	1	-	-	5	-	-	6
		Lap joint	3	-	-	-	-	-	3
		Bevel/chamfer	-	6	-	-	-	-	6
		Rebate	-	-	-	-	-	-	-
		Groove	-	-	-	1	-	-	1
		Notch	1	1	-	-	-	-	2
		Nail hole	-	-	-	-	-	-	-
		Nail	-	-	-	-	-	1	1
		Scribe marked	-	-	-	-	-	-	-
	Other	Burnt	-	-	-	-	-	-	-
		Reused	1	1+2?	-	-	-	1?	2+3?
Brine tank [1182]	**Species**	*Quercus*	13	25	22	7	-	1	68
		Pinus	-	-	-	-	-		-
		Betula	-	-	-	-	-	-	-
	Carpentry evidence	Saw	3	6	8	1	-	1	19
		Axe/adze	5	15+1?	14	5	-	1	40+1?
		Chisel	2	-	-	-	-	-	2
		Mortice	7	-	-	-	-	-	7
		Tenon	1	-	22	7	-	-	30
		Lap joint	6	-	-	-	-	-	6
		Bevel/chamfer	1	2	2	-	-	-	5
		Rebate	-	2	-	-	-	-	2
		Groove	-	1	1	-	-	-	2
		Notch	2	-	-	-	-	1	3

		Nail hole	-	1	-	-	-	-	1
		Nail	-	-	-	-	-	-	-
		Scribe marked	-	-	-	-	-	-	-
	Other	Burnt	-	-	-	-	-	-	-
		Reused	1	1+1?	1	-	-	1	4+1?
Trough [1264]	**Species**	*Quercus* sp.	-	17	7	2	1	-	27
		Pinus sp.	-	-	-	-	-	-	-
		Betula sp.	-	2	-	1	-	-	3
	Carpentry evidence	Saw	-	4	1	-	-	-	5
		Axe/adze	-	3	4+1?	2	-	-	9+1?
		Chisel	-	-	-	-	-	-	-
		Mortice	-	-	-	-	-	-	-
		Tenon	-	-	-	-	-	-	-
		Lap joint	-	1	-	-	-	-	1
		Bevel/chamfer	-	2	-	-	-	-	2
		Rebate	-	-	-	-	-	-	-
		Groove	-	-	-	-	-	-	-
		Notch	-	-	-	-	-	-	-
		Nail hole	-	1	-	-	-	-	1
		Nail	-	-	-	-	-	-	-
		Scribe marked	-	--	-	-	-	-	-
	Other	Burnt	-	3	-	-	-	-	3
		Reused	-	1+3?	-	-	-	-	1+3?

Conversion of the Timbers from the Parent Logs

The main types of timber showed a variety of carpentry techniques for converting the parent log. The wood used seems to have varied in age from *c* 20 to *c* 300 years.

The roundwood timbers had between 25 and 40 rings. They were cut from whole logs, usually with a significant amount of sapwood remaining and often with the bark left on, eg timbers (1556), (1565), (1615), (2012), (2541) and (3017) (Figs 16.2, 16.5 & 16.7). Their carpentry mainly comprised the addition of tenons, reflecting their principal use as posts within the two brine tanks.

The squared timbers, which were mostly beams and posts, were boxheart, box halved (squared from half a roundwood) or box quartered (squared from a quarter of a roundwood), eg timbers (1505), (1510), (1569), (1611), (1612), (1619), (1622), (2027), (2029), (2058), (2061), (2501, unstratified), (2527) and (2548) (Figs 16.1-3 & 16.5-6). They had generally between 30 and 50 rings.

The assemblage of posts included some radially cleft stakes, eg timbers (2515), (2517) and (3523) (Fig 16.6). These had 20+ rings, with sapwood sometimes being present.

The planks were typically tangentially faced timbers (formed by dividing a log lengthwise, at an angle to its growth rings), eg timbers (1541), (1584), (1652), (2535), (2545) and (2573) (Figs 16.3-4). Generally, the planks had more than 75 rings. At least eighteen had more than 100 and four more than 200, the largest having 293 (Table 17.1).

In terms of the tools used to fashion and prepare the timbers, saw marks were observed on all timber types, eg timbers (1565), (2027), (2515), (2535), (2545) and (2573) (Figs 16.3-7). Axe marks or adze marks were likewise found on all types, eg timbers (1584), (1590), (1612), (1652), (2027), (2029), (2069), (2535), (2541) and (2548) (Figs 16.1-5). These marks can be difficult to tell apart; adze blades tend to be thinner and lighter than axe blades and, since less force is required in striking an adze, the blade is not deeply buried within the wood (Ulrich 2007, 18). Two iron adze-hammers and an iron axe head were recovered from the fills of brine tank [1207] and represent the kinds of tool which would have been used in its construction (see Chapter 7, 4.28-30).

A small number of timbers of most types showed tool marks consistent with the use of a chisel. Evidence of splitting was observed on a few planks. Two timbers, (1569) (Fig 16.6) and (2548), carried scribe marks, which in both cases lined up with a mortice.

The Source of the Timber

Ian Tyers in his dendrochronological analysis notes a close correlation between the sequences from the oak timbers at Kingsley Fields and sequences from elsewhere within Nantwich and from Middlewich, a few kilometres to the north (see Chapter 17). There were also good matches with timbers from London, Carlisle and Droitwich but the strongest correlations were with the Middlewich and other Nantwich samples, suggesting that a supply of suitable raw material was locally available.

It is possible that at least some of this timber was derived from managed oak woodland, in the form of stands or coppices (cf Goodburn 1991, 90). Evidence for local woodland management, particularly of hazel, may be provided by analysis of the roundwood used in wicker-lined pits [1053], [1054], [1072] and [1099] (see Chapter 18). However, with the possible exception of [1099], these are known to be later than the brine tanks and other main structural features, being dated to the late 3rd to 4th century.

Reused Timbers

A minority of in-situ timbers showed evidence of certain or possible reuse in the form of carpentry features such as redundant tenons or mortices, nail holes and nails, chamfers, grooves and rebates. Some also had evidence of burning, again possibly indicative of reuse. The largest number of certain or possible reused timbers were planks (16+10?), followed by beams (6+2?), squared posts (4), braces (4), roundwood posts (3), and wedges (2+1?). Most examples were located in brine tank [1207] (14+4?) and chamber [1347] (13+2?) (Tables 16.1 & 16.2).

In the case of trough [1264], in addition to a small number of timbers (1+3?) which showed signs of reuse in their carpentry, the trough's irregular construction with poorly fitting timbers hinted that it may, as a whole, have been constructed with material salvaged from elsewhere.

Planks formed the majority of timbers with chamfering (33+1) or rebating (11), eg timbers (1541), (1584), (1590) and (2535) (Figs 16.3 & 16.4). In some cases these elements were possibly tailored to help fit individual timbers into the structure in which they were excavated and may be evidence that some final finishing of the timbers took place on site.

Very few timbers included nail holes (8) and/or nails (5+1?), eg timbers (1565), (2517), (2535) and (2575) (Figs 16.4, 16.6 & 16.7), with most examples being planks. From the pattern of nailing and other redundant features, some of these timbers were clearly reused. The infrequency of nails and nail holes is such that all instances may be the result of reuse.

Twelve timbers showed evidence of burning in antiquity, the majority of which were located in brine tank [1207]. Burning was specific to individual timbers rather than localised areas of the structural features, indicating that it took place prior to the feature's construction.

The majority of reused timbers appear to have originally served a structural purpose, rather than being derived from other wooden objects. Three, reused as planks, are of an uncertain function. Two of these, (2538) and (2573), were trapezoidal with two curving sides, while the third (1652) was more semicircular (Fig 16.4).

Other timbers illustrated here were reused as beams, (1611), (1612) and (1619); posts, (1565), (2517) and (2541); planks, (1590), (2069), (2527), (2535) and (2575); and braces, (1505) and (1569) (Figs 16.2-4 & 6-7).

Timber (2501) (Fig 16.3) was an unstratified timber found in the fill of chamber [1347]. It had three evenly spaced blind mortices on one face and a lap joint at both ends. Comparable timbers excavated, for example, in Carlisle and London suggest that it was part of a sill beam or wall plate of a timber-framed building, with tenoned posts set into the mortices (Goodburn 1991; Darrah 2009, 794-5). An alternative possibility for consideration is that it was part of an original top plate of the side walling of [1347]. However, this can be discounted on the grounds that the spacing between the mortices of timber (2501) was much smaller than that between the tenoned posts of [1347].

Timber (1611) (Fig 16.3) had a central groove running the length of one face, to either side of a slightly deeper rebate. The rebate continued to one edge of the timber and contained one rectangular and one circular socket. The timber had been reused as a cross-timber in the base of brine tank [1207] and the lap joint at either end was also related to a previous use. It is possible that this timber formed the sill beam of an open-ended building and that shuttering was inserted into the groove to close off the open end when required. Similarly grooved timbers have been excavated at Carlisle and have been likened to stone grooved sills at Pompeii, occasionally found with the impression of the timber shuttering (Darrah 2009, 796-7). One example at Carlisle included a rebate which could have been used to feed the shuttering into the grove; this rebate also contained a circular socket, interpreted as the seating for an end shutter which could be pivoted shut. In the case of timber (1611), the rebate could have held a harr-hung door, pivoting within the circular socket, while the rectangular socket may have held a peg, securing the door from the inside.

Timber (2575) (Fig 16.4), from its pattern of nails and nail holes, had originally been nailed as a plank to an arrangement of regularly spaced timber posts. Roman structures of nailed post and plank construction are known in a variety of forms and include the use of planks as cladding for timber-famed buildings.

Beam (1612) (Fig 16.3), which was positioned centrally within the base of brine tank [1207], had the most complex carpentry of all the timbers. Its row of evenly spaced blind mortices suggests that it was originally a sill beam, supporting a timber post wall. Its ends were later modified, either in connection with its role in the brine tank or with an intermediate reuse. A rebate and a groove which ran the length of opposing sides of the timber may also be original features, designed to accommodate timber planking.

Plank (2535) (Fig 16.4), a fairly thick piece of timber with a rebate at each end, one with a nail hole, may have originally served as a cover. A comparison may be drawn with the timber covering of drains excavated at Carlisle and Vindolanda (Darrah 2009, 797-8).

Brace (1569) (Fig 16.6) has a distinctive series of slots recessed at an angle into one face, suggesting that this timber may originally have been a stringer, or side support, for a flight of steps. This brace was located at the south-east of [1207] where the other two braces, (1574) and (1577), included similar angled slots.

Post (2541) (Fig 16.2), incorporated within the roughly built south-west elevation of chamber [1347], may have been a reused section of brine-tank baseplate.

The Timber Construction of the Brine Tanks

The two brine tanks, [1182] and [1207], were the most complete examples from the Roman period to have been excavated and clearly displayed common elements in their timber structure. Furthermore, while the two tanks also showed some variations in their construction, a number of these are likely to be alterations to an original common design. That design is most evident in the case of tank [1182] which appears to be free from such modifications. It involved a fairly simple box-frame construction utilising horizontal and vertical timbers, linked by relatively simple joints. A base-frame was formed using beams for the sides and ends, either butted end to end or jointed by lap/scarf joints. Mortices were cut into the outer edges of these baseplates at regular intervals for the tenons of vertical posts which were positioned along the beams. To give greater rigidity to the base-frame, transverse timbers extended between the side baseplates but were not jointed to them. In one case in [1182] and more frequently in [1207] these cross-timbers did not span the full width of the tank and additional timbers were used as chocks. To complete the walls, horizontal planks were laid, either edge to edge or overlapping, between the posts and the tank's outer clay lining. Although the uppermost parts of the tanks had not survived, it can be assumed that the tops of the posts were tenoned into an upper arrangement of beams, or top plates, reflecting the end and side baseplates below. It is also likely that this superstructure was strengthened by upper cross-beams, jointed into the top plates.

In the case of [1207] it is probable that wall [1348] and chamber [1347] were originally intended to form the ends of the tank but both failed during construction, resulting in the shortening of the tank and the modified structural arrangement at its new ends (see Chapter 3). Here the posts were now set between a double line of beams, only one of which (1619) was jointed, via mortices, into the side baseplates (see timber (1622) (Fig 16.1)). The new end walls were given additional stability by the use of inclined braces, eg (1505) and (1569) (Fig 16.6), jointed into cross-timbers within the tank.

The two tanks also showed a number of differences which may have been more general variations on the overall design. The tenoned posts in [1182] were of a smaller width than in [1207], in keeping with the tank's smaller size, but also included a greater proportion of squared timbers. In [1182] the base timbers were embedded within the tank's clay base, whereas in [1207] they sat proud of the clay.

The failed wall [1348] suggests the process by which the timber and clay sides of the tanks were constructed. Following its rupture, [1348] was faced with planking against which clay was placed to form the lining of the new end wall of the tank. This implies a method of construction by which a single course of planking was laid on its edge against the upright posts, clay was built up against this and the process was then repeated for a second course. If the same procedure was used throughout the construction of the tanks, the planking and clay lining were built up together against the vertical posts (rather than the clay lining being built first and the timberwork built up against it). This method would also be consistent with the large construction cuts which seem to have been dug for each tank. As well as being a possible quarry for clay, these would have created a working space around the outside of the timber frame.

The closest known parallel to the two tanks at Kingsley Fields is provided by a timber structure excavated at the salt-making site at Upwich, in Droitwich. The remains comprised two opposing timber-built walls of a tank measuring *c* 10m long, *c* 5m wide and at least 1.55m deep. The walls were of a post and plank construction, the posts in this case seemingly being set directly into the ground. The planks were set on edge and, as at Kingsley Fields, were not fixed to the posts but were held in place by the infill of a construction cut for the tank. The structure was located adjacent to brine springs and is believed to have been probably used for the storage of brine, as in the case of the Nantwich tanks. Dendrochronological analysis of timbers from the structure suggested a felling date range of AD 61-5. However, some of the posts showed evidence of reuse and the pottery from the fill of the construction cut points to a date no earlier than the Hadrianic period, making the Upwich tank roughly contemporary with those at Kingsley Fields (Hurst & Hemingway 1997, 9-16).

Catalogue of Illustrated Timbers

A total of 32 timbers have been selected for illustration as examples of the main structural types, forms of carpentry, and evidence of reuse.

1505 Inclined in-situ brace in [1207] (Fig 16.6). Oak (*Quercus* sp.). A boxheart converted timber with a degraded broken upper end and charring at the lower end. Cut into one face of the timber were five blind mortices, four complete and one damaged, with a maximum length of 90mm, a width of 50mm and depth of 50mm. Chisel? marks were also present and there was a bird's mouth joint on the lower end. Together with (1507) and (1509), the timber was a structural support for the north-eastern end elevation of brine tank [1207] with the lower end jointed into morticed cross-timber (1604). Length 1.8m, width 180mm, thickness 130mm. The redundant mortices indicate reuse.

1510 Vertical in-situ post in [1207] (Fig 16.5). Oak (*Quercus* sp.). A box halved converted timber with fragmented ends, one with the remains of a tenon, damaged by movement and collapse during excavation. A post-excavation longitudinal split and possible faint tool marks were present along the flat face of the timber. Originally set between baseplates (1599) and (1600) and supported by inclined brace (1505), the timber was part of the north-eastern end elevation of brine tank [1207]. Length 1.25m, width 200mm and thickness 120mm.

1541 Horizontal in-situ plank in [1207] (Fig 16.3). Oak (*Quercus* sp.). A tangentially faced converted timber with a chamfered edge to the end of one face. The plank was secured by posts (1542), (1543) and (1544) as part of the north-western side elevation of brine tank [1207]. Length 3.8m, width 310mm, thickness 40mm.

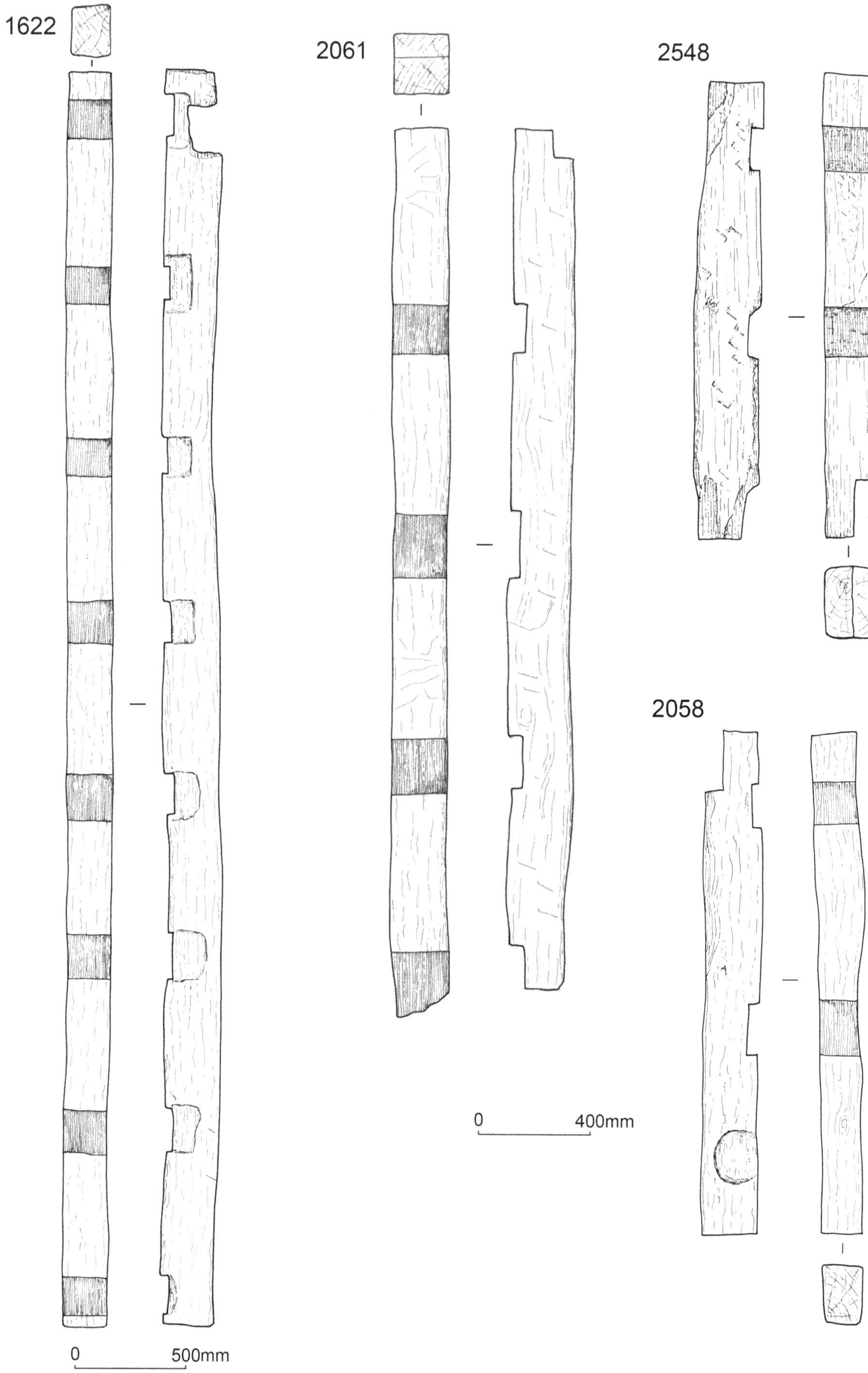

Figure 16.1: Structural timbers, beams.

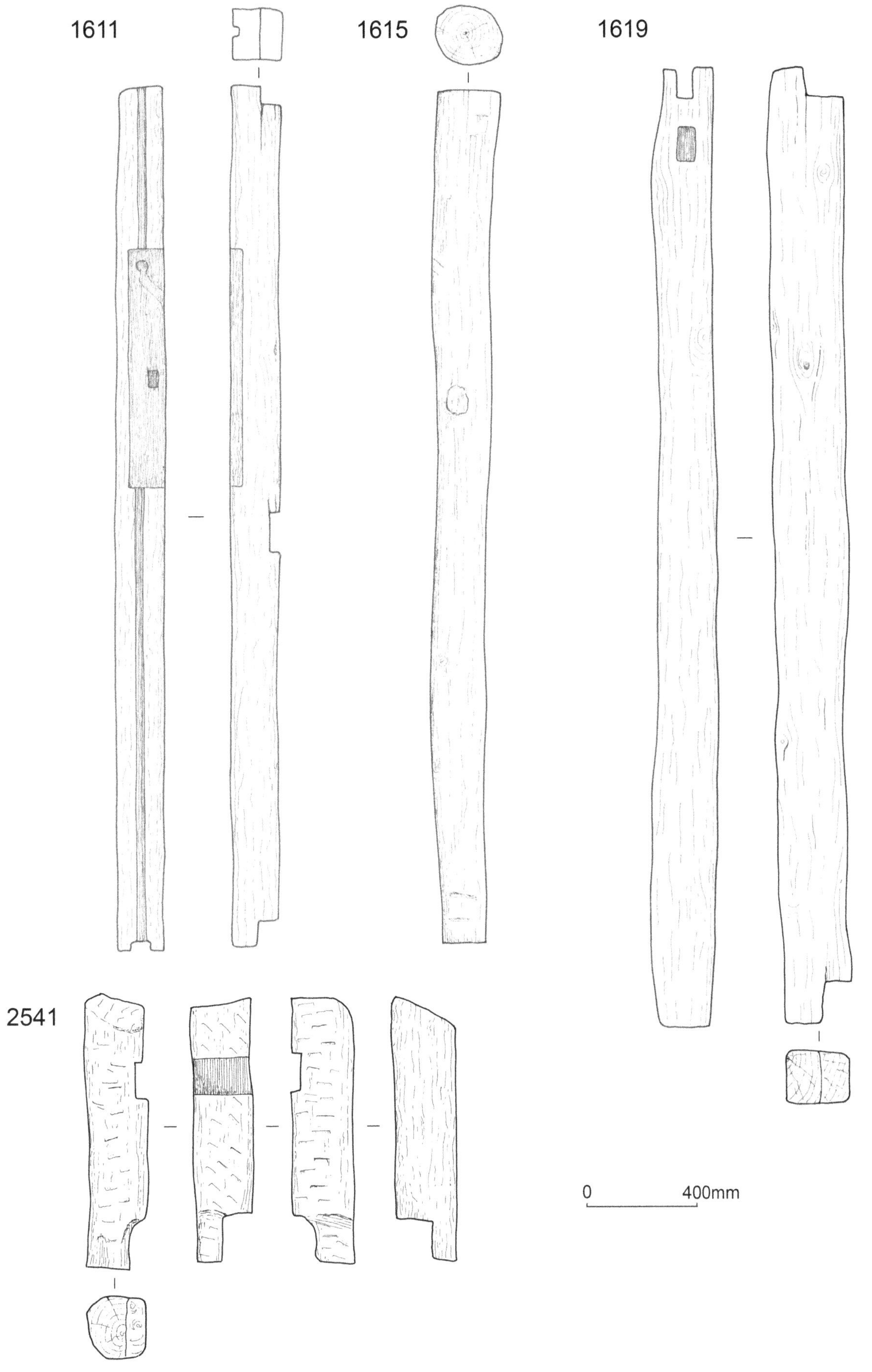

Figure 16.2: Structural timbers, beams.

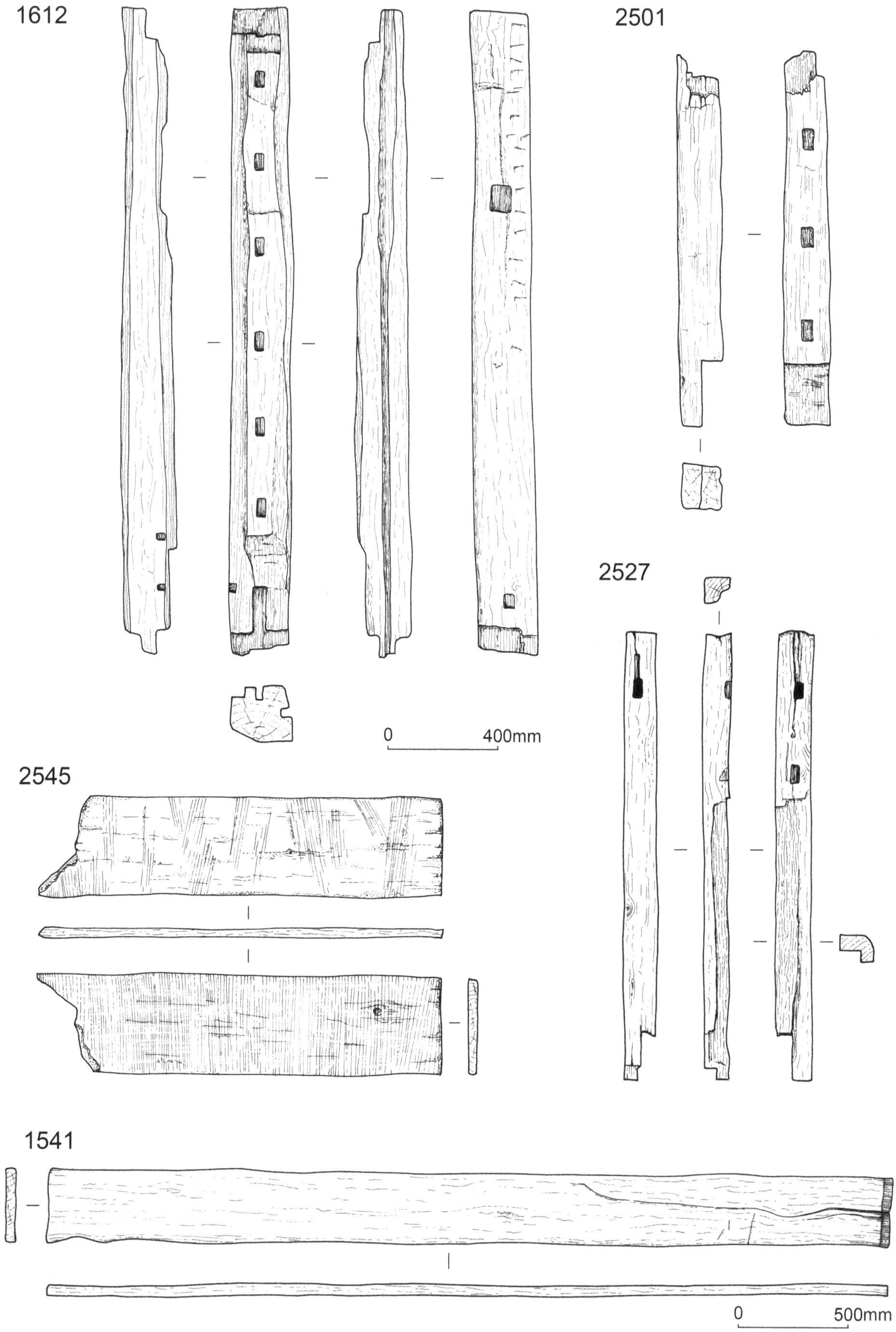

Figure 16.3: Structural timbers, beams and planks.

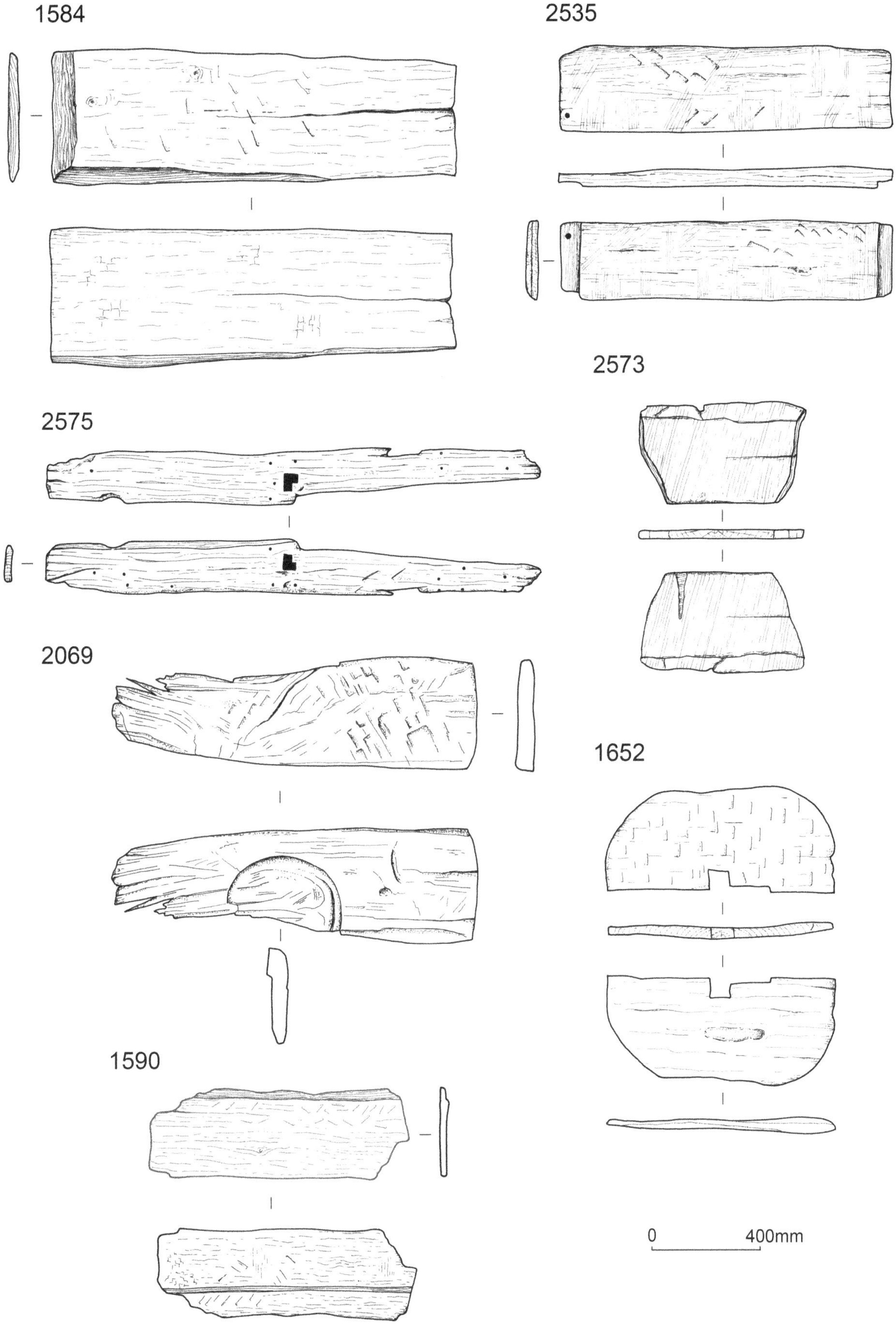

Figure 16.4: Structural timbers, planks.

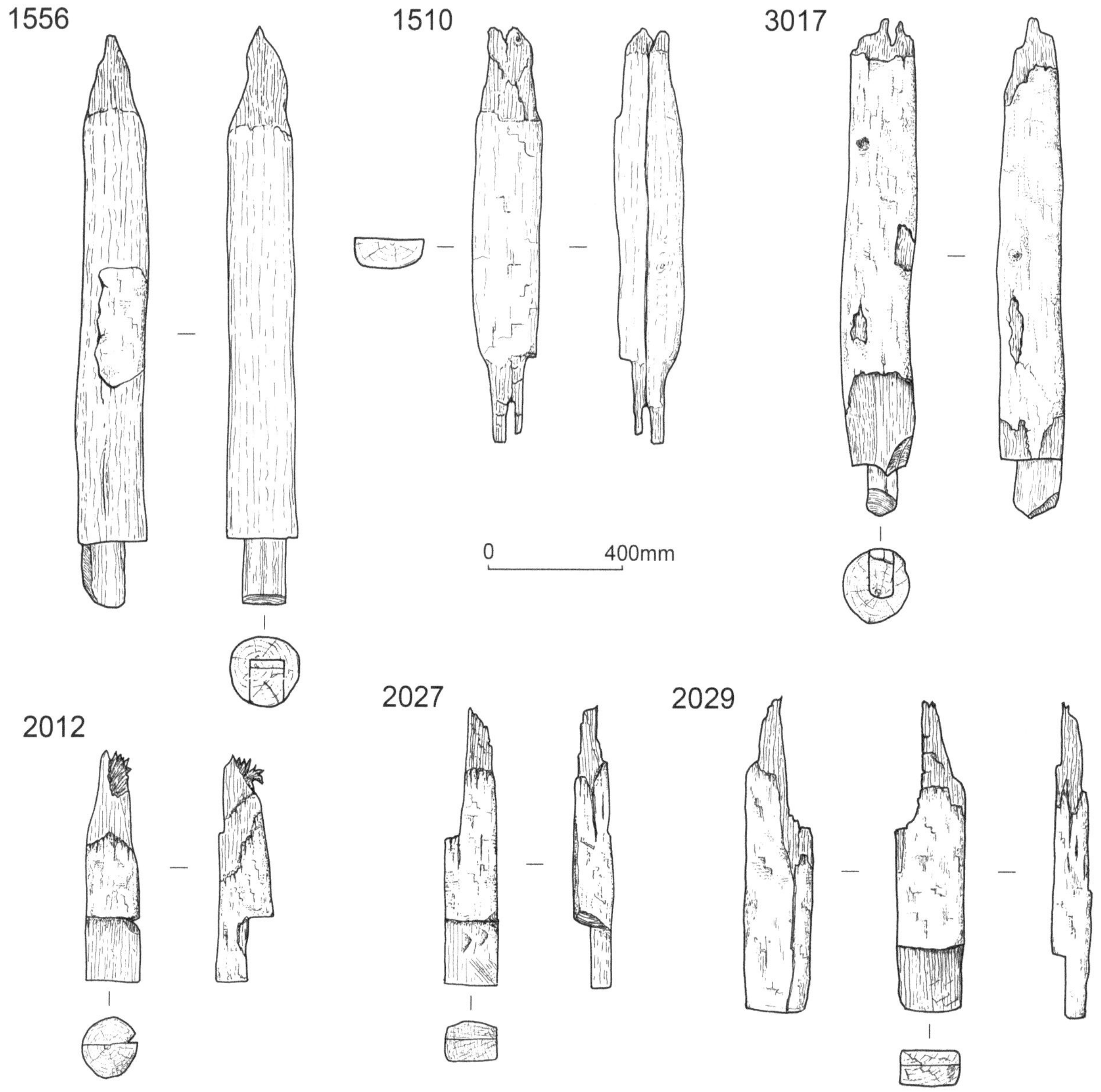

Figure 16.5: Structural timbers, posts.

1556 Vertical in-situ post in [1207] (Fig 16.5). Oak (*Quercus* sp.). A whole timber with a degraded broken upper end, localised covering of bark, and a bare-faced tenon at the lower end with a length of 200mm. The tenon has a bevelled side and end, with axe/adze marks. Part of the north-western side elevation of brine tank [1207]. Length 1.65m, diameter 210mm.

1565 Vertical in-situ post in [1207] (Fig 16.7). Oak (*Quercus* sp.). A whole timber with a charred lower end, partial covering of bark and a partially sawn flat face with ripple marks. Cut into the worked face was a squint lap joint with a depth of 20mm, compression marks from former use, and a centrally positioned square nail hole. Set between baseplates (1619) and (1620), the timber was part of the south-western end elevation of brine tank [1207]. Length 1.2m, maximum diameter 120mm. The redundant joint and nail hole indicate reuse.

1569 Inclined in-situ brace in [1207] (Fig 16.6). Oak (*Quercus* sp.). The upper fragment of a boxheart converted timber with a decayed and burnt lower end which broke into two pieces during removal. The squared upper fragment had four mortices, three complete and one damaged, cut into one face, with a maximum depth of 200mm. Scribe marks were present on one other face in line with a mortice. Length 1.8m, width 180mm, thickness 140mm. The adjoining decayed and burnt lower fragment (not illustrated), with a recorded length of 0.65m, had scribe marks in line with the lower mortice. The fragments had axe/adze marks and possible chisel marks. Together with similar inclined braces (1574) and (1577), the timber provided additional support for the south-western end elevation of brine tank [1207]. The redundant mortices indicate reuse.

1584 Horizontal in-situ plank in [1207] (Fig 16.4). Oak (*Quercus* sp.). A tangentially faced timber with localised splitting and decay. The plank had a chamfered end, a partially chamfered edge along one face, and on the opposing edge of the reverse face a damaged rebate with a maximum width of 30mm and depth of 15mm. The plank had very faint ripple marks on both faces and

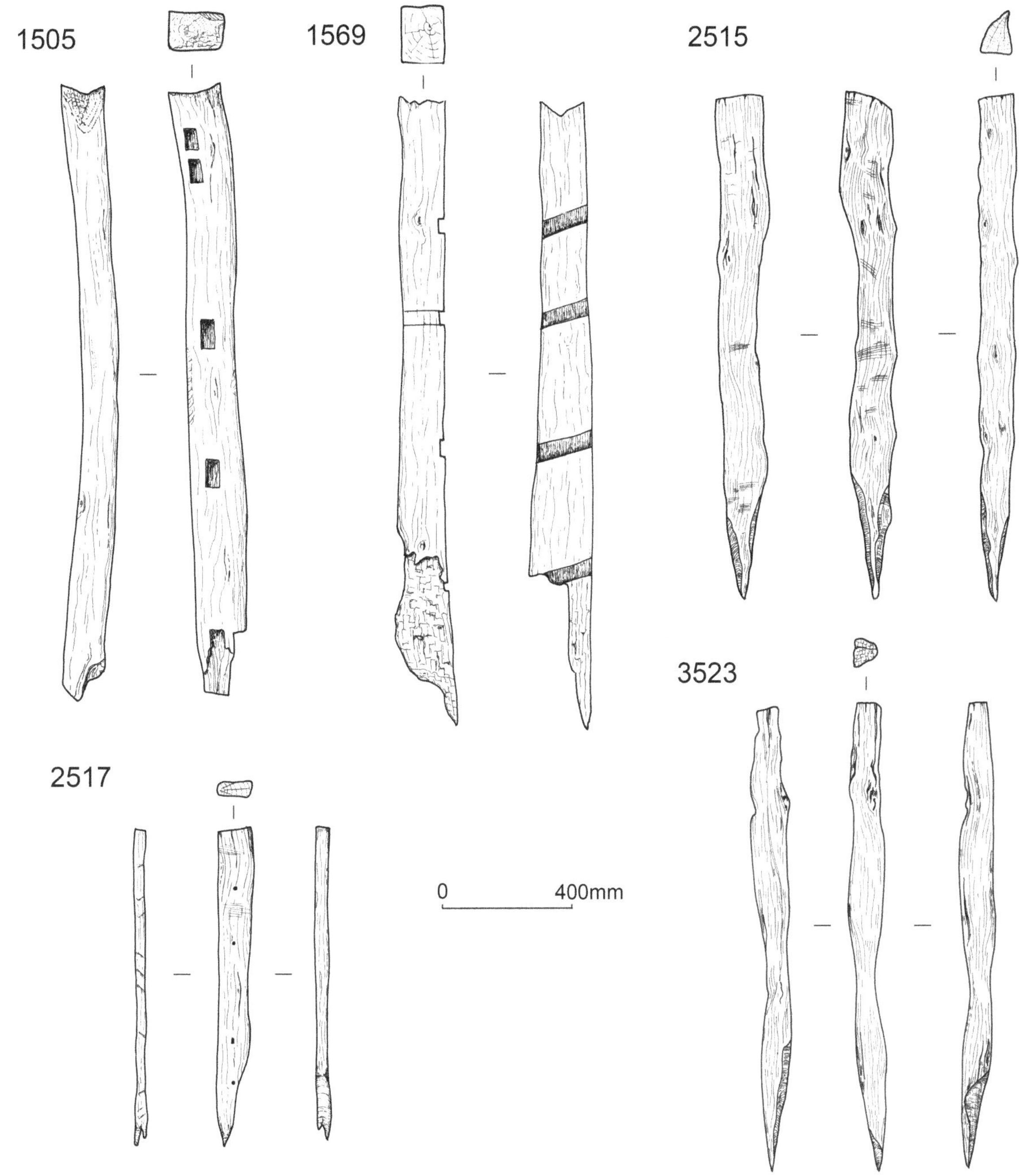

Figure 16.6: Structural timbers, braces and stakes.

occasional axe/adze marks. Part of the south-west end elevation of brine tank [1207]. Length 1.43m, maximum width 470mm and thickness 130mm. The rebate and chamfering indicates possible reuse.

1590 Horizontal in-situ plank in [1207] (Fig 16.4). Oak (*Quercus* sp.). Broken in antiquity and decayed with laminar splitting. One face had a rebated edge with a maximum width of 50mm and depth of 10mm, with axe/adze marks both on and adjacent to the rebate. The reverse face had a longitudinal groove, 15mm in width and with a maximum depth of 10mm, together with axe/adze marks and abraded ripple marks. Part of the south-western end elevation of brine tank [1207]. Length 0.92m, width 320mm, thickness 35mm. Originally joined with plank (1571). The redundant groove and rebate indicate reuse.

1611 Horizontal in-situ beam in [1207] (Fig 16.2). Oak (*Quercus* sp.). A box quartered well-preserved timber with a lap joint on each end and a centrally positioned mortice or housing for a lap joint on one face. The opposing face had a longitudinal groove 30mm in width and 20-40mm in depth, extending to either side of a rebate which was positioned off-centre and had a length of 0.83m, width of 130mm and depth of 50mm. The rebate contained a central rectangular mortice or socket with a length of 60mm, width of 30mm and depth of 40mm, and a circular socket with a diameter of 40mm and a maximum depth of 40mm, positioned at one end of the rebate and aligned with the groove. Length 3m, width 180mm. Used as a cross-timber in the base of brine tank [1207]. The redundant lap joints and other carpentry features indicate reuse.

1612 Horizontal in-situ beam in [1207] (Fig 16.3). Oak (*Quercus* sp.). A box halved converted timber with a degraded end and well-defined axe/adze marks on one face. This extensively worked timber had a complex arrangement of carpentry features. These included, on the upper face as excavated, a linear arrangement of six similarly sized, evenly spaced and centrally positioned blind mortices 220-225mm apart, 60mm in length, 40mm in width and with a maximum depth of 40mm. The timber also had an end tenon? and lap joint, a central groove running the full length of one side, a rebate and recessed mortice/peg holes on the other side, and two square blind mortices on the underside, which was also partly chamfered. Centrally positioned within brine tank [1207] and aligned on the tank's long axis, the timber rested on cross-timbers (1611) and (1614) and a short similarly aligned timber (1613). Its function in [1207] is unknown but the complex carpentry is clearly indicative of reuse, possibly of more than one phase. Length 2.25m, width 240mm, thickness 200mm.

1615 Horizontal in-situ beam in [1207] (Fig 16.2). Oak (*Quercus* sp.). A well-preserved whole timber with patches of bark, possibly notched and with possible saw marks. Used as a cross-timber in the base of brine tank [1207]. Length 2.95m, maximum diameter of 200mm.

1619 Horizontal in-situ beam (Fig 16.2). Oak (*Quercus* sp.). A well-preserved box quartered timber, with a lap joint on each end and with possible saw marks. The upper face as excavated had a blind mortice with a length of 130mm, width of 60mm and depth of 50-55mm and a 110mm long notch cut into one end. Used as a cross-timber in the base of brine tank [1207], jointed into mortices in side baseplates (1621) and (1622) (see below). Length 3.3m, width 200mm, thickness 260mm. The redundant mortice and notch indicate reuse.

1622 Horizontal in-situ beam in [1207] (Fig 16.1). Oak (*Quercus* sp.). A well-preserved box quartered timber with eight evenly spaced mortices for tenoned posts along one edge. Depressions on the upper face indicated the original position of the posts. On the opposite edge was a single mortice forming a joint with cross-timber (1619). The timber was one of four similar sized baseplates which formed the two long sides of brine tank [1207]. At its north-eastern end it butted side baseplate (1603). Length 5.46m, width 250mm, thickness 180mm.

1652 Horizontal in-situ plank in [1207] (Fig 16.4). Oak (*Quercus* sp.). A tangentially faced timber with an irregular shape, possibly warped in antiquity. The timber had one straight edge with a centrally positioned recessed and damaged mortice? with a length of 70mm, width of 40mm and depth of 70mm. The remaining curved edges were chamfered to create a blunt rounded profile. Abraded axe/adze marks and possible chisel incisions were present on both faces, with saw marks on the straight edge. One face had a broken nail and the reverse face had ancient compression marks with a maximum length of 230mm, width of 50mm and depth of 3mm. The timber was used as a plank in the south-western end elevation of brine tank [1207]. Length 0.83m, width 370mm, thickness 50mm. The shape and redundant joint indicate reuse.

2012 Vertical in-situ post in [1182] (Fig 16.5). Oak (*Quercus* sp.). A whole timber with a broken, splintered and decayed heartwood and sapwood upper end, a partial covering of bark, and a bare-faced tenon on the lower end with a length of 180mm. The lower part of the post was in a better condition but had decay along one side and on the tenon. Housed within a mortice in baseplate (2062); part of the south-eastern side elevation of brine tank [1182]. Length 0.65m, diameter 160mm.

2027 Vertical in-situ post in [1182] (Fig 16.5). Oak (*Quercus* sp.). A box quartered timber with a broken and decayed heartwood and sapwood upper end, and a bare-faced tenon on the lower end with axe/adze marks, saw marks and a length of 180mm. Housed within a mortice in baseplate (2059); part of the north-western side elevation of brine tank [1182]. Length 0.82m, width 165mm, thickness 130mm.

2029 Vertical in-situ post in [1182] (Fig 16.5). Oak (*Quercus* sp.). A box quartered timber with a broken and decayed heartwood upper end, split sapwood, and a bare-faced tenon on the lower end 180mm in length. Axe/adze marks on the tenon. Housed within a mortice in baseplate (2059); part of the north-western side elevation of brine tank [1182]. Length 0.92m, width 180mm, thickness 120mm.

2058 Horizontal in-situ beam in [1182] (Fig 16.1). Oak (*Quercus* sp.). A well-preserved box quartered timber with a lap joint on one end, mortices for two tenoned posts on one edge and a circular compression mark on the upper face. A baseplate in the north-western side elevation of brine tank [1182], jointed to side baseplate (2059) at its south-western end and butted against end baseplate (2057) at its north-eastern end. Length 1.76m, width 210mm, thickness 150mm.

2061 Horizontal in-situ beam in [1182] (Fig 16.1). Oak (*Quercus* sp.). A well-preserved box quartered timber fragment with a lap joint on one end, four mortices (three complete and one damaged) for tenoned posts along one edge and axe/adze marks on one face. A baseplate in the north-western side elevation of brine tank [1182], jointed to side baseplate (2056) at its south-western end. The north-east end of the timber (not illustrated) was joined by a lap/scarf joint to similarly aligned baseplate timber (2059). Length of fragment 3.2m, width 220mm, thickness 200mm. The complete timber had a length of 4.4m.

2069 Horizontal in-situ plank in [1182] (Fig 16.4). Oak (*Quercus* sp.). A damaged timber fragment with a broken end, longitudinal splitting and decay. One face had a subcircular rebate extending from the edge, with a depth of 20mm, and a short poorly defined abraded curved depression. The reverse face had a random arrangement of well-defined axe/adze marks. Part of a larger plank positioned to the rear of plank (2067) in the north-eastern end elevation of brine tank [1182]. Length of fragment 1.32m, maximum width 390mm, thickness 60mm. Redundant features indicate reuse.

2501 Beam unstratified in fill of [1347] (Fig 16.3). Oak (*Quercus* sp.). A box quartered timber, with two end lap joints, one badly decayed and damaged during removal. The other, well-preserved, lap joint had faint saw marks and abraded axe/adze and chisel marks. Three similarly sized, centrally positioned, evenly spaced blind mortices 250-270mm apart, 70-80mm in length, 40mm in width and 80mm in depth were cut into one face of the timber. Length 1.32m, width 180mm, thickness 160mm.

2515 Vertical in-situ post in [1347] (Fig 16.6). Oak (*Quercus* sp.). An irregular, radially cleft stake, axe-hewn to a point. Two of the three faces had faint saw marks. The sharpened stake was driven directly into the clay to provide additional support for the south-western elevation of chamber [1347]. Length 1.52m, maximum width 150mm.

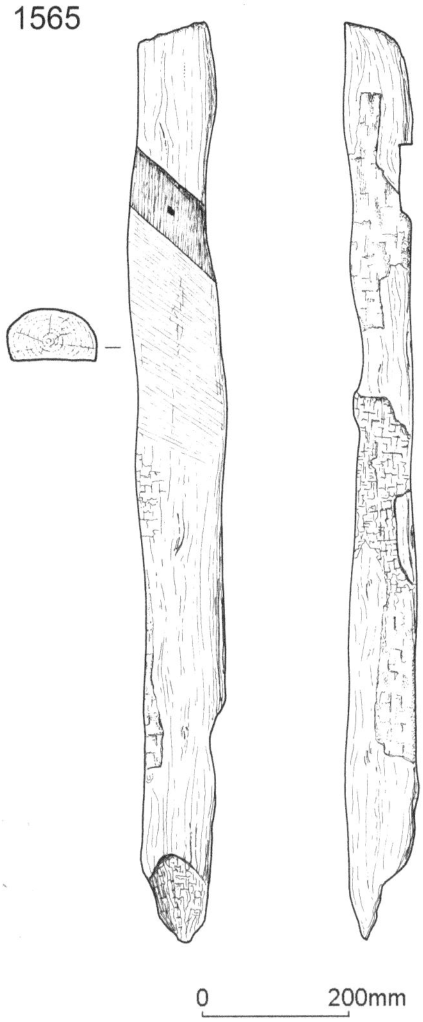

Figure 16.7: Structural timbers, post 1565.

2517 Vertical in-situ post in [1347] (Fig 16.6). Oak (*Quercus* sp.). A radially cleft stake with flat faces and a decayed lower end hewn to a point. One face had a centrally aligned linear arrangement of two iron nails and two nail holes, with faint saw marks and axe/adze marks on a side face and the sharpened end. Driven directly into the clay, the stake functioned as additional support for the south-western elevation of chamber [1347]. Length 0.95m, maximum width 100mm, thickness 40mm. The redundant nails and nail holes indicate reuse.

2527 Horizontal in-situ plank in [1347] (Fig 16.3). Oak (*Quercus* sp.). A moderately well-preserved box halved, rebated beam reused as a plank in the south-western elevation of chamber [1347]. The timber had abraded axe/adze marks, a longitudinal rebate, two mortices 230mm apart (one blind, one through) and a decayed lap joint on one end. Length 1.56m, width 120mm, thickness 85mm.

2535 Horizontal in-situ plank in [1347] (Fig 16.4). Oak (*Quercus* sp.). A tangentially faced converted timber with localised decay and damage. The plank had clearly defined saw and axe/adze marks on both faces and was unusually thick compared to the majority of planks recovered from the site. Each end of the timber had a rebate extending across the width of the plank, with one rebate having a nail hole. Part of the south-western elevation of chamber [1347]. Length 1.2m, width 300mm, thickness 70mm. The rebates and nail hole indicate reuse.

2541 Vertical in-situ post in [1347] (Fig 16.2). Oak (*Quercus* sp.). A whole timber with a mortice, a roughly hewn end and a bare-faced tenon. The timber had clearly defined axe/adze marks on several faces and one end, and axe/adze and saw marks on the other end. Part of the south-western elevation of chamber [1347]. Redundant joints indicate reuse. Although truncated the timber was similar in form to the baseplates used within [1207] and [1182]. Length 0.94m, width 100mm.

2545 Horizontal in-situ plank in [1347] (Fig 16.3). Oak (*Quercus* sp.). A tangentially faced timber with decay and splitting at one end which snapped during removal. The timber had two clearly defined, differently aligned sets of saw marks on one face, one on the reverse. Part of the south-eastern elevation of chamber [1347]. Length 1.45m, width 350mm, thickness 40mm.

2548 Horizontal in-situ beam in [1347] (Fig 16.1). Oak (*Quercus* sp.). A boxheart timber with a decayed lap joint on one end, mortices on one edge for two tenoned posts and abraded and well-defined axe/adze marks. The timber also had faint saw marks and two saw-made scribe marks on the underside, as excavated (not illustrated), in line with one mortice. Baseplate of the south-eastern elevation of chamber [1347], lap-jointed at the north-eastern end with transverse baseplate (2055). Length 1.6m, width 250mm, thickness 190mm.

2573 Horizontal in-situ plank (Fig 16.4). Oak (*Quercus* sp.). A tangentially converted board which split into four pieces following removal. The timber had chamfered edges, clearly defined saw marks on both faces and on one face a 170mm long saw scar. It also had woodworm damage. This timber was similar in shape to timber (2538), which measured 0.31-0.54m long and 0.27m wide but lacked chamfering. Part of the south-western elevation of chamber [1347]. Length 0.61m, width 350mm, thickness 35mm. The shape and position of the timber indicate reuse.

2575 Horizontal in-situ plank in [1347] (Fig 16.4). Oak (*Quercus* sp.). A tangentially faced? timber with broken ends, a centrally positioned L-shaped hole, nails and nail holes. Part of the south-western elevation of chamber [1347]. Length 1.75m, maximum width 180mm, thickness 20mm. The redundant hole, nails and nail holes indicate reuse.

3017 Vertical in-situ post in [1348] (Fig 16.5). Oak (*Quercus* sp.). A whole timber with a broken degraded upper end, a covering of bark, an axed/adzed lower end and a chamfered tenon with a length of 170mm. Housed within a mortice in horizontal beam (3008), the timber was part of failed wall [1348]. Length 1.41m, diameter 200mm.

3523 Vertical in-situ post in [1264] (Fig 16.6). Oak (*Quercus* sp.) A well-preserved irregular, knotty, sub-triangular stake with a compressed, split, degraded upper end and a lower end hewn to a point. The stake was driven deep into the clay in the north-west corner of trough [1264] providing support for the plank walling. Length 1.36m, maximum width 125mm.

17. Dendrochronological Samples of Structural Timbers

Ian Tyers

Following the recording and preliminary assessment of the structural timbers by UMAU, a total of 41 samples were sent to ARCUS for dendrochronological assessment and where appropriate dendrochronological analysis. Seven samples were collected and analysed in November 2002 as part of an initial assessment of the potential of the material. A further 34 samples were submitted for analysis in May 2004 as part of the post-excavation phase of the project. After preparation of the samples, standard dendrochronological methods (see for example English Heritage 1998) were applied to 27 of these samples (Table 17.1). The remaining fourteen samples were found to be either not oak, to contain too few rings or to contain a variety of anatomical anomalies which prevented their reliable analysis. The tree-ring sequences from 24 of the measurable samples were found to cross-match with either each other or with reference sequences and these yield a range of different types of results depending on the completeness of the section of the tree from which they are derived (Table 17.1).

Three types of dating result are usually obtained by dendrochronological analysis of oaks. Firstly, where a sample is complete to bark-edge a precise year of felling is obtained directly from the date of the last ring on the sample; where there is good survival of this outer ring it is sometimes possible to assign seasons to the felling period, the principal distinctions being between early spring, early summer and winter. Where a sample has some sapwood but is not complete to the bark-edge, a felling date range is obtained by applying the maximum and minimum numbers of rings of sapwood normally seen in oaks in the relevant region to the end dates of the samples. The range 10-46 has been used in this report. Finally, where no sapwood survives a *terminus post quem* (*tpq*) date is obtained by adding the minimum number of sapwood rings likely to have been lost to the date of the latest surviving ring. This type of date is less useful than the other two types since a great number of rings could have been lost either through ancient carpentry practice or poor site preservation, and thus the felling date of such material may be considerably later than the tree-ring date.

Results

A summary of the findings is presented in Table 17.1 and Appendix 17.1. None of the datable samples definitely include bark-edge. Somewhat frustratingly many of the unsuitable timbers and even some of those suitable for analysis but which were not found to be datable do retain bark-edge. The results can be combined to provide dates for the structures as follows. Brine tank [1182] dates to AD 130-7. Tank [1207] is dated post-AD 94, with its north-east chamber [1347] dated post-AD 99 and failed wall [1348] possibly dated to AD 114-42. Trough [1264] is dated post-AD 96. The absence of datable bark-edge effectively prevents the dendrochronology from determining whether all of this material is precisely contemporaneous or is the product of a number of construction events. Although most of the material analysed was not obviously reused it must be emphasised that if the dated material included unrecognised reused elements this will affect the validity of the interpretation of these results.

The tree-ring correlations (Appendix 17.1) and visual similarity of the samples indicate that the following timbers are derived from common parent trees: (2027) and (2029) are from one tree; (2042) (both samples) and (2049) are from another single tree; and (2550) and (3001) are from yet another single tree. The latter two come respectively from the plank facing of the failed wall [1348], at the south-west end of brine tank [1207], and chamber [1347] at the tank's north-east end. If the timbers are not reused, the result indicates that these are contemporary structures.

The samples included several which exhibited bands of narrow rings, with wider rings in between, indicating some sort of cyclical growth pattern. These patterns occur both before and during the period of the Roman occupation. The cause of these cycles may be natural or anthropogenic in origin.

The tree exploited for timbers (2027) and (2029) was damaged in AD 107 and this resulted in an unusual feature where there were two sapwood bands in the timber. One of these occurred under the area of the damaged bark when the damage prevented the cambial layer from converting the sapwood to heartwood. After a few years the cambium had recovered from this damage and there was a narrow band of heartwood and then the real sapwood. This damage may be natural or anthropogenic in origin. It is worth pointing out that, although this is a relatively rare occurrence, if sapwood estimates had been applied to the first band (for example if it had not recovered at all or the additional heartwood had not survived the usage and excavation of the timber) then dendrochronology would have created a 'felling date range' which was too early for this structure.

The material was sub-selected for tree-ring analysis on the basis of the tree-ring assessments undertaken during the post-excavation recording of the timbers. It was evident from the selected material that all the large roundwood posts contained too few rings for analysis. The success rate with the planking was, however, quite remarkably high. It should be noted that no material with sapwood has been dated from tank [1207], trough [1264] or chamber [1347], and only a single piece with sapwood has been dated from failed wall [1348].

The sequences from the 24 dated samples can be combined to create a site tree-ring sequence that is both well replicated and unusually long (198 BC - AD 130). This sequence is a useful addition to the regional tree-ring series since it replicates an earlier Nantwich sequence (Howard *et al*, pers comm) and a recent series from Middlewich (author unpublished). This new series matches these local series extremely strongly (Appendix 17.2) which indicates that they are all probably local in origin. The new series also matches well with data sets from Carlisle and London (Appendix 17.2). This strong internal matching indicates that the Nantwich and Middlewich data and the data from Droitwich represent a useful cluster of Roman 1st- and early 2nd-century tree-ring data from the English western-midlands counties.

Timber/Structure/Sample	Rings	Sap or bark	ARW (mm)	Date of sequence	Interpretation
0031/1207/106	-	-	-	Unsuitable for analysis	-
1511/1207/103	84	-	2.16	23 BC-AD 61	After AD 71
1535/1207/116	116	-	1.69	74 BC-AD 42	After AD 52
15571207102	-	-	-	Unsuitable for analysis	-
1571/1207/101[1]	110	-	2.32	168 BC-59 BC	After 49 BC
1582/1207/125[1?]	185	-	1.68	101 BC-AD 84	After AD 94
1583/1207/107	-	-	-	Unsuitable for analysis	-
1584/1207/105[1?]	243	-	1.89	198 BC-AD 45	After AD 55
1592/1207/104	-	-	-	Unsuitable for analysis	-
1594/1207/127	-	-	-	Unsuitable for analysis	-
2007/1182/109	93		1.71	143 BC-51 BC	After 41 BC
2027/1182/132	114	16	1.50	AD 14-AD 127	AD 127-57
2029/1182/100	132	21	1.54	2 BC-AD 130	AD 130-55
2037/1182/131	169	-	1.49	91 BC-AD 78	After AD 88
2042/1182/123	260	h/s+*33s*	1.23	168 BC-AD 92	AD 125-40
2042/1182/146	178+*22*	-	1.25	113 BC-AD 65	After AD 97
2044/1182/120	67	-	4.02	AD 6-AD 72	After AD 82
2048/1182/121[1?]	134	14	1.95	14 BC-AD 120	AD 120-52
2049/1182/111	246	27	1.38	128 BC-AD 118	AD 118-37
2515/1347/143	100	-	1.60	88 BC-AD 12	After AD 22
2548/1347/151	-	-	-	Unsuitable for analysis	-
2550/1347/137	191	-	1.40	102 BC-AD 89	After AD 99
2551/1347/124	81	-	1.20	10 BC-AD 71	After AD 81
2552/1347/142	98	-	1.27	60 BC-AD 37	After AD 47
2554/1347/156	-	-	-	Unsuitable for analysis	-
2564/1347/126	-	-	-	Unsuitable for analysis	-
2565/1347/129	-	-	-	Unsuitable for analysis	-
2571/1347/139	-	-	-	= *Betula*?	-
3001/1348/133	162	-	1.02	96 BC-AD 66	After AD 76
3012/1348/149[1?]	98	h/s+*18s*	2.56	2 BC-AD 96	AD 114-42
3016/1348/152	-	-	-	Unsuitable for analysis	-
3017/1348/150	-	-	-	Unsuitable for analysis	-
3019/1348/158	-	-	-	Unsuitable for analysis	-
3021/1348/136	157	-	1.14	131 BC-AD 26	After AD 36
3029/1348/128	114	-	1.46	127 BC-14 BC	After 4 BC
3501/1264/155	85	22+Bw	1.27	Undated	-
3513/1264/138	198+*6*	-	1.41	118 BC-AD 80	After AD 96
3527/1264/137	37	15+Bw	3.00	Undated	-
3531/1264/153	123	-	1.13	52 BC-AD 71	After AD 81
3533/1264/140	84	37+Bw	1.01	Undated	-
u/s/1182/157	-	-	-	Unsuitable for analysis	-

*Table 17.1: Dendrochronological sample details. ARW: average ring width (mm/year). +Bw: winter felled bark-edge. Unless otherwise noted the material is oak (*Quercus *spp.).* [1] *reused.*

18. Plant Macrofossils, Wood, Diatoms, Faunal Remains and Insect Analysis

Charlotte O'Brien, Lorne Elliott, Nigel Cameron, Louisa Gidney & Steve Davis

This chapter presents the results of analyses by Archaeological Services Durham University of the plant macrofossil, wood, diatom, faunal and beetle remains recovered during the excavation. Except in the case of the beetle remains, each of these followed an earlier assessment (Archaeological Services 2004).

Plant Macrofossil Analysis

Charlotte O'Brien

Methods

A total of 49 bulk samples were assessed for plant macrofossil remains. In the case of non-waterlogged contexts, 5 litre sub-samples were manually floated and sieved through a 500 μm mesh. The residues were retained and described. The flots were dried slowly and then scanned for botanical remains at up to x40 magnification. 1 litre samples from waterlogged contexts were wet-sieved though a 500 and 250 μm mesh and both fractions were scanned at up to x40 magnification. Following assessment, twelve contexts were recommended for further work, and all remaining material from these contexts was processed. Plant macrofossils were identified by comparison with modern reference material held in the Environmental Laboratory at Archaeological Services Durham University. Plant taxonomic nomenclature follows Stace (1997).

Results

All of the contexts examined were from Roman period features. Full plant macrofossil analysis was undertaken on five contexts from brine tank [1207], one context from brine tank [1182], three contexts from Building 1, one context from each of well [5109], pit [3359] to the south-west of the U-shaped enclosure and [2067], the ditch of the large enclosure. The assessment and analysis results are presented in Appendices 18.1-4.

Brine Tanks [1207] and [1182]

Uncharred seeds were abundant and well preserved in the brine tank fills as a result of the waterlogged conditions. Species from arable, heathland, ruderal, wetland and woodland habitats were represented. In brine tank [1207] the lower fills (1337), (1239) and (1304) contained a diverse range of charred plant remains. These were dominated by hulled barley grains, along with chaff of 6-row barley and spelt wheat. Oats, emmer wheat, rye and hazelnuts were also recorded at very low frequencies. Arable weed seeds included wild radish and scentless mayweed, and other weeds present were nipplewort, redshank, ribwort plantain, sedges, bedstraw, grasses, buttercups, dock, vetch and wild members of the cabbage family. By contrast, later fills (1303) and (1295) contained fewer charred plant remains which included a few weed seeds and a barley rachis internode. The only charred remains in brine tank [1182] were a hazelnut shell fragment and a seed of ribwort plantain and grass.

Building 1

Full analysis was undertaken on three contexts from Building 1, (2157), (2159) and (2160), which formed the fills of posthole [2156]/[2518]/[2177]. Charcoal and charred plant remains were abundant in these contexts. Cereal grains were dominated by wheat and indeterminate cereals, with only a few barley grains present. Many of the grains were fused together in clumps. Low amounts of chaff occurred relative to the number of grains and were all of spelt wheat. Very few weed seeds were recorded. Peas and beans were present in high numbers.

Other Features

Waterlogged conditions in basal fill (5108) of well [5109] and fill (2095) of ditch [2067] allowed the preservation of a diverse range of uncharred seeds. These were dominated by arable, ruderal and damp ground taxa, particularly fat-hen, lesser stitchwort, pale persicaria and rushes. A few charred wheat grains and fragments of spelt chaff occurred in (2095) in addition to a few charred seeds of sedges, grasses, knotweed, buttercups, dock, hazelnut and sloe. Waterlogged remains of spelt chaff were also present in (5108). The only charred plant remain in fill (3370) of pit [3359] was an indeterminate cereal grain and few uncharred seeds were preserved.

Discussion

Economic Taxa and Agricultural Practices

The charred plant remains indicate that the dominant cereals used at the site were spelt wheat and hulled 6-row barley. This is not an unusual assemblage, as by the Roman period spelt had taken over from emmer as the most commonly used wheat throughout Britain (Caseldine 1990; Huntley & Stallibrass 1995). In addition, hulled barley had superseded the naked form in the Bronze Age and the 2-row species was not common until the medieval period. Spelt wheat was also recorded from the salt production site at the Old Bowling Green, Droitwich (Colledge & Greig 1992), while large amounts of spelt, together with some rye and barley, occurred at the Roman villa at Bays Meadow, Droitwich (Straker 1979).

Peas and beans also formed a part of the diet at Kingsley Fields and the occurrence of charred hazelnuts and sloe suggests that gathered wild nuts and berries were an important additional food source. Uncharred fruitstones of hawthorn, blackberry, raspberry and elderberry indicate that these taxa were growing locally and are likely to have been exploited for their fruits.

A few oat grains were recorded but in the absence of oat chaff it is not possible to ascertain if this is the cultivated or wild species. Oats were not commonly cultivated during the Roman period and, considering the low numbers recorded, they were probably only present here as a weed. A rye grain and an emmer glume base were present in (1337) at the base of brine tank [1207] and are also likely to represent sporadic weed occurrences.

Culm nodes were present in the lower fills of brine tank [1207] and reflect debris from an early stage of the threshing process. In addition, a range of weed seeds and smaller chaff were recorded which together suggest that all stages of crop processing were undertaken at the site. It is also more likely from this that the crops were grown locally rather than brought to the site from elsewhere.

Charred and uncharred seeds of several taxa associated with agricultural ecosystems, including pastoral farming, occurred in moderate to high frequencies. These include ribwort plantain, buttercups and fat-hen (Behre 1986) and henbane may indicate the presence of manured soils (Stace 1997). These herbaceous plants, in combination with the faunal remains of cattle, sheep and pig, suggest that a mixed regime of arable and pastoral farming was practised at the site.

The Brine Tanks

The fills of both brine tanks were dominated by waterlogged plant remains including seeds, fruits, buds, wood and vegetative plant material. Thus, a proportion of the infilling of the tanks will have resulted from the natural accumulation of undecomposed plant material from vegetation growing in and around the tanks. In tank [1207] the lower fills (1337), (1239) and (1304) contained a range of charred plant remains including barley and wheat grains and numerous weed seeds. Charcoal was abundant and fragments of pottery and mussel shells occurred in (1239). These results suggest that some domestic waste may have been thrown into [1207] during this time. Charred heather and ferns may have derived from bedding or roofing material. The presence of chaff (including culm nodes) and arable weeds indicates that some crop-processing waste may also have been disposed of in the tank. However, as they occur in relatively low numbers, these remains may have accumulated in the tank fill inadvertently, due to the presence of waste lying around the site. By contrast, later fills (1303) and (1295) had very few charred plant remains, which may indicate that the tank was not continually used for waste disposal. In addition, botanical food waste was absent from tank [1182], which suggests that it was not used for domestic waste disposal. There is little evidence of religious or ceremonial deposition of plant remains into the tanks but it is possible that some of the cereals, hazelnuts and other wild fruits represent votive offerings.

None of the plant species which occurred in the tanks are obligate halophytes, although a number of the taxa recorded have tolerances of brackish or saline conditions. For example, members of the Chenopodiaceae (Goosefoot family) were abundant in most of the contexts, including oraches, fat-hen and red goosefoot. These all occur on waste and cultivated ground, but oraches and red goosefoot often also grow in brackish habitats inland or by the sea (Stace 1997).

The uppermost fills from the tanks, including (1229) and (1208) in [1207] and (1220) in [1182], contained few or no plant remains. This is likely to be due to periodic drying out of the upper sediments as the water levels fluctuated.

Building 1

Charred cereal remains were abundant in fills (2157), (2159) and (2160) from posthole [2156]/[2518]/[2177] in Building 1. These were dominated by grains of wheat and cerealia indeterminate. It is likely that all of the wheat was spelt, as the grains had a dorsal ridge typical of this species and the only chaff present was spikelet forks and glume bases of spelt. An oat grain and a few barley grains occurred but the low numbers suggest that these were growing as contaminants of the wheat crop rather than as a maslin crop. Peas and beans were present in relatively high numbers.

The predominance of cereal grains and legumes in the flots suggests that Building 1 may have been used as a granary. Charcoal was abundant in the flots, often with many large pieces present. In addition, numerous clumps of charred cereal grains occurred. This fusing together of grains can result when stored crops are burnt in situ. It suggests that the building may have been damaged or destroyed by fire resulting in the charred crops and burnt wood within these contexts. It is unlikely that the fire was a deliberate act to wipe out beetle or pest infestations, as few insect remains were recorded in the flots. The low amount of chaff recovered suggests that the cereals were fully processed before being stored.

Local Environment

The uncharred seeds are dominated by wetland plants, reflecting the waterlogged conditions in and around the tanks and in other areas of the site. These include sedges, spike-rush, members of the carrot family, rushes, gypsywort, crowfoot, blinks, meadowsweet, bur-marigolds and many other herbaceous plants which grow on damp ground. Buds and nut fragments of hazel suggest that woodland or scrub vegetation grew locally. This also comprised birch, sloe and elder with bramble and raspberry forming the understorey. A single seed from a yew tree occurred in fill (1158) within pit [1157], indicating the presence of this tree also.

Ruderal taxa including knotgrass, sow-thistles, stitchworts, henbane, hemp-nettle and redshank were also abundant. These taxa would have grown in areas of open or waste ground and indicate that the local landscape was relatively open rather than being extensively wooded. Many other taxa occurred which occupy a range of habitats including grasses, nettles, docks, thistles, common knapweed, fairy flax and buttercups. Twigs and seeds of a number of heathers occurred which suggest that areas of heath also grew close to the site.

Conclusions

Spelt wheat, hulled 6-row barley, legumes and wild foods were the main plant food sources used. The cereals were grown and processed locally and a mixed regime of arable and pastoral farming was practised. Brine tank [1207] may have been used for the disposal of domestic waste at an early stage of its infilling. Uncharred seeds were abundant and well preserved in many of the contexts, and show that the local vegetation at the time of the infilling of the tanks was largely open with wetland and ruderal habitats and some nearby woodland. Charred remains from a posthole suggest that Building 1 was a granary and had been damaged or destroyed by fire.

Wood Analysis

Lorne Elliott

Introduction

This study involved the analysis of a large amount of small roundwood from the excavation. The wood derived from wicker-lined pits, the lining of which contained vertical staves interwoven with withies. The aim of this study was to gain insight into woodland management during the Roman occupation of the site, in particular to look for evidence of coppicing or age/size selection, or the use of specific tree species.

Methods

Four of the five pits excavated contained more than 100 pieces of wood. These were sampled in the field with a section taken from each piece of wood for laboratory analysis. Samples of 100 roundwood pieces were measured and identified for contexts (1073), (1100) and (1222) from pits [1072], [1099] and [1053]. Samples of 238 roundwood pieces were measured for (1221) from [1054], with 119 of these randomly selected for identification. Due to the level of decomposition, each piece of wood was frozen prior to identification. Thin transverse, radial and tangential sections were taken for age (ring counting) and species determination. The diameter of the samples was also recorded. Reference material and guides (Hather 2000) were used for identification purposes.

Results

Nine species were present in the 419 samples identified: *Corylus* (hazel), *Quercus* (oak), *Alnus* (alder), *Betula* (birch), Salicaceae (willow/poplar), *Prunus avium* (wild cherry), *Prunus spinosa* (blackthorn), *Fraxinus* (ash) and Maloideae (which includes hawthorn, apple, pear and rowan). Ages and diameters of the various species were used to produce scatter plots and histograms for each context.

	(1073)	(1100)	(1221)	(1222)
Alnus (Alder)	14	-	1	1
Betula (Birch)	2	-	-	2
Corylus (Hazel)	67	90	106	57
Fraxinus (Ash)	1	-	5	9
Maloideae (Hawthorn, Apple, Pear, Rowan)	4	-	-	4
Prunus avium (Wild Cherry)	-	-	1	-
Prunus spinosa (Blackthorn)	-	-	1	-
Quercus (Oak)	11	10	2	14
Salicaceae (Willow/Poplar)	1	-	3	13
Total no of samples	**100**	**100**	**119**	**100**

Table 18.1: Number of wood species by context.

The number of the wood species for all four contexts is shown in Table 18.1 and their relative frequency in Appendix 18.5. All four contexts show that hazel is the dominant species used, particularly so in context (1100) where 90% of the samples were hazel. In the remaining contexts, hazel ranges from 57% to 88% of the samples. Oak is the next most abundant species overall, with the remaining species in various small amounts. Context (1100) contains only two species of wood (hazel and oak) in comparison to the other contexts each of which has seven species. Ages and diameters of the total wood are shown as histograms in Appendices 18.6 and 18.7. Scatter plots of total wood age and diameter are presented in Appendices 18.8-11. Age and diameter of hazel are shown as histograms in Appendices 18.12 and 18.13 and scatter plots in Appendix 18.14. Ages and diameters of the other main species (those with at least five samples present) are presented in Appendices 18.15-18. Contexts (1073) and (1222) show a wide age distribution with samples ranging from 2 to 25 years. Context (1221) has two clusters, a large concentration of samples in the 5-7 years age range and a smaller concentration at 11-14 years. Context (1100) has a narrow age distribution with a particularly large cluster at 7 years. All four contexts show a relatively broad distribution of diameter sizes.

Contexts (1073), (1221) and (1222) are predominantly between 10mm and 20mm (averaging 16.5mm, 16mm and 18.1mm respectively), while context (1100) tends to have slightly larger diameters concentrated between 19mm and 27mm (average 20.4mm).

Hazel

The patterns of age and diameter distribution for hazel mirror those of total wood due to its abundance. Linear regression lines are shown for hazel in the four contexts in order to establish the relationship between age and diameter (Appendix 18.14). While diameters and age are linked to a certain degree, a strong concentration in age, and a wider range of diameters would be expected if coppicing were taking place. The r^2 values range from 0.0392 to 0.2057 showing a very weak relationship between age and diameter, with a particularly low value for (1100).

In contexts (1100) and (1221) hazel has a concentration around 5-7 years with a wide size distribution, particularly so in (1100) where 41 samples are 7-year old hazel but have a size range of 15-27mm. Appendix 18.12 shows that in (1100) and (1221) 55% and 50% respectively are either 5-year old or 7-year old hazel. There also appears to be a smaller concentration of hazel of around 11-14 years in (1221). In contrast, hazel has a wide age distribution (2-25 years) in (1073) and (1222). These two contexts also have a relatively wide size distribution. The average diameter of hazel is larger in (1100) than the other three contexts.

Oak

In context (1100) oak is clustered between 4 and 8 years and has a wide diameter distribution of 14-27mm. In (1073) oak is either between 5 and 9 years or 17 or 20 years old with sizes ranging from 10mm to 26mm. In (1222) oak has a wide age range from 8 to 17 years but a narrow size range of 17-26mm diameter. Only two oak samples occurred in (1221). Oak appears to have a wider diameter than other species at the same age.

Alder

The only context with significant numbers of alder was (1073). The samples are mainly between 8 and 11 years, but have a wide size range of 9-27mm.

Ash

Ash occurred in relatively small numbers. In (1221) the majority fall between 17 and 18 years, while in (1222) they are between 7 and 9 years with two samples at 14 years. Diameters range from 13mm to 21mm. Present-day coppicing of ash has a cycle of 15-30 years.

Willow/Poplar

In context (1222) thirteen pieces of willow/poplar occurred. These were all 3 or 4 years in age with a relatively narrow size of *c* 11-14mm.

Others

Maloideae, birch, blackthorn and wild cherry were not present in sufficient number to observe patterns of age or size distributions.

Discussion

There is evidence that woodland management has been undertaken in Britain since prehistoric times (Rackham 2003), and the Roman interest in baths, bricks, hypocausts, corn-driers and salt production would have provided a strong incentive to maintain fuel supplies. For example, Cleere (1974) calculated that work on the Roman iron industry of the Weald in Sussex would have consumed about 75,000 acres or 9% of the Weald wildwood, with the need for some kind of stable coppice system being highly likely. Despite the likelihood of woodland management, few studies have been undertaken which have provided direct evidence for formal woodland management, in particular coppicing, in the Roman period. The large number of roundwood pieces from Nantwich has allowed a detailed study of woodland management to be undertaken.

In (1100) the concentration of samples of age 7 years, the relatively broad range of diameters and low species diversity (with only hazel and oak present) suggest that some form of coppicing, particularly of hazel, was undertaken rather than the wood samples having been selected randomly. The slightly wider average diameter of the samples from (1100) may also be the result of coppicing, as evidence suggests that there is often extremely rapid growth during the first two years following coppicing (Huntley 1991). A broadly similar pattern of results is shown for (1221), where the majority of hazel pieces are between 5 and 7 years, but a secondary cluster of hazel aged 11-14 years is also apparent. These age ranges appear to be similar to coppice cycles in eastern England during the Middle Ages (Rackham 2003). They were usually short cycles of 4-8 years, with 20% of the wood having a longer cycle of 12-20 years and nothing over 20 years. Present-day coppicing of hazel tends to range from 6 to 8 years to produce hoops and crate rods and from 12 to 14 years for hop poles. This is due to hazel losing suppleness after 10 years. The same may be evident in contexts (1100) and (1221), where the majority of the samples were harvested between 5 and 7 years, presumably for their suppleness, while a smaller amount was allowed to grow on to 11-14 years which would have provided more strength. Ash and oak, although low in numbers, also have narrow age ranges in these contexts. Herbs and shrubs which have been associated with managed woodlands, including bristle club-rush (*Isolepis setacea),* common hemp-nettle *(Galeopsis tetrahit)* and wild cherry (*Prunus avium*) (Rackham 2003), have been recorded from the site either as seeds or wood.

By contrast, the relatively broad age ranges and diversity

of species which occurred in contexts (1073) and (1222) suggest that the wood samples for these pit linings were selected on a more random basis and there is little evidence for coppicing. It is possible that these pits were dug and lined at an earlier stage than (1100) and (1221) before a change to a more managed wood supply was introduced. This is supported by the fact that pit [1072] associated with wicker lining (1073) appears to be cut by the edge of pit [1054] containing lining (1221) and therefore may be earlier.

Pollen diagrams from the region give an insight into the natural vegetation of the area prior to Roman occupation. Sites at White Moss, south Cheshire (Lageard *et al* 1999), and Crose Mere, Shropshire (Beales 1980), provide evidence for a predominantly wooded landscape, with trees and shrubs making up around 70-80% of the pollen for the period under consideration. Hazel is the most abundant species followed by alder, oak, birch and ash. Small amounts of blackthorn, wild cherry and willow were also present. This suggests the presence of two types of environment near the site. Oak and ash prefer drier soil conditions, whereas alder and willow would have grown on damp ground such as along the banks of the River Weaver. Hazel grows on a variety of soils but favours moist, base-rich conditions (Preston *et al* 2002).

As hazel is the most abundant species in both of the above pollen diagrams during the Roman period, this large naturally occurring supply may explain why it was the preferred material for the contexts mentioned above. It should be noted, however, that woodland management can result in over-representation of certain species in pollen diagrams (Rackham 2003). For example, in a mixed coppice wood, hazel flowers more rapidly after coppicing compared to species such as ash and alder, and oak is often allowed to grow as a timber tree, therefore producing a full pollen output. The preference for hazel is also likely to be in part due to its flexibility and thus the ease with which the withies could have been woven between the staves.

Woodland management during the Roman period has also been investigated at Castle Street, Carlisle, adjacent to the fort (Huntley 1987). Analysis of roundwood pieces showed that alder was the main species used with a size appropriate to the job being chosen. There was little evidence for formal management of the nearby woodlands, as a wide range of ages for any given size of material was recorded (*ibid*). At nearby Annetwell Street, within the fort at Carlisle, alder was again chosen during the early period of occupation with a wide range in ages. By the late 1st to early 2nd centuries more oak and hazel were used and alder, although still dominant, had a narrower age range. The author suggested that, once there had been local clearance of alder from the river valley, the Romans moved further afield to collect wood. With time, the local alder woodland regenerated and produced even-aged stands to be used again (Huntley 1989).

The mid 1st- to early 2nd-century salt production site at the Old Bowling Green, Droitwich, produced a large number of pieces of wood, mainly from boards lining four brine tanks or from debris within the tanks (Crone 1992). Oak was dominant, with lesser amounts of hazel, ash, maloideae, wild cherry, birch and willow. Studies of Iron Age and Romano-British wood from Prestatyn in North Wales found that half of the wood was oak, with the other half split between alder, blackthorn, ash and poplar/willow (Morgan 1989). At both of these sites significant quantities of roundwood were absent.

Conclusions

The main conclusion from this analysis is that hazel was the preferred material for all of the wicker-lined pits due to its suitability for the specific job and its apparent abundance at the time as illustrated by pollen evidence. There appears to be a change in wood-use activity from the random selection of diverse species to the predominant use of hazel of a specific age range. This may be the result of formal woodland management in the form of coppicing, although natural hazel coppicing cannot be ruled out.

Diatom Analysis

Nigel Cameron

Introduction

Following the assessment of four contexts from the excavation (Archaeological Services 2004), two were identified as having potential for further percentage diatom analysis and diatom-based salinity reconstruction: context (1283), a mid fill in brine tank [1182], and context (1335), a basal fill in brine tank [1207].

Methods

The slides analysed were those previously prepared for the diatom assessment. Diatom preparation followed standard techniques: the oxidation of organic sediment, removal of carbonate and some clay, concentration of diatom valves and washing with distilled water. Two coverslips of different concentrations of the cleaned solution were prepared from each sample and fixed in a mountant of suitable refractive index for diatoms (Naphrax). Slides were counted under phase contrast illumination at a magnification x1000.

A number of diatom floras and taxonomic publications were consulted to assist with diatom identification, including Hartley *et al* (1996), Hustedt (1930-1966), and Krammer and Lange-Bertalot (1986-1991). Diatom species' salinity preferences were classified using the halobian groups of Hustedt (1953; 1957). These are summarised below:

1. Mesohalobian: 0.2-30 g l^{-1} optimum in brackish water;
2. Oligohalobian - Halophilous: optimum in slightly brackish water;
3. Oligohalobian - Indifferent: optimum in freshwater but tolerant of slightly brackish water;

4. Halophobous: exclusively freshwater;
5. Unknown: taxa of unknown salinity preference.

The diatoms are grouped by their halobian classifications in Appendix 18.19, with the more saline groups on the left of the diagram. The principal source used for diatom ecological data was Denys (1992). Salinity reconstruction was carried out using the combined EDDI diatom-salinity training set (Battarbee *et al* 2000) and associated ERNIE software (Juggins unpublished). Harmonisation of diatom taxa from the archaeological samples with those in the EDDI combined salinity training set was also carried out using ERNIE software. The diatom diagram (Appendix 18.19) was plotted using Tilia, Tiliagraph, and Tgview programs. Conversion of data between formats was carried out with the Wintran program (Juggins unpublished).

Results and Discussion

A total of approximately 270 valves were counted in each sample. The samples contain moderately diverse diatom assemblages. Planktonic diatoms (eg the halophilous species *Cyclotella meneghiniana,* and a single occurrence of the mesohalobous taxon *Cyclotella striata*) are rare, while non-planktonic diatoms from shallow water habitats dominate the assemblages. A number of the diatoms present are aerophilous and are tolerant of desiccation. These aerophilous species are most common in context (1283) and include *Hantzschia amphioxys, Pinnularia borealis, Pinnularia major, Pinnularia subcapitata.* The aerophiles *Navicula mutica* and *Navicula cincta* are also halophilous diatoms. All the aerophilous diatoms may be derived from damp soil but it is also possible that they grew on the damp sides of the tanks or represent post-use deposits. Part of the sediment may have filled in the tanks after they had fallen out of use as brine tanks and therefore the overall assemblage may not represent as strong a saline environment as one might predict. The presence of undifferentiated chrysophte cysts found both in the assessment and analysis supports the idea that some of the deposits formed when there were low water levels in the tanks.

Inferring the values of past aquatic salinity, or other chemical or environmental variables, from fossil diatom remains preserved in water-lain sediments involves two stages (eg Cameron *et al* 1999). First the relationship between diatom abundances and the contemporary environmental variable of interest (here salinity or conductivity) is modelled using a modern training or calibration set of surface sediment diatom samples and associated water chemistry. These relationships are used to derive a transfer function. The second stage is to use the transfer function to infer past salinities from the fossil diatom assemblages. In this case the most comprehensive existing diatom training set, the combined EDDI salinity training set, was used. This training set is derived from a wide geographical range (Europe and North Africa) of approximately 300 water bodies (Battarbee *et al* 2000) and has been used to reconstruct the past conductivity of the water stored in the tanks.

A total of 35 taxa were present in context (1283) and 66 taxa in context (1335). After harmonisation a total of 35 taxa were present in context (1283) and 61 taxa in context (1335). Of these 28 and 56 taxa respectively were present in the EDDI combined salinity training set, representing 93% and 94% of the total number of diatoms present in contexts (1283) and (1335). There was therefore good representation of the fossil diatom assemblages in the modern diatom training set. The results of the diatom-based salinity reconstructions, from a simple weighted average method, and the associated standard error estimates are shown below (Table 18.2). Results from inverse and classical regression methods are very similar. These results indicate that the diatom assemblage in both tanks is derived predominantly from slightly brackish water with relatively high concentrations of dissolved salts.

Sample	Salinity g l^{-1}	Standard error
1283	3.37	0.47
1335	3.18	0.47

Table 18.2: Diatom-based salinity reconstructions for the brine tank samples.

Although there are significant numbers of freshwater taxa (oligohalobous indifferent group), halophiles such as *Nitzschia umbonata, Amphora veneta, Navicula veneta* in context (1283) and *Navicula cincta, Navicula veneta, Gomphonema olivaceum* and *Melosira varians* in context (1335) reflect the relatively high levels of dissolved salts. Despite some differences in the species composition of the two samples analysed, the salinity values reconstructed from diatoms for both tanks are very similar at just over 3 g/litre.

Conclusions

Diatom analysis and grouping of the species using a halobian classification showed that there was a mixture of diatom groups present. Although freshwater species were common, there were also relatively high numbers of brackish water and halophilous species in both samples. Some aerophilous diatoms were also common.

Quantitative salinity reconstruction carried out using a transfer function showed that salinity levels were relatively high (low end of brackish range at *c* 3 g l^{-1}). Although no direct comparison is made with surface or ground water at the site, diatom analysis supports the idea that water present in both tanks during their use or post-use was of relatively high conductivity. Furthermore the salinity of water in both tanks was similar, although the diatom species composition of each shows some differences.

Faunal Remains Analysis

Louisa Gidney

Introduction

The faunal remains under discussion were recovered from the fills of the two brine tanks, [1182] and [1207]. The waterlogged preservation of the bones within these features was generally excellent. However, some of the uppermost fills do show some surface degradation of the bones typical of repeated episodes of wetting and drying. Preservation of animal bones outside the special depositional environment of the brine tanks was extremely poor.

Results

Animal bones were not present in every context within the two brine tanks. In tank [1207] bones were concentrated in the lower and upper fills, with a large number of contexts in the middle devoid of animal bone. Quantities of animal bone were only present in a few particular contexts, including clay lining (1241)/(1290), lower fill (1239) and upper fill (1240). In tank [1182] animal bones were more evenly distributed throughout the fills though with one outstandingly rich context, lower fill (1285).

At the assessment stage, counts were made of the identifiable fragments present using the criteria normally adopted of the presence of a recognisable 'zone' or distinguishing non-repeatable, characteristic on an individual skeletal element (Archaeological Services 2004). During further analysis, it became apparent that the unusual contents of these tanks were not well represented by this methodology. Groups of articulating bones were certainly deposited in several contexts but recovery and recognition of these groupings has been made difficult by the particular circumstances of excavation (on which see Chapter 3). Table 18.3 indicates some of the differences obtained by counting recognisable groupings, such as a dog skeleton or cattle head and neck, as one unit of deposition compared to counting all the individual bones within the articulating group. This partly explains the difference between the quantity of bone recovered, filling fourteen boxes, and the comparatively small numbers of identified fragments in Table 18.3. This is still unsatisfactory as many of the sheep/goat fragments are parts of maxillae and mandibles which were probably deposited together but could not be matched up with confidence. It does, however, highlight that this is an atypical assemblage which does not neatly fit into a standard recording procedure. Given the size of the brine tanks, even fourteen boxes of bones represent a very small proportion of the total fills. It would appear that the deposition of animal bones was not an ongoing disposal of general refuse, but short episodes of dumping of particular waste. Table 18.3a-d further demonstrates that the fills of the two brine tanks are not homogenous either within or between the features, despite the fact that they are broadly chronologically contemporary.

	Groups of articulating bones counted as one specimen					
	(1241) & (1290)	(1239)	(1240)	(1225)	(1208)	u/s
Cattle	2	51	32	4	1	12
Cattle size	-	9	2	1	-	-
Sheep/goat	-	19	10	1	-	4
Sheep	-	4	1	-	-	1
Sheep size	-	2	1	1	-	-
Pig	-	8	6	-	-	-
Horse	-	6	3	-	-	1
Dog	-	3	1	1	-	-
Red deer	-	5	3	-	-	-
Domestic fowl	-	2	-	-	-	-
Raven	-	-	-	-	-	1
Mussel	-	3	-	-	-	-
Total	**2**	**112**	**59**	**8**	**1**	**19**

	Individual fragment counts					
	(1241) & (1290)	(1239)	(1240)	(1225)	(1208)	u/s
Cattle	10	52	37	9	1	12
Cattle size	3	24	3	2	-	-
Sheep/goat	-	19	10	1	-	4
Sheep	-	4	1	-	-	1
Sheep size	-	6	1	1	-	-
Pig	-	13	6	-	-	-
Horse	-	9	3	-	-	1
Dog	-	*c* 45	58	6	-	-
Red deer	-	6	3	-	-	-
Domestic fowl	-	2	-	-	-	-
Raven	-	-	-	-	-	1
Mussel	-	3	-	-	-	-
Total	**13**	**138**	**122**	**19**	**1**	**19**

Table 18.3a & b: Species fragment counts for the fills in tank [1207].

	Groups of articulating bones counted as one specimen							
	(1286)	**(1284) & (1285)**	**(1283)**	**(1280) - (1282)**	**(1279)**	**(1213) & (1215)**	**Other**	**u/s**
Cattle	3	47	13	8	5	14	7	2
Cattle size	-	2	1	1	-	1	3	-
Sheep/goat	2	25	17	10	4	5	2	1
Sheep	-	4	3	-	-	-	-	-
Sheep size	-	-	2	-	-	-	-	-
Pig	-	3	1	-	1	1	-	-
Horse	-	8	1	1	-	6	-	-
Dog	-	1	-	1	-	1	-	1
Red deer	-	1	-	-	1	-	-	-
Domestic fowl	-	1	-	-	-	-	-	-
Raven	1	3	-	1	-	-	-	1
Mussel	-	2	1		-	-	-	-
Total	**6**	**97**	**39**	**22**	**11**	**28**	**12**	**5**

	Individual fragment counts							
	(1286)	**(1284) & (1285)**	**(1283)**	**(1280) - (1282)**	**(1279)**	**(1213) & (1215)**	**Other**	**u/s**
Cattle	3	47	63	8	17	15	7	2
Cattle size	-	3	1	1	16	1	3	-
Sheep/goat	2	26	21	11	4	5	2	1
Sheep	-	4	3	-	-	-	-	-
Sheep size	-	-	2	-	-	-	-	-
Pig	-	3	1	-	1	1	-	-
Horse	-	8	8	1	-	8	-	-
Dog	-	1	-	1	-	11	-	1
Red deer	-	1	-	-	1	-	-	-
Domestic fowl	-	2	-	-	-	-	-	-
Raven	1	6	-	1	-	-	-	7
Mussel	-	2	1	-	-	-	-	
Total	**6**	**103**	**103**	**23**	**39**	**41**	**12**	**11**

Tables 18.3c & d: Species fragment counts for the fills in tank [1182].

The excellent condition of the bones has clearly preserved butchery marks. There is no doubt that in some contexts the cattle bones have been severely modified by the typical, heavy-handed, use of the cleaver by the Roman butcher. Context (1239) in [1207] was noted for an unusual concentration of shattered long bone shaft fragments. These appeared to derive from the rapid deposition of refuse from short-lived incidents of activity, so the opportunity was taken to try to reassemble individual bones. It proved possible to reassemble one almost complete tibia from eight fragments. Other bones had fewer joining pieces. Some of the unstratified finds almost certainly derive from this context and smaller fragments may have been missed under the difficult circumstances of excavation. This exercise demonstrates that whole bones were being shattered and all the fragments were being deposited together in a fresh state.

Less usual is the find from context (1240) in [1207] of a cattle mandible, charred and holed midway along the horizontal ramus, and of a chopped and charred fragment from context (1239). This technique is recorded from other Roman sites, for example Lincoln (Dobney *et al* 1996), although the reason for this practice remains obscure.

Slightly different is a complete mandible from context (1239), which appears to have been laid on a fire, rather than charred midshaft to produce a weak spot to break the bone. This may be related to the charred mandible from a complete head discussed below.

A more ubiquitous practice, both here and on a plethora of other Roman sites, is the cattle scapula variously trimmed and with a suspension hook hole through the blade. This is usually interpreted as a brined and smoked beef product, singularly appropriate for a site with a commercial interest in brine.

Cattle

It is clear from Table 18.3 that cattle bones far outnumber those of any other species in both the tanks. The greater size and bulk of cattle bones compared to those of the smaller species means that the greatest volume of bone deposited and recovered comprises cattle bones. This is not a product of biased recovery given the extremely good survival of bones from juvenile animals and smaller species.

As noted above, raw fragment counts were not thought to be the most appropriate guide to quantification. Table 18.4 presents the counts of individual, non-repeatable, anatomical features for each element. This gives a guide to the number of individual bones deposited. Even this method fails to give adequate representation as it does not count horn cores detached from the frontal, which are therefore particularly under-represented in tank [1182], or ribs and vertebrae. The skull and mandible are well represented in all the major contexts in both tanks. Scapula and humerus are also well represented, being absent only from the minor fills of tank [1207]. Radius, ulna and metacarpal are present only in context (1239) in [1207] but are in all the major fills of [1182]. In particular, mid fill (1279) in [1182] produced what had probably been a complete forelimb, now represented by humerus, radius and ulna, proximal metacarpal - the distal end had been chopped off. The absence of the lower forelimb elements from context (1240) in [1207] is particularly striking given the presence of both scapula and humerus. It suggests that the forelimb was divided at the junction of the humerus and radius/ulna. In context (1239) there was at least one association of ribs and vertebrae with forelimb bones, suggesting deposition not just of a forelimb but probably also of the associated thorax since the vertebrae were not split or butchered.

Table 18.4: Cattle anatomical zones (after Rackham 1987).

		[1207]			[1182]		
		(1239)	(1240)	Other	(1284)/(1285)	(1283)	Other
Skull	Zone						
	1	1	2	2	-	1	1
	2	4	4	2	-	2	3
	3	1	3	2	5	1	2
	4	5	2	2	-	2	3
	5	6	2	3	7	4	2
	6	2	2	2	-	-	-
	7	-	2	2	-	-	-
	8	5	2	4	1	3	4
	9	2	2	4	1	3	2
	0	4	-	2	6	4	2
Jaw	1	1	6	3	-	-	3
	2	3	5	3	-	-	3
	3	1	6	3	-	-	3
	4	1	4	3	2	-	1
	5	1	4	3	2	-	2
	6	-	7	3	2	3	2
	7	-	4	3	3	2	4
	8	-	6	3	2	2	3
Scapula	1	2	3	-	3	2	1
	2	5	3	-	7	2	2
	3	5	3	-	9	3	3
	4	2	2	-	6	1	2
	5	5	3	-	8	2	3
	6	-	1	-	2	-	-
	7	-	1	-	1	1	-
Humerus	1	-	1	-	1	-	-
	2	-	-	-	-	-	-
	3	-	-	-	-	-	-
	4	-	-	-	-	-	-
	5	2	1	-	-	-	2
	6	1	2	-	2	2	3
	7	5	1	-	1	2	2
	8	-	1	-	1	1	2
	9	2	2	-	2	2	3
Radius	1	3	-	-	1	1	4
	2	1	-	-	1	1	3
	3	1	-	-	2	1	2
	4	1	-	-	-	1	1
	5	-	-	-	-	1	1
	6	-	-	-	1	1	2
Ulna	1	-	-	-	-	-	-
	2	3	-	-	-	1	2
	3	3	-	-	1	1	2
	4	-	-	-	-	-	-
Metacarpal	1	1	-	-	2	-	2
	2	1	-	-	3	-	2
	3	-	-	-	2	-	1
	4	-	-	-	2	-	1
	5	-	-	-	2	-	2

Innominate	1	-	-	-	-	-	-
	2	1	-	2	1	2	-
	3	-	-	2	1	2	-
	4	-	1	2	1	-	-
	5	-	1	2	1	-	1
	6	-	-	2	-	-	1
	7	-	1	2	2	1	1
	8	-	-	-	-	-	-
	9	-	1	2	1	1	1
	0	-	-	2	1	-	1
Femur	1	-	-	1	-	1	-
	2	1	-	-	-	-	-
	3	1	1	1	-	1	1
	4	3	2	1	2	-	2
	5	1	-	1	-	-	-
	6	-	1	1	-	-	-
	7	1	1	1	-	-	-
	8	-	-	1	-	-	-
Tibia	1	-	-	2	-	-	-
	2	1	-	2	-	-	-
	3	-	-	2	-	-	-
	4	1	3	1	-	-	-
	5	2	3	-	1	1	3
	6	5	2	-	1	1	3
	7	1	3	-	1	1	-
Calcaneum	1	-	-	-	1	-	-
	2	-	-	-	-	-	1
	3	-	-	-	-	-	1
Metatarsal	1	1	1	-	-	1	1
	2	1	-	-	1	1	1
	3	1	-	-	-	2	2
	4	1	-	-	-	2	1
	5	1	-	-	1	1	1
Phalanx 1	1	3	1	-	3	-	-
	2	3	1	-	3	-	-
Astragalus	1	-	-	-	-	-	1
Centro-quartal	-	-	-	-	1	-	-

Compared to the scapula and humerus, representation of all parts of the hindlimb is much lower in both tanks. Parts of the pelvis are infrequent in tank [1207], with the exception of context (1225). This was one of the upper fills with poorer preservation, probably caused by repeated wetting and drying. It was possible to reassemble an articulating group of pelves, sacrum, femur and both proximal tibiae; the lower leg had been detached. One femur is missing and may have decayed beyond recognition. The sex was not clear from the pubes but was more masculine than feminine. This one find contributes all the zones for the hindlimb for the other contexts in tank [1207]. Tibia finds from [1207] otherwise approach the numbers seen for the scapula and humerus, but the remainder of the hindlimb is nearly as scarce as the lower forelimb. Tank [1182] has more even representation of hindlimb in the major fills but fewer legs appear to be present than the forelimb. There is an interesting anomaly among the metapodials with no metacarpals present in mid fill (1283) in [1182], only metatarsals, whereas the other fills have more metacarpals than metatarsals. Overall, the zone data indicate deposition of heads and upper forelimbs from more animals than are represented by the remaining limb bones. This would appear to be deliberate human selection rather than the product of taphonomic factors.

The cattle heads are an unusual collection, showing a range of different treatments. Tank [1207] had unequivocal foundation deposits in contexts (1241) and (1290), the clay lining of the tank, of cattle heads with attached cervical vertebrae. The skull from (1241) is now fragmentary but was probably intact when deposited. Both mandibles are present, together with all of the cervical vertebrae. The horn cores are missing and it is not clear whether these had been deliberately detached. The animal was a mature adult with all permanent teeth present and at an advanced wear stage (Mandibular Wear Stage 45, Appendix 18.20). There are no clear chop marks so it is not obvious how the animal was killed or the head and neck detached from the body. The presence of the complete head and neck may indicate a method of sacrifice analogous to that depicted on the Mithraic tauroctony.

The skull from context (1290) is in better condition. It had clearly not been poleaxed, but the horn cores had been removed with part of the frontal. The atlas vertebra is present but only the anterior articulation of the axis, which lies within the ring of the atlas. The head had clearly been chopped off through the junction of the atlas and axis, a very different method of slaughter to the companion animal in (1241). This could suggest different or complementary rites. One articulating mandible was recovered. The ventral aspect of the horizontal ramus is charred and has been broken in two. The teeth suggest a similar age stage (MWS 43) to the animal in context (1241). The charring of the mandible perhaps suggests that the entire head was placed on an altar fire as part of the ceremony prior to deposition. These cattle skulls are without parallel at either Droitwich (Locker 1992) or Upwich (Meddens 1997).

A further notable skull was recovered from context (1239) in tank [1207]. This find has the whole cranial area surviving, while the facial area has been deliberately removed. The right frontal has five, mostly square-sectioned, projectile point holes (Plate 18.1). None of these have penetrated through the frontal sinus into the brain cavity. A larger square-sectioned hole is also present in the left temporal. This has penetrated the brain cavity and could have proved fatal, if the animal was alive and not merely a head mounted as a target. Finds of such skulls are not numerous but other examples have been found on Roman forts, such as Vindolanda and Wallsend. The horn cores of this specimen appear masculine and have been smoothed off on the caudal aspect to expose the internal sinuses. This modification enables the back of the skull to fit smoothly against a flat surface. A similar trimming has been observed on the Vindolanda examples (Dr D Bennett, pers comm). Such skulls may have been mounted for target practice or displayed as trophies of successful achievement. This find is evidence to support the hypothesis of a Roman military presence at Nantwich. The possibility that the military contributed to the infill of the brine pits has important connotations for the interpretation of this whole assemblage.

Another largely complete skull may also come from context (1239). This skull has been clearly poleaxed and the horn cores have been chopped off together with the posterior section of the frontals. Repeated chop marks on the left frontal and orbit suggest a failed attempt to detach the facial area. The occipital condyles show chop marks, indicating decapitation at the junction of the occipital condyles with the atlas vertebra. This was an adult animal with all the permanent teeth present and in wear. Also present in this context is a calf skull. The prominent horn buds suggest this was several months old at death.

Context (1240) produced a skull with mandibles, atlas and axis vertebrae from an immature animal. The horn cores appear masculine. The third molars were yet to erupt, giving a MWS of 19. There was no evidence for cause of death or method of detachment from the body.

Brine tank [1182] does not appear to have had a cattle foundation deposit comparable to tank [1207]. Instead there may have been equivalent deposits within the fills. Context (1279), one of the upper fills, had rather mixed preservation, probably caused by wetting and drying. Only during the re-examination of this context did it become apparent that parts of probably three cattle heads and articulating necks are represented. The most complete head was at the same developmental stage as that from context (1240) in [1207], with a MWS of 19. Two older animals are indicated, one at MWS 41, with advanced wear on all permanent teeth, and another incomplete tooth row from an adult animal with less wear on the teeth. It was not possible to accurately correlate the neck vertebrae with the heads but there are clearly one larger and two smaller animals. The neck of the larger animal included four of the thoracic vertebrae. One of the smaller necks appears to have been severed through the cervical vertebrae and one other neck chopped through

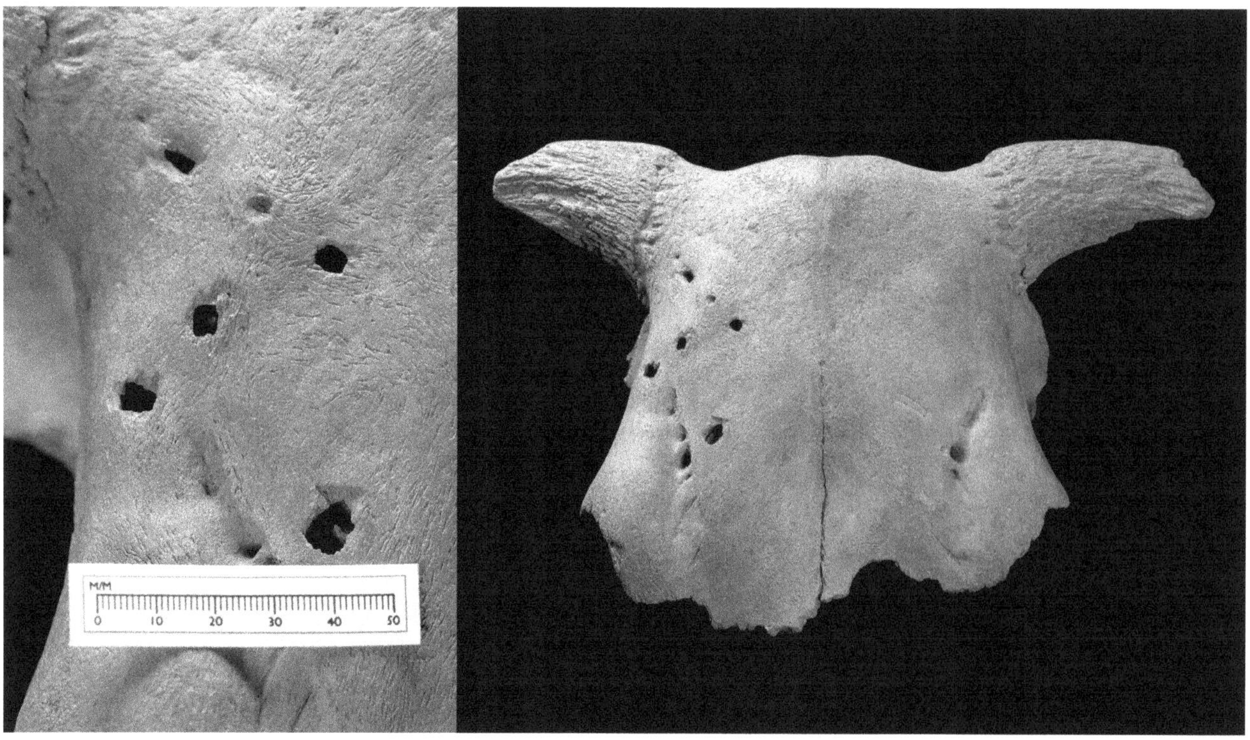

Plate 18.1: Skull from context (1239), brine tank [1207] showing square-sectioned, projectile point holes in the right frontal.

the thoracic vertebrae. Despite the presence of thoracic vertebrae and a forelimb, there were no ribs recovered. This, together with the butchery marks, strongly suggests that detached body parts rather than complete torsos were deposited. Context (1283) also appears to have had at least two, probably three, heads and necks and a forelimb. One head appears to have been female with 'celtic shorthorn' horn type. The facial area had been chopped. The animal had not been poleaxed. A juvenile head had been deliberately chopped up but most parts have been recovered, so all the pieces were deposited together. A further head appears to have been chopped up and dumped but fewer pieces have been retrieved. Three atlas vertebrae are present and part of one axis, suggesting that these heads were detached through the atlas/axis joint. The forelimb is indicated by articulating humerus, radius and ulna and possibly an associated scapula. Also of note is the newborn calf found in this context. Context (1285) produced abundant fragments of skulls. The horn cores indicate parts of at least three female and two male or castrate heads. The only complete head appears to have been that of an infant calf.

Some horn cores appear to have been detached from the skull, indicating the removal of the horn sheath for craft working. These finds contrast with the entire frontals or skulls deposited with the horn cores attached. This difference in treatment could reflect the contrast between sacred/sacrificial and mundane waste or between the parts dedicated to the divine and the priest's portion, depending on the nature of the sacrifice.

In tank [1207] loose horn cores were found in context (1239) (three), (1240) (two and a joining pair) and (1208) (one). The types include adult 'celtic shorthorn' female and juvenile, probably male. The majority of loose horn cores were recovered from context (1285) in tank [1182] but include one pair of severed horn cores with attached frontals which reunite, suggesting rapid disposal of refuse after dismemberment. Appendix 18.21 plots the basal diameters of the horn cores from both brine tanks. There appear to be three main groupings but the principal distinction is between those less than 60mm and those more than 65mm on the Y axis. There was visual evidence during cataloguing for large and robust, probably male, horn cores and distinctive small, probably female, 'celtic shorthorn' horn cores which may equate with the top and bottom of the range. The mid range certainly includes examples of immature, porous horn cores thought to derive from either bulls or castrates. It may also include cows with a different horn shape to the 'celtic shorthorn'.

The superb preservational conditions in these two tanks has led to the recovery of bones from very young calves from contexts (1239), (1249) and (1225) in tank [1207] and context (1285) in tank [1182]. The most complete find was in context (1283) in tank [1182], which produced most of the major limb bones, the mandibles and some ribs and vertebrae of an infant calf. This was almost certainly deposited as a complete articulated body and the missing elements not recovered. Measurements of the limb bones at first suggested this was a foetal calf of about 240 days gestation, using the data from Simmental and Fribourg cattle (Prummel 1989). However, these bones were visually similar to those from a full-term still-born Dexter bull calf in the author's reference collection and the measurements of this archaeological find and the Dexter are practically identical (Table 18.5).

	Dexter	[1182] (1283)	[1207] (1225)
Humerus	97	97	-
Radius	101	101	-
Metacarpal	96	-	98
Femur	119	120	-
Tibia	131	129	-
Metatarsal	114	115	-

Table 18.5: Comparisons of calf bone sizes (length of diaphysis in mm).

It seems much more probable that both these archaeological finds derive from neonatal calves from cattle more comparable in size to the Dexter rather than the Simmental. It should be noted that the specimen from context (1283), though the same length as the Dexter, is more gracile in build. Plate 18.2 compares the humerus from this archaeological calf with the modern Dexter and a still-born continental cross bred calf. The Romano-British specimens from the brine tanks also fall within the range for the medieval calf bones from Lindisfarne, thought to derive from vellum production (Scott 2000), further indicating that the modern data are not useful comparanda for archaeological specimens.

The high representation of cattle heads has provided a reasonable sample of complete mandibular tooth rows from which Mandible Wear Stages (MWS) may be calculated (Grant 1982). It can be seen from Appendix 18.20 that the deposition of cattle heads in both tanks

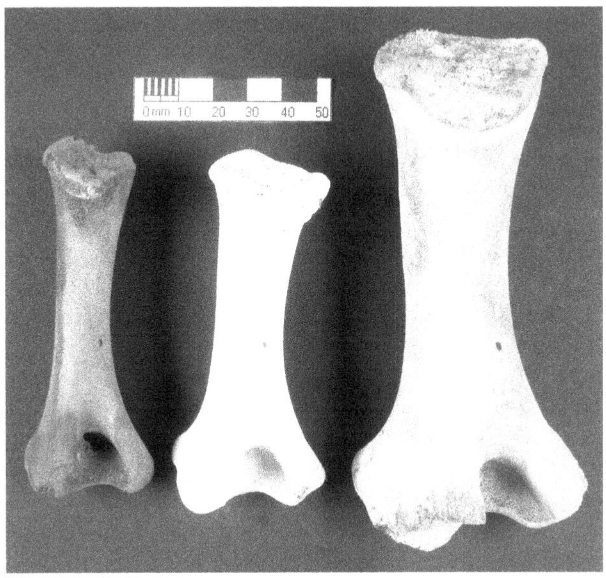

Plate 18.2: Comparison of the calf humerus from context (1283), brine tank [1182] (left), a modern Dexter (centre) and a still-born continental cross bred calf (right).

			[1207]			[1182]		
			U	S/W	H/W	U	S/W	H/W
Cattle	5-6m	M1	2	1	14	-	4	15
	15-18m	M2	2	4	9	-	7	12
	24-30m	P2	5	3	4	1	1	2
	18-30m	P3	5	-	7	5	1	2
	24-30m	M3	4	1	10	4	3	11
	28-36m	P4	5	-	9	6	5	4
Sheep/goat	3-5m	M1	-	2	16	-	5	19
	9-12m	M2	2	9	7	3	14	8
	21-24m	P2	1	2	-	3	3	-
	21-24m	P3	2	2	1	8	3	-
	18-24m	M3	13	4	2	9	6	7
	21-24m	P4	7	3	5	11	4	6
Pig	4-6m	M1	4	-	-	-	-	-
	7-13m	M2	4	1	-	-	-	-
	12-16m	P2	4	-	-	-	-	-
	12-16m	P3	5	-	-	-	-	-
	12-16m	P4	5	-	-	-	-	-
	17-22m	M3	4	-	-	-	-	-

Table 18.6: Tooth eruption and wear. U = unerupted/deciduous; S/W = slight wear; H/W = heavy wear. Approximate ages of eruption after Silver (1969).

		[1207]			[1182]		
		Fused	Just fused	Unfused	Fused	Just fused	Unfused
By 18 months	Scapula tubercle	5	-	-	5	-	1
	Acetabulum symphysis	3	-	-	1	-	-
	Proximal radius	3	-	-	6	-	-
	Distal humerus	6	-	-	5	-	-
	Proximal phalanx 2	-	-	-	-	-	-
	Proximal phalanx 1	4	-	-	3	-	-
By 2-3 years	Distal tibia	8	-	1	5	-	-
	Distal metacarpal	2	-	-	-	-	-
	Distal metatarsal	1	-	-	4	-	-
By 3.5-4 years	Proximal calcaneum	-	-	-	1	-	-
	Proximal femur	-	1	-	1	-	-
	Distal radius	1		-	2	1	1
	Proximal humerus	1	1	-	1	-	1
	Proximal tibia	1	2	1	-	-	-
	Distal femur	3	1	-	-	-	-
	Proximal & distal ulna	-	-	-	-	-	-
By >5 years	Anterior vertebral	1	1	4	4	5	12
	Posterior vertebral	-	-	7	2	3	20

Table 18.7: Cattle epiphyses in approximate order of fusion. Ages of fusion after Silver (1969).

derived from animals at comparable stages of tooth eruption and wear. It is of note that no more than two jaws, or one animal, are represented at any one MWS in each tank. The calves noted previously are indicated at MWS 0 and 7. The heads at MWS 19 have also been discussed. Young adults are indicated at MWS 34-36 and mature or aged animals at MWS 41-45. Comparison with the author's reference collection of Dexter cows suggests a calendar age of 3 years for MWS 35 and 11-17 years for MWS 45-8 (Gidney 1999). The overall pattern in Appendix 18.20 suggests the sourcing of at least four pairs of animals of matching developmental stage, with at least the heads from one of each pair deposited in each tank. Heads from a further two young adult animals were deposited in [1182] and a calf in [1207].

The data in Table 18.6, which includes loose and maxillary teeth as well as mandibular rows, show the same general pattern with similar numbers of teeth at comparable wear stages present in both tanks. The epiphysial data in Table 18.7 do not include the infant calf from context (1283), discussed above. Fusion data are less abundant than tooth data due to the unusual composition of this assemblage. Unfused bones in the earlier fusing categories are scarce. Some culling of animals younger than roughly four years is indicated but the vertebrae suggest that the majority of cattle did not survive to an advanced age and full skeletal maturity. These scant data complement the information from the teeth.

Sheep/Goat

Though the generic term sheep/goat is used, there was no positive evidence for the presence of goat, whereas all the distinguishable skull fragments derive from sheep, and this appears to have been the economically important animal. Table 18.3 shows that remains of sheep were far less abundant than those of cattle, although sheep bones were the most common finds after cattle. Table 18.8 suggests there has been even more selection of the body parts deposited than was the case with cattle. In both tank [1207] and tank [1182], the majority of elements found derive from the head, particularly the mandible. Detached heads far outnumber post-cranial elements. Tank [1207] attracted the deposition of fewer body parts than tank [1182]. The absence of scapula from [1207] may be a mere chance of recovery since there appears to be only one or two examples of the other bones of the forelimb. The absence of pelvis and femur from [1182] is more noticeable since there are several examples of all the other major limb bones, suggesting that this part of the leg was disposed of elsewhere.

Like the cattle heads, the sheep heads show a variety of treatments. Both polled and horned heads are represented. Several of the surviving primitive sheep breeds also have both horned and polled variants within the phenotype. The polled variety is less common with only two examples from context (1239) in tank [1207]. The majority of heads have had the horn cores chopped off but there are examples with the horn cores left intact. The majority of skulls from tank [1182] had been split in half along the sagittal plane. Some examples of this practice were also present in tank [1207] but there were also relatively intact heads and an example of a failed attempt to detach the facial area across the frontals. The motivation for these different treatments is unclear.

The horn cores had been detached with vigour leading to damage both on the base of the detached horn cores and the stump on the frontal, which has generated a smaller than anticipated sample for metrical analysis. Appendix 18.22 shows examples of larger and smaller basal diameters which equate with the general robust, masculine and gracile, feminine categories observed during recording.

The predominance of heads has generated a good sample for ageing from tooth eruption and wear. The MWS data, in Appendix 18.23, show very clear age-related patterning in the slaughter population, with the finds from both tanks falling into the same broad categories. There is a small group of first year lambs at MWS 6-11; then a large group, of nearly half the jaws found, at a tight grouping of MWS 18-22 from second year animals; and finally a spread of animals in their third year and older at MWS 29-40. This is broadly comparable with the cull pattern seen at Leicester (Gidney 1999) and suggests a general sheep-keeping strategy for this midland region based on prime meat production. The inclusion of the data from the maxillary teeth in Table 18.6 amplifies this pattern and confirms that the maxillae and mandibles derive from the same heads, although none could be confidently articulated and several jaws had clearly been detached from the head at the mandibular hinge. Deciduous teeth and permanent teeth in the early stages of wear are abundant, whereas permanent teeth in the later stages of wear are scarce.

It can be seen from Table 18.9 that there are virtually no epiphysial data for ageing from tank [1207] and little from tank [1182]. The post-cranial elements are again mostly from younger animals with unfused epiphyses, with few examples of fused bones among those fusing later in life.

		[1207]	[1182]
		All contexts	All contexts
Skull	Zone		
	1	2	4
	2	2	5
	3	1	4
	4	2	5
	5	5	3
	8	1	7
	9	5	11
	0	6	8
Jaw	1	6	11
	2	10	19
	3	7	16
	4	7	10
	5	8	14
	6	6	13
	7	9	15
	8	8	13
Scapula	2	-	3
	3	-	3
	4	-	1
	5	-	4
	6	-	2
	7	-	2
Humerus	5	-	2
	6	1	2
	7	-	1
	8	-	1
	9	1	2
Radius	1	-	4
	2	-	4
	3	1	4
	6	1	4
Ulna	1	1	1
	2	2	2
	3	2	3
Metacarpal	1	1	3
	2	1	5
	5	-	3
Innominate	4	1	-
	5	1	-
	7	1	-
	9	1	-
Femur	1	2	-
	2	1	-
	3	2	-
	4	1	-
	5	1	-
	6	1	-
	7	1	-
Tibia	4	1	7
	5	-	2
	6	-	1
	7	1	7
Metatarsal	1	-	1
	2	-	1
	3	-	2
	4	-	2
	5	1	3

Table 18.8: Sheep/goat anatomical zones (after Rackham 1987).

Pig

Pig remains were infrequent finds. Two concentrations of pig bones were recovered from tank [1207], from contexts (1239) and (1240) respectively. Context (1239) appears to have contained a largely complete body of a juvenile animal, of which only parts have been recovered, and two disassociated bones. Context (1240) may originally have contained half a body of a second juvenile animal at the same developmental stage as that in context (1239) but fewer bones have been recovered. The zone data in Table 18.10 show that head and shoulder blade survive most intact and that three examples of these elements are present. Ribs and vertebrae also survive, which suggest deposition in articulation. By contrast, tank [1182] produced no such concentrations, only finds of one or two bones from four contexts. The teeth in Table 18.6 principally derive from the four tooth rows of the head in context (1239) and the maxilla in context (1240). The deciduous teeth had little wear and none of the permanent teeth had erupted. One older female animal is indicated by a jaw, also from context (1239), from an older animal, with molar 2 and canine present. The epiphysial fusion data in Table 18.11 show that none of the major limb bones were fused and only one vertebrae in the process of fusing indicates the presence of an older animal. While the ageing data clearly show that the two pigs deposited in tank [1207] were only a few months old, they were certainly of a size to have been weaned and were beyond the stage of sucking pig.

		[1207	1182]
		All contexts	All contexts
Skull	Zone		
	4	2	1
	5	2	-
	8	2	1
	9	3	-
	0	1	-
Jaw	1	3	-
	2	3	-
	3	3	-
	4	1	-
	5	2	-
	6	3	-
	7	3	-
	8	3	-
Scapula	2	3	-
	3	3	1
	4	2	1
	5	3	-
	6	2	-
	7	3	-
Humerus	5	2	-
	6	2	-
	9	2	-
Ulna	2	1	1
	3	1	1
Femur	3	2	-
	4	2	-
	5	-	1
	6	-	1
	7	-	1
Tibia	4	2	-
	7	1	-
Calcaneum	2	1	-
	3	1	-

Table 18.10: Pig anatomical zones (after Rackham 1987).

		[1207]			[1182]		
		Fused	Just fused	Unfused	Fused	Just fused	Unfused
By 1 year	Distal humerus	-	-	-	1	-	1
	Proximal radius	-	-	-	2	2	-
	Scapula tubercle	-	-	-	-	-	3
	Acetabulum symphysis	1	-	-	-	-	-
By 1-2 years	Distal tibia	-	-	-	1	-	6
	Distal metacarpal	-	-	-	1	-	1
	Distal metatarsal	-	-	-	1	1	1
By 2.5-3.5 years	Proximal femur	2	-	-	1	-	-
	Distal femur	1	-	-	-	-	-
	Proximal tibia	-	-	-	-	-	2
	Distal radius	-	-	-	-	-	3
	Proximal humerus	-	-	-	-	-	1
	Proximal & distal ulna	-	1	-	-	1	-
By >5 years	Anterior vertebral	-	-	1	-	-	3
	Posterior vertebral	-	-	1	1	-	2

Table 18.9: Sheep/goat epiphyses in approximate order of fusion. Ages of fusion after Silver (1969).

		[1207]			[1182]		
		Fused	Just fused	Unfused	Fused	Just fused	Unfused
By 1 year	Scapula tubercle	-	-	2	-	-	-
	Distal humerus	-	-	2	-	-	-
By 1-2 years	Distal tibia	-	-	1	-	-	-
	Proximal calcaneum	-	-	1	-	-	-
By 2.5-3.5 years	Proximal & distal ulna	-	-	2	-	-	-
	Proximal tibia	-	-	1	-	-	-
	Proximal humerus	-	-	2	-	-	-
	Proximal & distal femur	-	-	6	-	-	1
By >5 years	Anterior vertebral	-	1	1	-	-	-
	Posterior vertebral	-	-	2	-	-	-

Table 18.11: Pig epiphyses in approximate order of fusion. Ages of fusion after Silver (1969).

Dog

Dog was present in both tanks. The conditions of excavation were not conducive to the recognition of complete skeletons in situ but the recovery of groups of associated bones indicates that whole bodies had been deposited. The following discussion concentrates on the ageing and height data from the major limb bones but all parts of the body are represented. The skulls and mandibles were generally too incomplete for measurement. Withers heights have been calculated using the factors of Harcourt (1974) and age from epiphysial fusion after Schmid (1972).

Parts of three corpses were recovered from context (1239) in tank [1207]. One complete radius, with part of the associated ulna, indicates an animal over 18 months old and about 54cm high. A juvenile between 6 and 12 months old is represented by several elements with all epiphysial ends unfused, including humerus and femur. An adult short, bandy-legged, animal is represented by bones from both forelimbs, indicating a height of 30-32cm. Parts of possibly a further three dogs are also represented in context (1240), though one tibia from an animal over 18 months old and about 54cm high could be part of the same animal represented in context (1239) by a complete radius. A puppy, less than 6 months old, is indicated by a group of unfused bones including humerus, radius and femur. Parts of a young male animal include the os penis, limb bones with one fusion line still clear and vertebrae unfused or just fusing. This dog would have been about 18 months old and 46-47cm high.

Tank [1182] did not produce such concentrations of dog bones, with a few individual elements in several contexts but only one partial skeleton, from context (1213). All the epiphyses were fused, indicating an age in excess of 18 months. Two femora and one tibia indicate a height of 44-45cm.

The ubiquity of dogs is better demonstrated by finds of gnawed bones than actual dog skeletons. Gnawed bones were recorded in contexts (1239), (1240) and (1225) in tank [1207], and contexts (1279), (1283), (1285), (1315), (1213), (1215) and (1277) in tank [1182]. Overall numbers of gnawed bones were similar for both tanks (Table 18.12).

	[1207]	[1182]
Cattle	19	19
Sheep/goat	7	4
Pig	1	-
Horse	6	7
Red Deer	1	-

Table 18.12: Bones gnawed by dogs.

The proportion of gnawed horse bones is particularly high and is discussed further below. Dog gnawing otherwise suggests some scavenging by, or deliberate feeding to, dogs before final deposition of the bones in the tanks but this would not appear to have had a major impact on the composition of the assemblage.

Horse

Horse bones appear to have been deposited as discrete concentrations in tank [1207], with most finds recovered from context (1239) and a few more in context (1240). Radius and ulna, with fused epiphyses, from three different adult animals were recovered from context (1239). The olecranon process of the ulna had been chewed by dogs on all three specimens. The radii could be measured and indicate estimated withers heights of 1.31m, 1.36m and 1.48m. These equate to pony-sized animals of about 14-15 hands. The presence of one humerus and one scapula suggests that entire forelimbs may have been deposited and that the other elements were not recovered. In support of this contention is the presence of a more complete hindlimb, comprising articulating tibia, metatarsal, calcaneum and astragalus. This limb too had gnawing marks on the proximal tibia and calcaneum. The metatarsal indicates a withers height of 1.42m. Two complete metacarpals and one third phalanx were found in context (1240). The withers heights of 1.19m and 1.34m estimated from the metacarpals suggest that these do not derive from the same forelimbs as the radii in context (1239).

Rather more remains of horse were recovered from tank [1182], concentrated in contexts (1281), (1283), (1284) and particularly (1285). One third phalanx, or hoof bone, from context (1281) may belong with the otherwise largely complete hindleg in context (1283) with femur, tibia, all tarsals and metatarsal present. Femur, tibia and calcaneum had been gnawed. A withers height of 1.40m is indicated by the metatarsal. The only horse skull present was found in context (1284). This was probably intact when found but has broken up since it has dried out. The presence of the canine teeth indicates that this was a male animal and the advanced wear on the teeth suggests that it was elderly at death. The concentration of horse elements in context (1285) includes a mandible which probably belongs with this skull. Other finds are two partial pelves, which have been gnawed so that it is not clear whether they are a matching pair, and at least two forelimbs represented by a scapula, radius and ulna, metacarpal and third phalanx. The scapula and ulna have been gnawed. The withers heights of 1.26m and 1.17m, estimated from the radius and metacarpal respectively, suggest that more than one animal is represented. One further grouping of horse bones was found in contexts (1213) and (1215). A matching pair of pelves and a femur were found in (1213), while a tibia, metatarsal and articulating first phalanx were present in (1215), together with articulating first and second phalanges from another, larger, animal. The tibia gives an estimated height of 1.23m while that for the metatarsal is 1.16m. Taken with the disparate size of the phalanges, this suggests that parts of two legs from separate animals are represented.

The horse bones have not suffered the heavy-handed butchery seen on the cattle bones, suggesting that these were not part of the human food chain. Articulated legs appear to have been chosen for dumping in these tanks but the ubiquitous presence of gnawing marks shows that dogs had access to these carcase parts prior to deposition. Individual limbs from a range of animals would appear to have been selected for disposal in the tanks. The range of heights estimated, spanning 1.16-1.48m or roughly 12-15 hands, suggests that radii from four individuals, metacarpals and metatarsals from three individuals each and a tibia from a further individual are represented. In contrast there is only one head and there are no ribs and vertebrae.

The explanation for this pattern is obscure. Whole carrion carcases were certainly not dumped. The gnawing marks might suggest either disposal of refuse from feeding dogs or scavenging of exposed limbs at, for example, a knacker's or tanner's yard. This does beg the question of where were the rest of the bodies disposed of, and why bring, mostly, only one leg per animal to deposit in the brine tanks. Bones from sacrificial animals were not normally permitted to be scavenged by dogs. However, some sort of offering principally of horse legs could explain this unusual collection.

Red Deer

The majority of red deer elements were recovered from contexts (1239) and (1240) in tank [1207]. Both contexts contained the cranial part of a female skull with the facial area clearly chopped off. That from context (1239) had chop marks on the occipital condyles indicative of decapitation. A matching pair of maxillae from (1239) may belong to the same head, as may a chopped fragment of mandible. Also from (1239) were a shed antler from a stag of five points, with no sign of use, and a complete metatarsal from an adult animal. Further elements from context (1240) were a scapula with gnaw marks and a charred fragment of tibia. Tank [1182] only produced two finds of red deer. These are a poorly preserved shed antler from context (1279) and a complete metatarsal from an immature animal.

No parallels for the two hind heads are currently known but it would seem that they are part of the pattern of deposition of, principally, cattle and sheep heads within these features. Only the tibia and scapula from context (1240) suggest any use of venison. The heads and metapodials suggest either primary butchery or hides, while the shed antlers are independent of the carcase.

Water Vole

The two water vole bones from context (1285) in tank [1182] are articulating elements and possibly suggest the presence of a complete body which was otherwise missed in the less than ideal conditions of excavation. This could be a natural mortality of an animal which became trapped in the tank.

Birds

Despite the superb preservational conditions, bird bones are rare finds. This situation is paralleled at Droitwich and Upwich. Only two domestic fowl bones were found, both from context (1239) in tank [1207]. The most frequent bird bones were of raven and the concentration of raven bones in tank [1182] currently appears to be unique. The stratified bones were concentrated in context (1285), with parts of one juvenile and two adult birds represented, with other raven bones present in contexts (1281) and (1286). These, together with further unstratified finds within this tank, all appear to derive from the same bodies. Tank [1207] produced only one unstratified raven bone. Raven is one of the most commonly found wild bird species on Roman sites, often occurring as more or less complete skeletons (Parker 1988). Various suggestions of symbolic or religious association (*ibid*) failed to consider the strong connection between the raven and the cult of Mithras. The raven was both the messenger of the sun god and the title of the first grade of initiate. Mithraism was a popular cult with the military, so these raven bones may possibly indicate a military and religious link with the infill of tank [1182] in particular.

Mussel Shells

Mussel shells are the only evidence for any procurement of seafood. Hand-recovered finds were sparse with three shells from context (1239) in tank [1207] and single shells from contexts (1283), (1284) and (1285) in tank [1182]. However, the soil sample from context (1239) contained abundant fragments of mussel shell. This suggests that mussels were originally more numerous than the hand-recovered finds suggest. It seems probable that the mussels were traded up the River Severn from the Severn estuary. No oyster shells were recovered from the brine tanks, whereas small numbers of oyster shells were recovered from the Friar Street (Hurst 1992) and Upwich (Hurst 1997) sites at Droitwich. The finds from the brine tanks are marine mussel shells, in contrast to the freshwater mussel shell from post-Roman contexts at Friar Street and Upwich.

Discussion

The unusual composition of this faunal assemblage would appear to be an example of the type of special deposit which Fulford (2001) suggested would invite comment, in his discussion of structured deposition in wells, shafts and pits. Some recurrent features of such structured deposits are articulated animal remains, frequently dogs, and certain parts of the skeleton only, such as cattle skulls. Both of these criteria are present in this collection. Fulford (*ibid*) noted that a preliminary survey indicated that such deposits were pervasive in the countryside. The exceptional preservation of the finds from the brine tanks has produced an assemblage to complement the urban and military finds considered by Fulford and provides a type site for the region.

Cattle heads can be associated with 'opening' deposits, as exemplified by the two heads from the clay lining of tank [1207]. Fulford (2001) similarly notes two ox skulls from the base of a shaft in Verulamium but offers no further details on deposits of cattle skulls or heads, so it is of particular interest that the cattle heads from these brine tanks fall into several groupings.

Firstly, complete heads with attached neck, with either no evidence of severance from the body or separation between cervical or thoracic vertebrae. These examples were not poleaxed, which may suggest that either the decapitation through the neck was fatal or death was caused by a stab wound, as depicted on the Mithraic tauroctony. Such complete bull's heads appear, for example, in Mithraic iconography as an attribute of Cautes, variously interpreted as the personification of day or the spring equinox (Ulansey 1989). The *taurobolium* is a further ceremony about which little is known of the practicalities. While not yet attested by epigraphy in Britain (Duthoy 1969), the rite of the *taurobolium* appears to have been added to the cult of Cybele by the emperor Antoninus Pius and became a public ceremony for the well-being of the emperor. Some offerings appear to have been made by the military. While the exact nature of the 2nd-century ceremony is unknown, it appears to have differed substantially from the later rite, where the blood of the sacrificed animal soaked the votive in a subterranean chamber. Bulls' heads are also part of the iconography of the 2nd- to 3rd-century *taurobolium* inscriptions (Rutter 1968).

Secondly, heads again not poleaxed but decapitated through the atlas and axis joint, although it is not clear if this was the method of slaughter. The facial area may be removed. The latter could be considered as bucrania, depicted as adornments of temples in sacrificial scenes in Roman art. A recent example of a Dexter cow's head, decapitated in this manner and exposed for several years, demonstrates that curated heads can remain in complete articulation, despite exposure to the elements (Archaeological Services 2007, Plate 3.5). It is therefore possible that some of the cattle heads from the brine tanks represent disposal of exhibited relics from a previous ceremony, when succeeded by the current offering.

Only one example of a poleaxed head with the horn cores removed was found, whereas several heads had been extensively dismembered and the horns chopped off. The latter certainly appear to have been deposited fresh, as there are at least two sets of conjoining frontals.

While there is no clear interpretation of any of these finds, it is apparent that these are not the mundane disposal of domestic or commercial refuse. The deposition of at least two, apparently complete, forelimbs may also have sacrificial connotations as a bull's shoulder appears, for example, in Mithraic iconography (Ulansey 1989). Fulford (2001) also notes the deposition of an articulated ox forelimb at Neatham.

In contrast to the cattle heads of the 'opening' deposit, a 'closing' deposit in tank [1207] may be represented by the articulating group of cattle pelves, femur and proximal tibia in upper fill (1225). These are the bones usually associated with the deity's portion of the sacrifice in the classical tradition (Detienne & Vernant 1989).

While some tentative links have been suggested with sacrificial deposition and Mithraic iconography, it is important to note that these brine tanks produced no evidence for the deposition of table refuse from the ritual meals, or feasts, associated with such cults (Martens 2004; Ulbert *et al* 2004). The remains of sacrificial offerings and the remains of communal dining may have been viewed as requiring separate repositories.

Three sites associated with salt production in Droitwich have produced assemblages of animal bone which have been published (Locker 1992; Meddens 1997). Several of the contexts are directly comparable to Nantwich, being the fills of abandoned brine tanks. The assemblage from Upwich (Meddens 1997) shows several similarities to Nantwich. The animal bones were hand-picked under difficult conditions of excavation; the waterlogged bones were well preserved and darkly stained; only a small number of identifiable bones were present in the Roman deposits; the sheep were mostly culled by their third year; other species present include dog, horse and red deer. The differences are that sheep and pig are proportionally more abundant and there is no suggestion that anything other than domestic refuse was deposited.

Several phases of Romano-British activity were identified at the Old Bowling Green site in Droitwich (Locker 1992). Similarities to Nantwich include a high proportion of cattle heads, split and horn cores chopped off; chopping up of the major limb bones; trimming of scapulae, splitting of sheep heads and removal of horn cores; comparable proportions of cattle and sheep, in a mid 3rd- to 4th-century phase; deposition of articulated dogs in disuse fills of barrels used for the storage and settling of brine; and overall stature and cull patterns. The main differences are the greater quantity of animal bones deposited in the brine barrels and the lack of discussion of the noted increase in cattle skulls together with complete dogs in these features.

The Friar Street site in Droitwich (Locker 1992) produced very small groups from the Roman phases, with most finds recovered from a well shaft. Two skeletons of fox from this feature are believed to indicate a ritual purpose.

Overall the assemblages from Droitwich indicate that these finds from Nantwich derive from the same types of livestock kept under comparable regimes of husbandry to supply meat. The differences lie in components such as the presence of infant calves, the preponderance of sheep heads and the non-food species horse and raven. Apart from the dogs at the Old Bowling Green site and the foxes at Friar Street, the Droitwich faunal remains have all given the appearance to experienced analysts of being normal domestic household refuse. It has become apparent that the Nantwich finds include a major component of waste which derives from activities other than food preparation and consumption and which may reflect structured deposition of a votive, religious or ritual nature.

Insect Remains Analysis

Steve Davis

Methods

Coleopteran remains were isolated following a standard paraffin flotation technique as described by Kenward *et al* (1980). Samples for insects were washed over a 300 µm sieve and the residues mixed well with paraffin. Following addition of cold water, decanting and washing with hot water and detergent, the flots were examined for insect remains. These were stored in denatured ethanol and identified with reference to the collections housed in the Royal Albert Museum, Exeter, and standard entomological literature. Taxonomy follows that of Lucht (1987). Valuable ecological information was derived through the computer package BugdCEP (Buckland & Buckland 2006). The distribution of coleopteran taxa within their ecological groupings is presented in Appendix 18.24 and the raw counts are listed in Appendix 18.26.

Results

Brine Tank [1207]

The four samples from this feature exhibit clear faunal differences between lower contexts (1239) and (1337) and upper contexts (1295) and (1303).

The sample from context (1337) comprised a rich assemblage, dominated by generalist decomposer taxa characteristic of rotting plant matter (eg *Anotylus rugosus* F., *Platystethus arenarius* Fourc., *Cercyon* spp.). The assemblage also incorporated significant proportions of dung taxa, synanthropic taxa (ie those with an intimate association with human activity) and members of the Lathridiidae (mould beetles). Lathridiidae were most abundant in the basal context, as were synanthropes. The synanthropic taxa present included low frequencies of two of the most frequently encountered taxa of infested grain, the 'saw-toothed grain beetle' *Oryzaephilus surinamensis* (L.) and the 'grain weevil' *Sitophilus granarius* (L.). Both of these taxa have a long history of association with man, and are recorded from a wide range of storage contexts of Roman age and later in Britain (eg Roman-period deposits at Rougier Street, York (Hall & Kenward 1990)). In addition, several taxa belonging to the 'house fauna' group (*sensu* Hall & Kenward 1990) were recorded. These included two species of ptinid ('spider beetles'), *Ptinus fur* (L.) and *Tipnus unicolor* (Pill.) as well as the blind colydid *Aglenus brunneus* (Gyll.), characteristic of manure heaps, tan-pits, corn-bins refuse and cellars (Fowler 1889). The 'house fauna' is considered typical of 'domestic, storage and stable buildings' (Kenward & Hall 1997) and, in conjunction with the numerous Lathridiidae, is suggestive of a relatively foul, mouldering substrate deposited within or proximal to a building. The low abundance of the grain taxa in addition to the presence of dung taxa suggests that these beetles have arrived in the assemblage post-consumption. By adopting an experimental approach, Osborne (1983) determined that such grain taxa suffered minimal damage other than disarticulation following human consumption and digestion. In addition, the woodworm *Anobium punctatum* (represented by three individuals within this context) as well as being a common structural pest of building timbers (eg Palm 1959) is also a ubiquitous find in such cesspit assemblages (Osborne 1983).

Context (1239) yielded a very similar assemblage to (1337) with the exception that mould taxa were less well represented. More aquatic taxa were present, suggesting the presence of some stagnant water, in addition to taxa characteristic of waterside environments (eg *Bembidion* spp. *Gymnetron beccabungae* L.). Once again, the primary indication is of an accumulation of decomposing vegetable matter with some mould, dung and synanthropic taxa and relatively few indications of surrounding environment. Synanthropes are dominated by *S. granarius* but also include the 'powder post beetle', *Lyctus linearis* (Steph.), a characteristic taxon of dry, starchy wood (Koch 1989), previously recorded from a wide range of Roman-period deposits in Britain (eg York (Hall & Kenward 1990) and The Lunt, Warwickshire (Osborne 1975)). Plants indicated include *Veronica beccabunga* (*G. beccabungae*), *Rumex* spp. (*Apion hydrolapathi* Marsh.), *Malva* spp. (*Apion rufirostre* F.) and *Lathyrus pratensis* (*Apion subulatum* Kirby, although this taxon also feeds on *Vicia* spp.). This assemblage is suggestive of a meadow environment and is also reminiscent of elements within the 'stable manure' indicator group of Kenward and Hall (1997).

Context (1303) was the most species poor of those examined, comprising only seventeen individuals. Of these, six were aquatic in nature (three of running water of indeterminate velocity, three of slow water). All other taxa were similar to those in the two lower contexts, suggesting refuse and mould. The only phytophagous taxon present was *Phyllotreta nigripes* F., which is characteristic of *Brassica* spp.

Context (1295) yielded an assemblage which was more aquatic in character than others recovered from this tank, being dominated by slow-water taxa, in particular *Helophorus* spp. The assemblage also included several individuals of the weevil *Tanysphyrus lemnae* (Payk.) which is oligophagus on duckweed (*Lemna* spp.) (Koch 1992), a plant characteristic of stagnant water rich in nutrients, and which grows best in water with high levels of nitrogen and phosphate. Refuse and dung taxa were again well represented whilst synanthropes were represented by a single individual of *Cryptolestes ferrugineus* (Steph.) which, while often found in grain and derived products (Munro 1966; Fogliazza & Pagani 1993), is not a direct pest but a feeder on mould and detritus (Horion 1960). This assemblage differed considerably from the two lower contexts, (1337) and (1239), which were far more terrestrial in nature. In combination with the sparseness of the previous context (1303), this may suggest a change or cessation in use of the feature.

Brine Tank [1182]

This assemblage from context (1284), being from an analogous structure to tank [1207], was expected to contain a similar assemblage. This proved to be the case, with the assemblage most closely allied to contexts (1239) and (1337) in [1207], although incorporating a stronger slow-water element. Numerically the sample was dominated by slow-water taxa (*Helophorus* spp.), Lathridiidae and dung beetles of the genus *Aphodius*. A diverse refuse fauna was present, indicative of rotting plant debris (eg *Onthophilus striatus* Forst., *Anotylus tetracarinatus* Block) and synanthropic taxa characteristic of stored grain were again present at low levels (*C. ferrugineus*, *O. surinamensis*). Few phytophages were recovered but these indicate a probable local pastoral environment. For example, the weevil *Alophus triguttatus* F., while polyphagous on a range of ruderal plant taxa, prefers *Plantago lanceolata*, while the nitulid *Brachypterus urticae* F. has a strong preference for stinging nettle, *Urtica dioica* L.

Well Fill (5108)

This context yielded an extremely rich assemblage (a subsample of 1/5 total flot produced 258 individuals from 81 taxa) numerically dominated by members of the Lathridiidae and the Ptiliid *Ptenidium* spp. These tiny beetles are of variable ecology, some being waterside taxa while others live on rotten wood or decaying plant material and dung (Koch 1989). Osborne (1983) found these to be abundant in samples taken from modern compost heaps as opposed to cesspit samples. The sample also included a strong refuse component (67 individuals of 21 taxa), a moderate dung component and six synanthropic taxa: *L. linearis*, *A. brunneus*, *O. surinamensis*, *P. fur* and *T. unicolor*. In combination with the scarabaeid *Oxyomus sylvestris* Scop., Robinson (unpublished) suggests a similar assemblage from a mid-Roman pit at Claydon Pike as typical of 'foul compost'. As can often be the case with well assemblages, aquatic taxa are relatively sparse. A number of taxa of arable/disturbed locales were also found within this context, including the xerophilic carabid taxa *Calathus fuscipes* (Goez.), *C. melanocephalus* (L.), *Harpalus rufipes* Dej., *H. affinis* (Schr.) and *Amara spreta* Dej. Three phytophagous weevil taxa were present within the assemblage, all characteristic of meadow environments; *Sitona hispidulus* (F.), *S. sulcifrons* (Thun.) and *Apion fulvipes* (Geoff.), all of which feed on clovers (*Trifolium* spp.).

Enclosure Ditch Fill (2095)

This context, from the ditch which defines the large enclosure, yielded a relatively species-poor assemblage (26 taxa) strongly dominated by dung and refuse taxa. Several phytophagous taxa were present, including three weevils of the genus *Ceutorhynchus*: *C. erysimi* (F.), which is polyphagous on Brassicaceae (especially Shepherd's Purse, *Capsella bursa-pastoris* L. (Morris 1991)), *C. assimilis* (Payk.), also polyphagous on Brassicaceae and *C. pollinarius* (Forst.) which lives on stinging nettle. The scolytid *Leperisinus varius* (F.) indicates the presence of dead wood of ash, *Fraxinus excelsior* (L.) and was found in large numbers in the Roman well assemblage at Dalton Parlours (Sudell 1990) whilst two relatively non-specific woodland taxa were also recovered: the woodworm *A. punctatum* and the elaterid *Prosternon tesselatum* (L.).

Pit Fills (1063), (1097), (1098), (1158) and (1270)

These five samples derived from pits [1054], [1096] and [1157] and trough [1264], which were thought to have been possibly associated with leather processing and finishing (Connelly & Power 2005, 39). As such they are dealt with here as a coherent unit.

Context (1063), from wicker-lined pit [1054], yielded a relatively sparse beetle assemblage (59 individuals), dominated by taxa of refuse, dung and slow-water environments and devoid of synanthropic taxa. As is indicated over much of the site, the feature appears to have been filled with foul, mouldering organic matter as evidenced by the presence of taxa such as *P. arenarius*, *A. rugosus* and *L. minutus* grp. in abundance. Few phytophagous taxa were present but these mirror those found previously, with *A. triguttatus* once again present and *Ceutorhynchus pollinarius* (Forst.) indicating the presence of stinging nettle. Also present was the weevil *Rhinoncus pericarpius* (L.) which lives upon large-leaved species of *Rumex* spp. (Koch 1992).

Context (1097), from pit [1096], yielded a rich assemblage dominated by taxa of slow-water and vegetable refuse. The latter were represented by generalist taxa such as *Megasternum boletophagum* Marsh. and *Cercyon haemorrhoidalis* (F.). Single individuals of two synanthropic taxa were present, *S. granarius* and *T. unicolor*. The context also included a number of phytophagous taxa including the nitulid *B. urticae* (characteristic of stinging nettle) and two weevils of the genus *Apion*, *A. subulatum* Kirby and *A. fulvipes* (Geoff.). The former of these lives almost exclusively on meadow vetchling, *Lathyrus pratensis* L. (although occasionally on *Vicia* spp.), whilst the latter is indicative of *Trifolium* spp. (Bullock 1993). *T. lemnae* is once again present as is the weevil *Notaris acridulus* (L.), a characteristic species of waterside vegetation, usually Cyperaceae, but also aquatic grasses, in addition to two other taxa indicative of similar waterside vegetation: *Cyphon coarctatus* Payk. and a fragment of a chrysomelid of the genus *Donacia*.

Context (1098) was also from pit [1096] and was similar in nature to (1097) with the exception that it yielded a greater element of phytophagous taxa with fewer taxa of refuse and slow-water environments. No synanthropic taxa were present, while woodland taxa were represented by a similar assemblage to that found in context (2095), including *P. tessellatum*, *A. punctatum* as well as the chrysomelid *Crepidodera fulvicornis* F. (strongly associated with willow, *Salix* spp.), *Dorytomus* spp. (on willow or poplar, *Populus* spp.) and the non-specific canopy species *Strophosoma melanogrammum* (Forst.). Other phytophagous taxa present indicate members of the Polygonaceae (*R. pericarpius* and the chrysomelids *Gastrophysa viridula* Deg. and *G. polygoni* L.), *Plantago* spp. (*Gymnetron labile* Hbst.) and stinging nettle (*B. urticae*).

Context (1158), from pit [1157], was similar in nature to context (1097), yielding a rich fauna dominated by taxa of rotting vegetable refuse and slow water, with dung taxa also well represented. Four synanthropic taxa were present, all indicative of stored grain. As well as *C. ferruginneus*, *S. granarius* and *O. Surinamensis*, the small-eyed flour beetle, *Palorus ratzbergi* (Wiss.) was present. This taxon feeds predominantly on the faeces of *Sitophilus* spp. but is also probably a predator of other grain pests (Pals & Hakbijl 1992). That this represents the most numerous synanthropic taxon (three individuals) may indicate that the remains found here were in effect 'sub-sampled' by means of animal consumption from a more severe infestation of grain pests elsewhere in the vicinity. Some taxa may be taken to indicate the presence of carrion, for example the staphylinid *Philonthus cephalotes* (Grav.) and the histerid *Hister impressus* F. However, this last taxon may be a later contaminant, as it has not previously been recorded in the British Isles in material dating prior to the 9th century AD (Buckland *et al* 1976). Two woodland taxa were recovered, *S. melanogrammum* and *Curculio pyrrhoceras* Marsh., the latter of which is found in galls of the oak gall wasp *Cynips quercusfolii* L. (Koch 1992). Other phytophagous taxa indicate the likely presence of *Plantago* spp. (*A. triguttatus*, *G. labile*, *Mecinus pyraster* Hbst.), nettle (*B. urticae*), *Rumex* spp. (*R. pericarpius*) and Cyperaceae (*Limnobaris t-album* L.).

Context (1270), from trough [1264], yielded a moderately rich assemblage (78 individuals of 37 taxa) dominated by taxa of dung and slow-water environments with fewer refuse taxa than in some previous contexts. Dung taxa include the Histerid *Pelarus bimaculatus* (L.) which is strongly associated with the dung of cattle (Skidmore 1991). Once again, synanthropic taxa were represented by low frequencies of the two grain taxa *P. ratzergi* and *O. surinamensis* and the ptinid *T. unicolor*, suggesting input of either household sweepings or faecal material. A wide variety of phytophagous taxa were recovered, several indicating the local presence of Brassicaceae (*Phyllotreta undulata* Kuts.), *Ceutorhynchus alliariae* Bris. (on garlic mustard, *Alliaria officinalis* L.) and *Psylliodes cuprea* (Koch). Also present were taxa indicative of clover (*A. fulvipes*), meadow vetchling (*A. subulatum*), stinging nettle (*B. urticae*, *C. pollinarius*) and polygonaceae (*Chaetocnema concinna* Marsh.) in addition to two generalist woodland canopy taxa, *Polydrusus cervinus* (L.), the larvae of which live upon the grass *Dactylis glomerata* L. (cocksfoot), and the Elaterid *P. tesselatum*.

Discussion

The Brine Tanks

None of the material analysed included taxa particularly indicative of the saline conditions which would be associated with the activity of salt production. The two lower contexts from tank [1207] and the context from tank [1182] were of similar character and indicate an accumulation of rotting organic matter. The presence of low numbers of grain taxa, while previously thought to indicate the disposal of spoiled grain as refuse, was reinterpreted by Osborne (1983) as more likely to result from consumption of infested grain resulting in deposition in animal or human faecal material. The assemblage recovered is relatively non-specific but clearly urban in nature, being most closely related to the 'house fauna'-like assemblage (Kenward & Hall 1995). This has been described by Carrot and Kenward (2001) as incorporating floor sweepings and being over-represented in and around buildings. It is possible that the lower context (1337) from [1207] represents single deposition of such material soon after the construction of the feature. This may explain the variation between this context and others from the same feature, in addition to its somewhat more terrestrial nature. Other contexts from within [1207] suggest that the tank held water (or at least very wet refuse) for some time and in its final phase was covered with a bloom of *Lemna* spp., further stressing its extremely eutrophic nature.

The Well

The well sample was largely comparable with other Roman well assemblages previously published, although some differences are evident. The material was extremely rich and may benefit from further analysis. However, the overriding impression is of an environment closely related to that indicated by the brine tank contexts. The structure was evidently filled with rather unpleasant mouldering compost-like material as is suggested by the abundance of *Ptenidium* spp. and *Lathridius minutus* grp. Once again, there is evidence of probable contamination with faecal material. The presence of a variety of xerophilic carabid taxa is comparable with the assemblage recovered from a well assemblage at the Roman villa site of Dalton Parlours (Sudell 1990) and implies that the feature was surrounded by dry, weedy or disturbed ground, probably with sandy soil. Several published Roman well assemblages are clearly less urban in nature. For example the sites at Dalton Parlours (Sudell 1990), Rudston Well (Buckland 1980) and Piddington (Simpson 2001) all lack the variety of synanthropic taxa found in the Nantwich material, synanthropic taxa being largely restricted to members of the Ptinidae in these sites, while levels of phytophagous taxa are comparatively elevated in the more rural sites. However, synanthrope levels do not reach those found in well assemblages from York (eg Hall *et al* 1980; Kenward *et al* 1986) into which waste material is considered to have been deliberately dumped. The most likely conclusion regarding the origin of the material is, as Simpson (2001) concludes in the case of the Piddington well, a slow infilling of the structure following its falling out of use.

Pit Fills (1063), (1097), (1098), (1158) and (1270)

These assemblages were surprisingly variable in nature but offer little in the way of direct evidence of leather production. In contrast to the material recovered from a leather-working site at Alcester (Osborne 1994) none of the sites produced the 'thick, greasy...soupy [smelling] scum' observed by Osborne during processing. Furthermore the Alcester assemblage includes a large number of carrion taxa (eg histerids) in addition to very high numbers of refuse taxa. As such the Nantwich contexts, of which (1158) is the most closely allied to the Alcester material, may represent extremely limited leather production activity but are more likely to be associated with leather working as opposed to actual leather production (ie tanning).

Enclosure Ditch

There are no clear indications of function associated with this feature. The lack of a strong aquatic element suggests that if it was merely a ditch, it was at least free-draining and never held water in any great quantity. The absence of *Ptenidium* spp. or Lathridiids seemingly precludes the presence of similar 'foul compost' type material as is present in the well. As such the feature is interpreted as a dryish ditch in which animal dung accumulated, surrounded by a range of ruderal plant taxa.

19. Conclusion

Peter Arrowsmith & David Power

The excavation at Kingsley Fields has provided the most detailed evidence to date for Roman Nantwich. It has revealed the existence of a previously unknown area of Roman occupation on the north side of Welsh Row, expanding upon the previous, more fragmentary, evidence of occupation to the south of that road and on the opposite, eastern, bank of the River Weaver. It has identified the line of a Roman road linking the settlement at Nantwich to the main road network to the west. It has also produced the fullest evidence recovered so far for salt production in the settlement, complementing the results of modern archaeological investigations in Roman Middlewich. The remains uncovered at Kingsley Fields include the two finest examples of timber-built brine tanks excavated from Roman Britain. The waterlogged conditions on the site have meant that these features in particular contained an exceptionally well-preserved and varied assemblage, including metalwork, wooden objects and animal bones, which throws significant light on the material culture of the site and its likely status.

Phasing of the Site

Two main phases of occupation are represented at Kingsley Fields, one dating from the 2nd century AD, the other from the 3rd. Between these phases, occupation was significantly reduced, if not brought to a temporary end.

The 2nd-Century Occupation

The first phase of occupation involved the creation and operation of an area of brine collection and salt production on the west side of the Roman road (Fig 19.1).

Within the area of excavation this included the construction of the two large clay and timber-lined tanks, [1182] and [1207], built for the storage of brine. At this phase brine was collected in a timber trough in pit [1096], dug on a natural spring line to the west of the tanks, and was probably transferred to the tanks by a channel or trough which had left only a scant trace on the site.

The two brine tanks were built to the same general design. A construction pit was first dug in the natural sand, in which the timber structure of the tanks was built along with a watertight clay lining. In each tank the timberwork included a base-frame, post and plank walls, and probably an upper frame to provide rigidity. In the case of the larger tank [1207], the original two end walls seem to have failed during construction. As a consequence, new end walls were erected to the front of the old, using a greater number of timbers in a more makeshift arrangement.

Immediately to the north of the brine tanks was an enclosure containing two large timber buildings, set at a right angle to the Roman road. Of these, post-built Building 1 was the more substantial and was possibly erected in two episodes of construction which saw the original structure being later doubled in length. Plant macrofossil analysis suggests that it contained a granary but a use for the storage of finished salt also seems likely. Building 2, which was partly of a foundation trench construction, was perhaps more shed-like. Its precise function is unclear but again is likely to be closely linked with salt production.

To the north of the large enclosure a small group of hearths, which survived as channels with evidence of burning, can also be dated to this phase. These were almost certainly used for heating the brine, in order to evaporate the water, leaving crystallised salt. A second, perhaps larger, group of hearths may have been situated between the two brine tanks and the road, outside the excavated area. The method of evaporation is suggested by the fired clay and the lead sheet and runoff found on the site. From these, the hearths used at Kingsley Fields probably combined sub-surface flues with above-ground clay-built walls and a suspended floor, possibly perforated, above which the brine was heated in lead salt pans (see Chapter 13). Kingsley Fields may provide the earliest known evidence from Roman Britain for the use of lead evaporating pans, rather than clay containers.

To the north of the hearths, on the west side of the Roman road, the area of industrial activity gave way to a zone containing numerous pits, some datable to the 2nd-century occupation. While the function of most of these pits is uncertain, three contained cremation burials. At least one, an urned burial [4003], probably falls within this first period of occupation, as may one of two unurned burials [4245].

Within the area of excavation, the eastern side of the Roman road showed comparatively little evidence of activity during this period. The chief feature was a roadside timber-lined well. It seems to have stood at the junction of the road and a trackway leading roughly north-eastward.

This first phase appears to have spanned a period from the Hadrianic to the late Antonine. The pottery evidence does not support occupation before the Hadrianic period, and dendrochronological analysis dates brine tank [1182] to AD 130-7. Activity on the site significantly increased during the Antonine, to which a number of other major features can be dated. These include brine tank [1207]. A

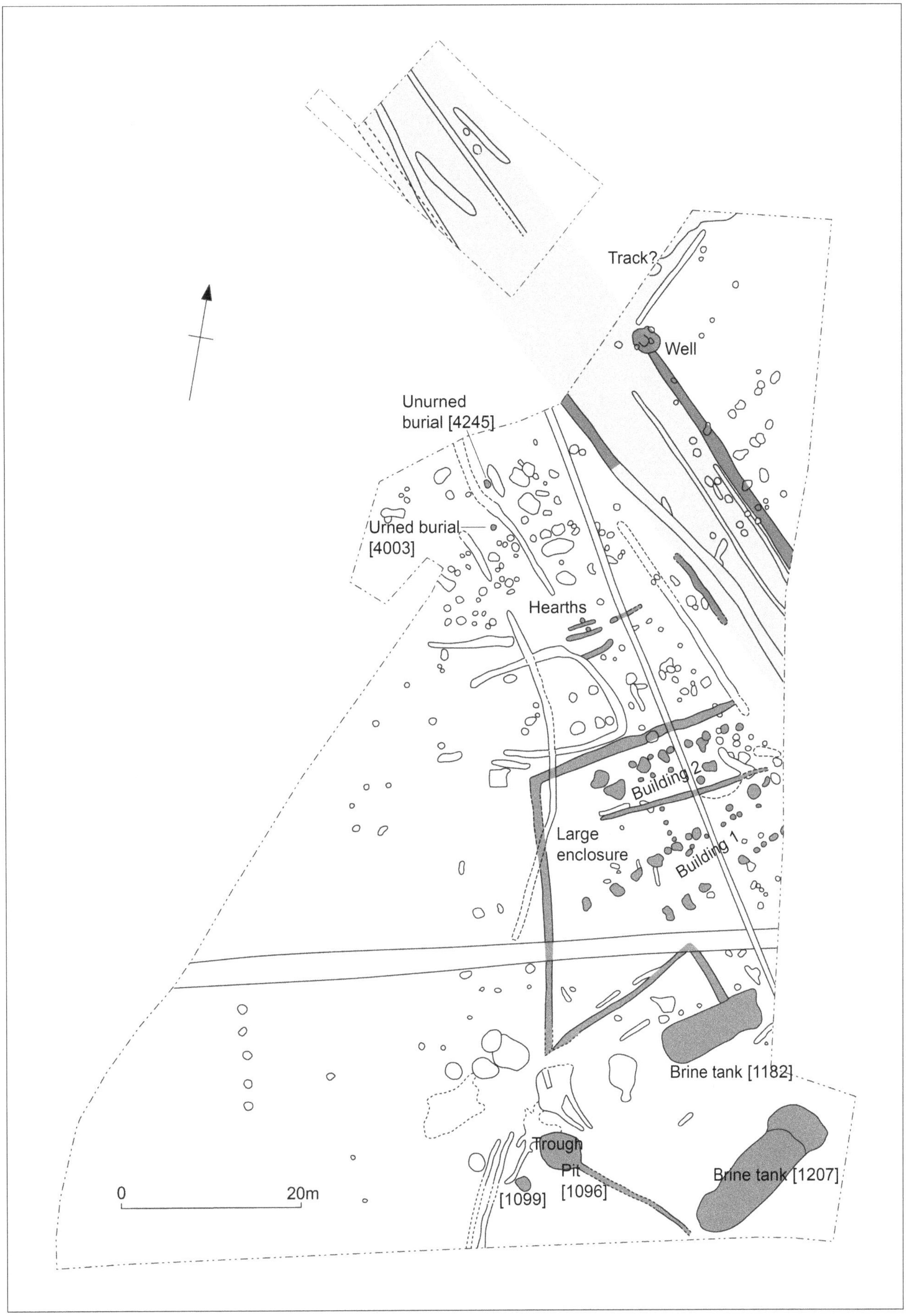

Figure 19.1: Plan of principal features in the 2nd-century phase of occupation.

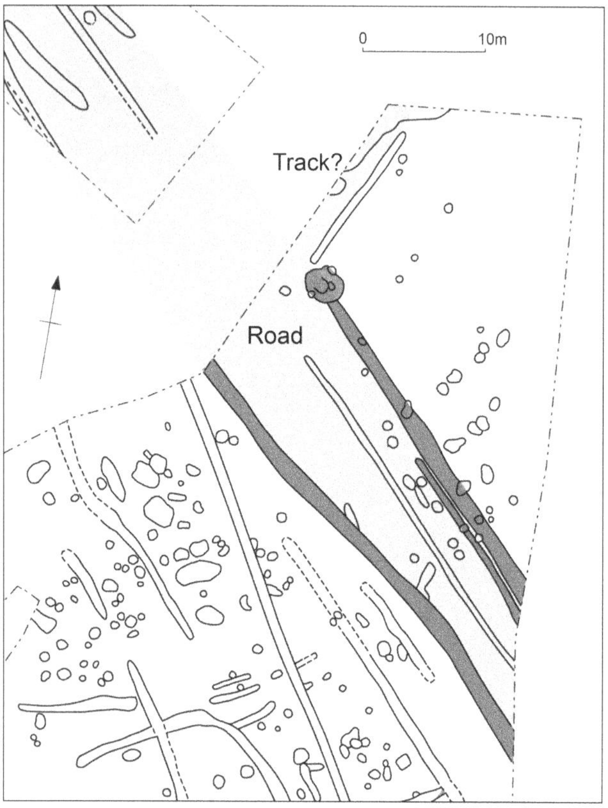

Figure 19.2: Plan of the narrowed road, late 2nd to early 3rd century.

dendrochronological date of AD 114-42 for this feature derives from a timber which had possibly been reused, and a sherd of decorated samian ware from the tank's clay lining gives a somewhat later *terminus post quem* of *c* AD 150. Building 1 can be dated to the mid to late Antonine and Building 2, the large enclosure and the hearths also to the Antonine.

The close of this first phase seems to have occurred in about the AD 180s, based on both the pottery evidence and the coinage which continues into the reign of Commodus. It was marked by the end of the large tanks being used for the managed storage of brine. The wooden trough in pit [1096] in which brine had been collected also went out of use, the hearths seem to have been dismantled and at least Building 2 was taken down.

In both brine tanks, the end of their industrial use was marked by what very much appears to have been a ritual deposit within a watery place. Although these deposits also contained some possible general waste, they were distinguished by the quantity, type and condition of much of the material. This included intact pottery vessels, animal remains encompassing a range of species and showing a degree of deliberate selection in their deposition, wooden objects which included a range of tools probably used in brine production, complete iron tools, an intact 'casserole', and matching upper and lower quernstones. Comparable ritual deposits have been identified at the military site of Newstead (Clarke 1997) and at a range of settlements in southern Britain (Fulford 2001).

In brine tank [1207] the Antonine ritual deposit seems to have fallen at the very end of this phase of occupation of the site and was followed by a period in which soils accumulated in the tank with very little evidence of human activity. By contrast, in brine tank [1182] an initial ritual deposit was followed by further episodes of similar deposition. These are characterised principally by animal remains and pottery vessels and seem to have taken place within, and up to the close of, the Antonine period of occupation. From this sequence, the industrial use of [1182] appears to have ended before that of [1207]. Although the construction of [1182] can also be placed at an earlier date, there is no indication that one tank was built to replace the other and on the pottery evidence the initial ritual deposit within [1182] may have occurred fairly close in time to the closure of [1207]. Several pottery vessel types are found exclusively in one tank or the other and imply different ritual elements.

The Late 2nd- to Early 3rd-Century Hiatus

The pottery record indicates a significant decline in the level of activity on the site at the end of the 2nd century. It was perhaps during this phase that the Roman road was narrowed where it ran past the area previously used for brine production (Fig 19.2). Occupation of the site, although at a much reduced level, may not have entirely ceased. The large enclosure seems to have continued in use and Building 1 perhaps remained standing until the early 3rd century, when it may have been destroyed by fire.

The Early to Mid 3rd-Century Occupation

The early to mid 3rd century saw a revival of the site, although not to the same level as in the mid Antonine. The road was restored to nearly its original width but on a line slightly further to the west (Fig 19.3). On its eastern side, Building 3 and its possible annex, Building 4, were constructed, partly over the back-filled ditches of the earlier road. On the west side of the road, the large enclosure was now vacant and was perhaps used as a paddock. To its north the U-shaped enclosure was added, which may have served as an animal pen and which seems to have contained the four-post Building 5. The north-west area of the main excavation site was again used as a burial ground. At least one of the two unurned burials [4250] dates from this period, which again also saw other pits being dug in this area.

The brine tanks were not reinstated. However, these remained watery places and received fresh ritual deposits during this period. The evidence for these primarily comprised animal remains and pottery vessels, although a few other artefacts, including one of the two matching pairs of quernstones (see Chapter 11, Q3 and Q4), may also belong to this later phase of ritual deposition.

Brine was collected during this period in wooden trough [1264] in pit [1157], situated on the natural spring line. There is evidence of a possible channel system leading from this trough to brine tank [1207], which may have

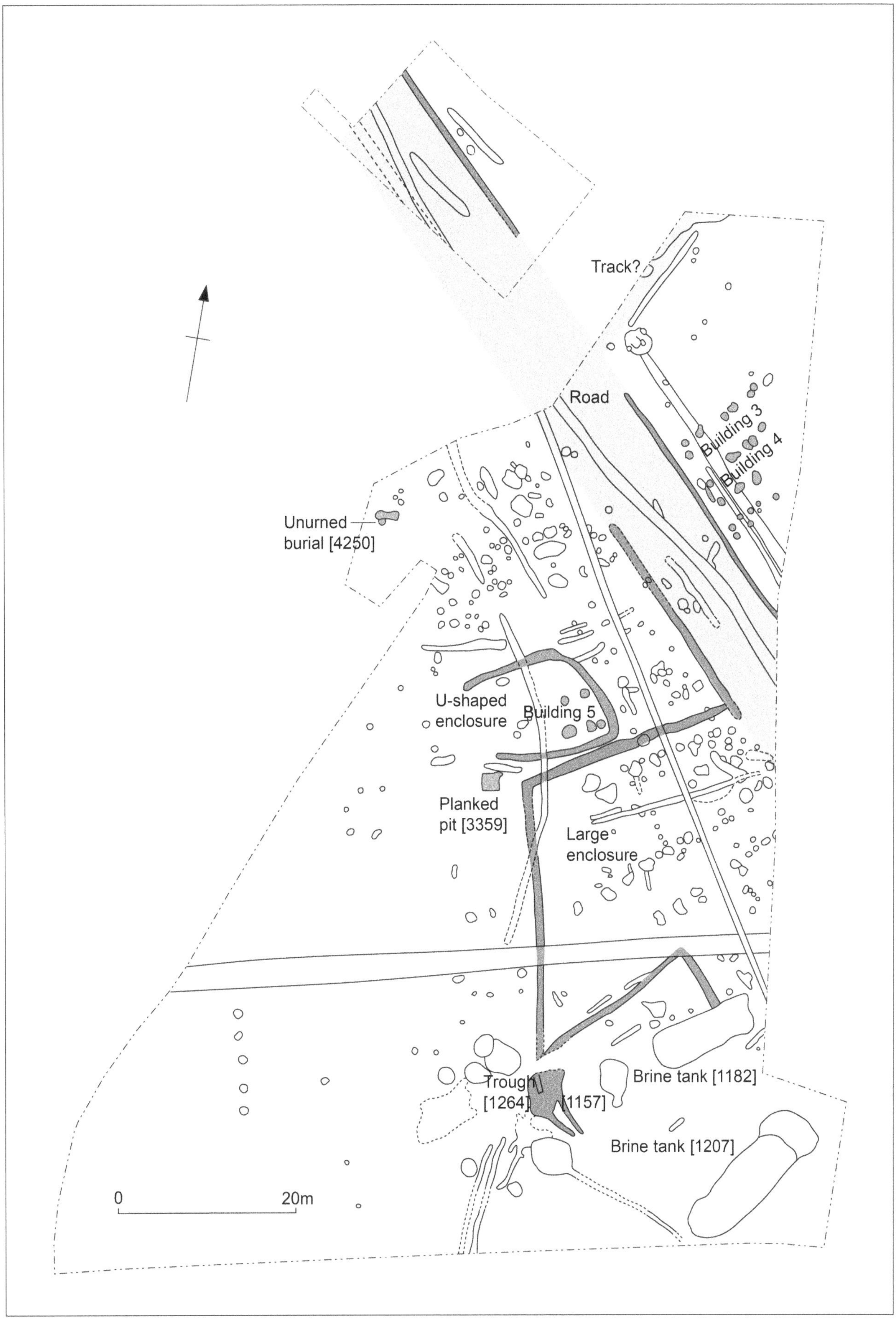

Figure 19.3: Plan of principal features in the early to mid 3rd-century phase of occupation.

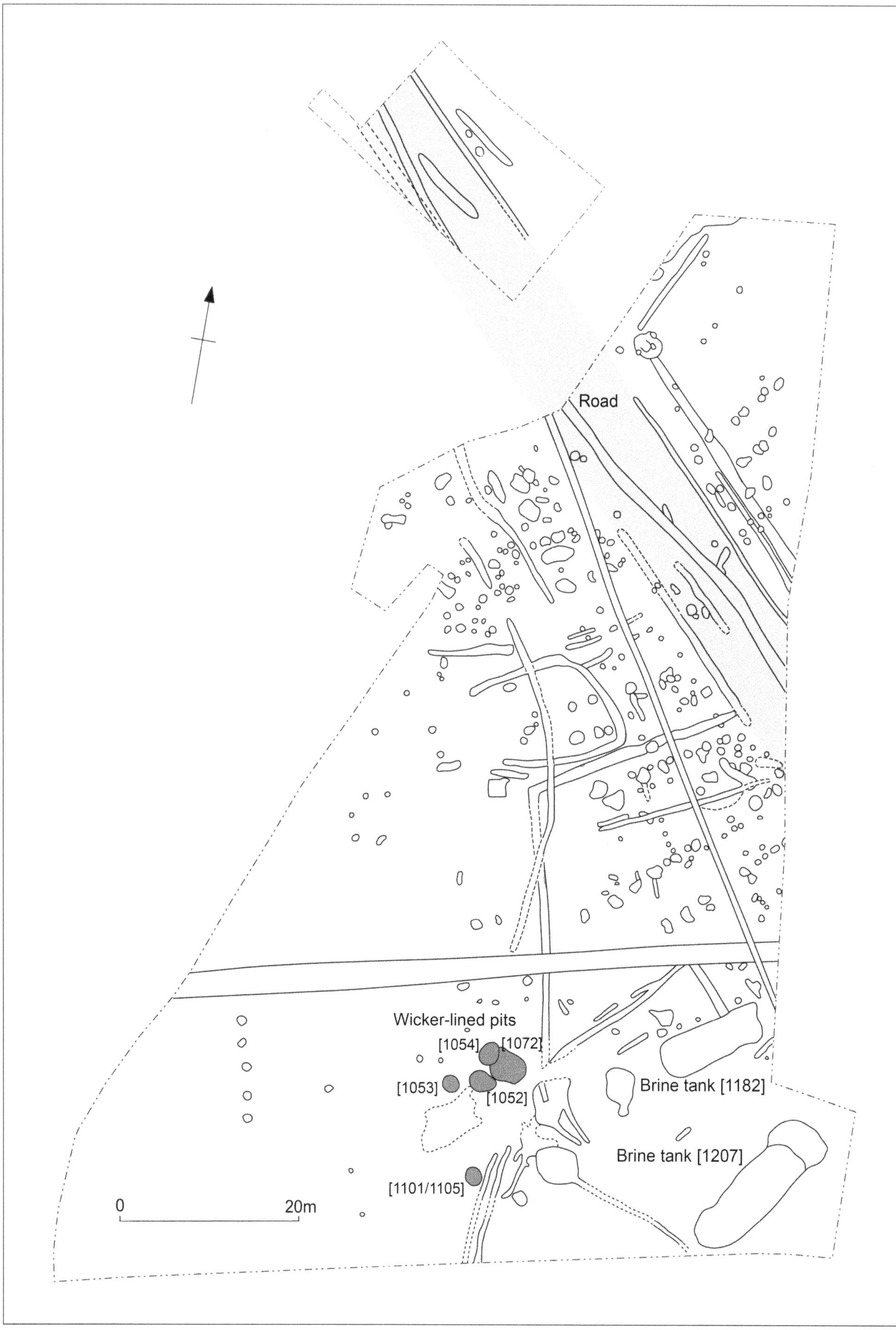

Figure 19.4: Plan of principal features in the late 3rd- to 4th-century phase of occupation.

been associated with the tank's continuing ritual use.

Other channels leading southward from the spring line imply the existence of a brine storage and processing area outside the area of excavation. The date of this channel system is uncertain but its alignment suggests that it may have been fed by trough [1264].

It was previously suggested that the 3rd century saw activity on the site expand into cattle processing, where the animals were corralled and slaughtered and brine was used for leather processing and finishing (Connolly & Power 2005, 39). This has not been supported by subsequent analysis. While both the large enclosure and the U-shaped enclosure perhaps served as animal paddocks, this may have been part of the settlement's general agricultural regime.

Mid 3rd Century Onwards

Occupation of the site largely ceased in about the mid 3rd century. In the late 3rd to 4th century a final phase of brine collection took place using wicker-lined pits [1052], [1053], [1054], [1072] and [1105], located on the spring line (Fig 19.4). Each of these pits probably had a short working life and they may represent successive episodes of brine collection on a relatively modest scale. Otherwise activity during this later period may have been largely associated with traffic along the road. However, the trackway leading north-eastwards from the road seems to have gone out of use during this period when pits were cut into it.

Post-Roman

In the post-Roman period, brine collection and processing on the site was not resumed. By the medieval period the land was given over to an agricultural use and remained as fields until the modern development which prompted the archaeological investigations. One long-lived survival of the Roman occupation was the road. A pair of probably late roadside ditches at Kingsley Fields contained medieval pottery. Subsequent investigations to the south-east imply that the road was still in use in the 11th or 12th century but was later replaced by Welsh Row, in existence by the late 13th (see below).

Kingsley Fields and Roman Nantwich

The Status of Roman Nantwich

The pottery assemblage from Kingsley Fields implies a military source of supply and a number of other finds point to an actual military presence. This evidence is strongest for the 2nd-century period of occupation, with such finds being particularly numerous in the case of the deposits in the brine tanks. For the early to mid 3rd-century occupation, the evidence is somewhat weaker but a military link is suggested by the bone veneer from the cremation burial in [4250]. The altar decorated with chip carving is one of the strongest indicators for a military presence but was unfortunately unstratified and might be of either an Antonine or 3rd-century date. The faunal remains in the 2nd-century deposits in the brine tanks, perhaps most particularly the presence of raven bones, hint at a connection with Mithraism, a cult especially popular with the army. The relatively early use of lead pans for the process of salt evaporation, suggested at Kingsley Fields, may also reflect a close military association.

Roman forts are known at the two other main Cheshire 'wiches', Northwich and Middlewich (Shaw & Clark 2003b, 3-5; Garner & Reid forthcoming), and it has been previously inferred from the finds at Kingsley Fields that one also existed in Nantwich (Connelly & Power 2005, 40). This requires reconsideration. Excavations at Northwich have not been fully published but the fort is known to have been established in the AD 70s. The fort at Middlewich has been dated to the beginning of that decade or slightly earlier. Both belonged to the process of campaigning and military consolidation in the North-West during and before the AD 70s. Roman Nantwich was of a later origin, at the earliest at the turn of the 1st and 2nd centuries (see below), and this may make it less likely that the settlement also started as a fort. Nor, had such a fort existed, is it likely to have continued in use into the late Antonine. The history of the forts at Northwich and Middlewich presents a number of uncertainties, not least because of the difficulty of supposing that two forts in such relative proximity, only 9km apart, were occupied at the same time (Webster 2005). At Northwich two main phases of occupation of the fort have been proposed, the first Flavian, the second Hadrianic, with a break in between. At Middlewich the coin evidence has suggested that following the establishment of the fort there was a fall, or possibly a break, in occupation until a revival in *c* AD 90. In both places it is believed that occupation of the fort continued no later than the Hadrianic or very early Antonine, when their abandonment can be attributed to the redeployment of troops to Hadrian's Wall and later to the Antonine Wall.

Rather than pointing to the existence of a Roman fort at Nantwich, the military source of supply and military presence at Kingsley Fields can be attributed to the site being part of a settlement whose primary function was as an 'industrial centre' serving the Roman army. As such it can be added to a group of such centres in the region which also includes Wilderspool and Walton-le-Dale (see Chapter 4). Salt was of vital importance to the army's needs, with uses which included the preservation of foodstuffs, tanning, and the dyeing of fabrics (Shotter 2005, 42). In Britain it can be assumed that the salt industry was brought into state ownership, as in other parts of the empire (Mattingly 2006, 362-3, 510). As such it was under the control of the province's procurator and his officials, who either managed production themselves or leased it out to private contractors. In Droitwich, a villa of notably high status was established in about the mid 2nd century and may have been the residence of a state official in charge of the local salt industry or of a wealthy contractor (Barfield 2006, 239). Droitwich was also the site of a Roman fort of the AD 50s and 60s (McAvoy 2006). It seems very probable that at Droitwich, Middlewich and Northwich the forts were directly involved in the takeover of the local brine springs by the Roman state. The same is perhaps true at Whitchurch, where excavation has revealed two phases of military occupation in the 1st century, as well as a later hearth suggestive of salt making (Jones & Webster 1968). At Nantwich the army may have had a hand in the layout and construction of the settlement, as has been suggested for other 'industrial centres' in the region (Philpott & Brennand 2007, 63). The subsequent military presence in the settlement, indicated by the finds from Kingsley Fields, may have carried out the day-to-day policing of salt production, storage and transportation.

The Extent of Roman Nantwich

From the excavation at Kingsley Fields and other modern investigations it is clear that the Roman settlement at Nantwich included land on both sides of the Weaver, although the full extent of the settlement remains unclear.

On the west side of the river, the main area of known activity is principally represented by the remains at Kingsley Fields and developed along the Roman road which linked the settlement with the road between Whitchurch and Middlewich.

The north-western extent of this roadside development seems to have roughly coincided with the extent of the main excavation. On the west side of the road, the cremation burials and pits of the Cremation Zone mark the periphery of the settlement. On the east side, the limit of development along the main road was possibly represented by the junction with the side routeway, adjacent to the well (Fig 19.1). It may have led to an area where the discovery of flue tiles, in areas F and G, implies the presence of a heated building. From its relative proximity to the Weaver this was possibly a bath house but alternatively the building may have been accommodation of a relatively high status. No other finds were made in this area to indicate its likely type and function.

The south-eastern extent of development along the main road is less clear. Archaeological trial trenching *c* 50m to the east at Kingsley Cottage in 2003 (Fig 2.1) found evidence of several features containing Roman material, some possibly residual, located on the north side of the Roman road (see Appendix A). However, the limited scope of the investigations means that the full extent and nature of activity here are uncertain.

In 2007 a watching brief towards the east end of Welsh Row revealed what was probably the continuation of the Roman road (Fig 1.2), in the form of a pebble surface at a depth of nearly 2.5m below the modern road. Five sherds of Roman pottery were recovered from the pebble layer, including two sherds of samian ware of a likely Hadrianic-Antonine date. A sondage into an underlying silt layer uncovered a rim sherd of Cheshire Plain ware, probably of the late 1st to early 2nd century (Earthworks 2008). The watching brief also produced evidence that the Roman road remained in use until the medieval period when it was replaced by Welsh Row. Overlying the Roman metalling was a brushwood surface which has been radiocarbon-dated between the early 11th and mid 12th century (Mark Leah, pers comm). Some 50m to the west an early surface of Welsh Row was uncovered, comprising timber planking which has been dated by dendrochronological analysis to the second half of the 13th century (Earthworks 2008).

Roman remains on the south side of Welsh Row are known to the west of St Anne's Lane (Fig 1.2). Prior to the investigations at Kingsley Fields, this area provided much of the evidence for Roman activity on the west side of the Weaver, in the form of an urned cremation burial, a plank-built tank and an east-west aligned ditch. In 2006 an excavation by Gifford to the north of these earlier finds revealed evidence of at least two phases of Roman activity (see Appendix B). The site was crossed by a substantial north-south channel, possibly manmade, with a metalled trackway along its west side. To their west was the base of a clay-lined tank, overlying which was a later urned cremation burial. In the east of the site was an east-west gully, bounded on the west by a line of posts which probably marked a fence line.

The various features discovered to the west of St Anne's Lane point to a band of activity running along the edge of the river's floodplain. It has been suggested that the east-west ditch uncovered in 1997 may have marked the southern boundary of the Nantwich settlement on this side of the river. The St Anne's Lane area was presumably linked to the main Roman road, possibly by a continuation of the metalled trackway discovered in 2006, but whether there was continuous development between the two is at present unknown.

On the east bank of the Weaver, Roman settlement was suggested by the discovery in the 1970s of over 200 residual sherds of Roman pottery found during the Crown car park excavation and by *c* 29 sherds at the National Westminster Bank, both sites being within the area of the medieval castle. Within the same area, trial trenching at Mill Street in 2003 produced a sherd of Mancetter-Hartshill mortarium of the 2nd or 3rd century and two body sherds probably from a Cheshire Plain ware vessel, again residual (Earthworks 2003, 23).

Firmer evidence of Roman settlement on the east side of the river was provided in 2004 at Snow Hill (Fig 1.2). Here groundworks for a new electricity substation revealed in-situ structural timbers and wattling, within a series of gravel, clay and sand deposits up to 1.75m thick overlying the natural boulder clay. These deposits also produced a small assemblage of finds, comprising five sherds of samian ware, all of Hadrianic-early Antonine date, a bone pin and a leather shoe, both of 2nd-century types, and two body sherds of Cheshire Plain ware, again probably of the 2nd century. Twelve unstratified timbers, comprising five posts, four planks, two posts or planks and a wedge-shaped timber, were removed from the site for dendrochronological analysis. This provided two dating sequences, one of which gave a probable felling date for two of the timbers of AD 119^{+}/-18 and AD 125^{+}/-18 (Reid *et al* 2004). The other sequence allowed a possible felling date for some timbers in the 1st century AD but given the other evidence from Nantwich this is probably too early.

The area of the town centre on the eastern side of the river has yet to produce any firm evidence of Roman salt production, although at a later date the town's brine well was situated on this bank, not far from the Snow Hill site. At present, therefore, it is unknown to what extent salt manufacture was carried out in the Roman period within both sides of the settlement or whether it was largely a feature of the west bank.

The Chronology of Roman Nantwich

The coin evidence from Roman Nantwich, reviewed in this volume by Shotter, does not support there being a settlement here earlier than the turn of the 1st and 2nd centuries AD. Other dating evidence from the town provides little indication of settlement before the 2nd century when development occurred on both sides of the River Weaver. This includes the sites at Kingsley Fields, Kingsley Cottage, St Anne's Lane and Snow Hill and finds made in the area of the medieval castle. More particularly, the evidence points to a burst of activity in the Hadrianic period. Not only did this see the beginning of occupation at Kingsley Fields, but also the Snow Hill site and the plank tank at St Anne's Lane, which produced a dendrochronological date of AD 132, can both be placed in this period. The coin evidence may allow the beginning of the settlement to be placed at a slightly earlier date. This is perhaps also supported by the inclusion of reused structural timbers in brine tanks [1207] and [1182], although these could have been brought from elsewhere.

The wider coin evidence from Nantwich corresponds with that from Kingsley Fields in showing a peak of activity in the Antonine. There are other indications that the two main phases found at Kingsley Fields are broadly representative of the settlement as a whole, with a significant reorganisation between the 2nd century and the 3rd. The dating evidence from the Snow Hill site belongs only to the 2nd century and possibly indicates an end to activity here as early as *c* AD 160 (Reid *et al* 2004). At the site at St Anne's Lane excavated by Gifford in 2006, the pottery extended into the 3rd century but the main period of activity was the 2nd. The clay-lined tank was sealed by a deposit containing mid to late 2nd-century pottery and after the tank's disuse an urned cremation burial was also placed there. Nearby, the dated pottery from the plank-built tank discovered in 1985 is also mostly of the 2nd century, with only a small proportion from the 3rd century or later. The tank itself seems to have undergone a change of use, having been perhaps built for the collection of brine before later becoming a cess pit (McNeil & Roberts 1987). It was located just to the north of where a cremation burial was discovered in a 2nd- or 3rd-century urn. The evidence from St Anne's Lane is thus consistent with a reduction in the level of activity between the 2nd century and the 3rd, with burials seemingly encroaching into what was formerly an industrial zone.

The assemblage of residual pottery recovered during the Crown car park excavation differs from the patterns found elsewhere in the town, dating predominantly to the 3rd and 4th centuries, and suggests that this area may have been the settlement's final core. The shrinkage of the settlement during this later period would tally with the evidence from Kingsley Fields, which shows that brine collection was still being carried out in the late 3rd-4th centuries but on a reduced scale.

Roman lead salt pans found to the north of Kingsley Fields and to the east of the town at Shavington carry inscriptions which may indicate that in the 4th-5th centuries salt production was managed by the Church (Shotter 2005). If these pans are of this late date, they may also suggest an industry which was still based on a settlement in Nantwich. Whether that industry was now collecting and processing brine at a more scattered range of sites than previously is difficult to conclude. At Shavington a small number of other finds are known which date from the 2nd century to the 4th and include *sestertii* of Hadrian and Antoninus Pius (Penny & Shotter 2000-1, 57, 60 n 4). The coins coincide with the 2nd-century phase of activity at Kingsley Fields and raise the possibility of an industry which from the beginning encompassed not only the brine springs and hearths in the Nantwich settlement but also others further afield.

Roman Nantwich and Salt Production in Cheshire

No evidence for the exploitation of the Nantwich brine springs prior to the Roman occupation has been identified as yet but may be revealed by future excavation. At present the existence of an Iron Age salt industry in Cheshire is largely known from the sherds of briquetage vessels which were used to dry and transport salt and which have a wide distribution outside the county.

The Roman settlement at Nantwich differs from Cheshire's other main 'wiches', Northwich and Middlewich, in the later date of its foundation and, related to this, the probable absence of a fort as the focus of the settlement. Instead, Nantwich appears to have been founded primarily for industrial purposes. The location of the settlement away from the existing road network, requiring a branch road to be built for access, implies that the main reason for its foundation was the natural resource of the local brine springs. The settlement may also have been involved in other industrial activity, as at Northwich and Middlewich which have both produced small-scale evidence of pottery production (Petch 1987, 224; Garner & Reid forthcoming), but such activity is likely to have been of secondary importance compared with the settlement's role as a centre of salt production.

The apparent burst of activity within the Nantwich settlement in the Hadrianic period is in keeping with its role in supplying the Roman army and probably reflects an increased demand for salt in connection with troop deployments for the construction and garrisoning of Hadrian's Wall. It is possible that the role of Nantwich was originally to supplement the supplies of salt from Northwich and Middlewich. However, in the Antonine period, when activity in Nantwich seems to have been at its peak, it may have been the main centre of the Cheshire salt industry. At Northwich, following the abandonment of the fort, an associated settlement seems to have continued only at a reduced scale, with relatively little evidence of activity after the mid 2nd century (Petch 1987, 202). At Middlewich, the associated settlement seems to have survived the demise of the fort rather better, flourishing in the mid 2nd century, and with occupation then continuing until at least the mid 4th. However, at present the main evidence for salt making at Middlewich dates from the late 1st and early 2nd century, although production continued into the Antonine and probably later (Garner & Reid forthcoming).

The end of the main phase of activity at Kingsley Fields in about the AD 180s may perhaps be related to the wider disturbances in Britain in that decade, which saw first an incursion into the province across the northern frontier and later mutiny within the Roman army (Salway 1997, 162-6). The resurgence of salt making at Kingsley Fields in the early 3rd century, probably as part of a wider revival and reorganization of the Nantwich settlement, was possibly prompted by the campaigning of Severus in Scotland. It may also have been linked to the subsequent redeployment of troops to the legionary fortress at Chester. For much of the 2nd century, beginning with the construction of Hadrian's Wall in the AD 120s, significant numbers of the Twentieth Legion were stationed away from their headquarters at Chester. Following the end of the Severan campaign in the north, most of the legion is believed to have returned to Chester, where the fortress now underwent a major rebuilding (Mason 2001, 155, 161-5). The decline in salt production at Kingsley Fields in the mid 3rd century seems to reflect a wider demise in the region's 'industrial centres', from which there was no further significant recovery in the Roman era.

Appendix A. Kingsley Cottage, Red Lion Lane, Nantwich, A Note on the Evaluation in 2003

Peter Arrowsmith

Introduction

In August 2003 the University of Manchester Archaeological Unit carried out an evaluation of a site at Kingsley Cottage, Red Lion Lane, Nantwich, on behalf of Black Sheep Properties Limited, ahead of a proposed development. The site was situated at NGR SJ 647 525, *c* 50m to the east of the Kingsley Fields site (Fig 2.1), and covered an area of *c* 0.1 hectare including Kingsley Cottage, outbuildings and, to the north of these, a garden.

The evaluation comprised the excavation of three trenches within the garden area. These were dug by machine and archaeological features were then excavated by hand. Trenches 1 and 3 were aligned roughly east-west and were situated at the north and south ends of the garden respectively, Trench 1 measuring 5.4m by 1.6m and Trench 3 12.2m by 1.6m. Trench 2 was located between the two, on an alignment running roughly north-east - south-west, and measured 22m by 1.6m.

Roman Activity

Archaeological features revealed in the trenches comprised pits, postholes and linears. Roman material was recovered from Trenches 1 and 3.

In Trench 1 (Fig A.1) pit [104], which was 0.8m long, at least 0.7m wide and 0.46m deep, contained an upper fill (103) with two sherds of mid 2nd-century pottery. Pit [108] was an irregularly shaped feature 2.42m in length and 0.18m deep, which on the south continued beyond the trench. It contained two fills, (107) and (121), of which (121), possibly the fill of a recut, produced eleven sherds of mid to late 2nd-century pottery. Pit [106], 0.57m long, 0.54m wide and 0.15m deep, contained a fragment of rooftile within its fill (105).

A linear cut [128] ran roughly east-west along the length of Trench 3 (Fig A.1). It had a typical width of 0.4-0.5m and a maximum depth of 0.26m, and contained two fills, (127) and (138). Upper fill (127) produced eight sherds of mid to late 2nd-century pottery, an unspecified silver coin, a fragment of rooftile, an iron nail and a fragment of industrial waste. At the west end of the trench, a linear cut feature [130], 0.8m wide and 0.6m deep, ran on a roughly north-west - south-east alignment and extended beyond the edges of the excavation. It contained a single fill (129) which included two sherds of samian ware. This feature was cut by linear [128], which also cut pits [140] and [144]. Pit [140] was 0.28m deep and [144] 0.18m deep. Neither produced any dating evidence.

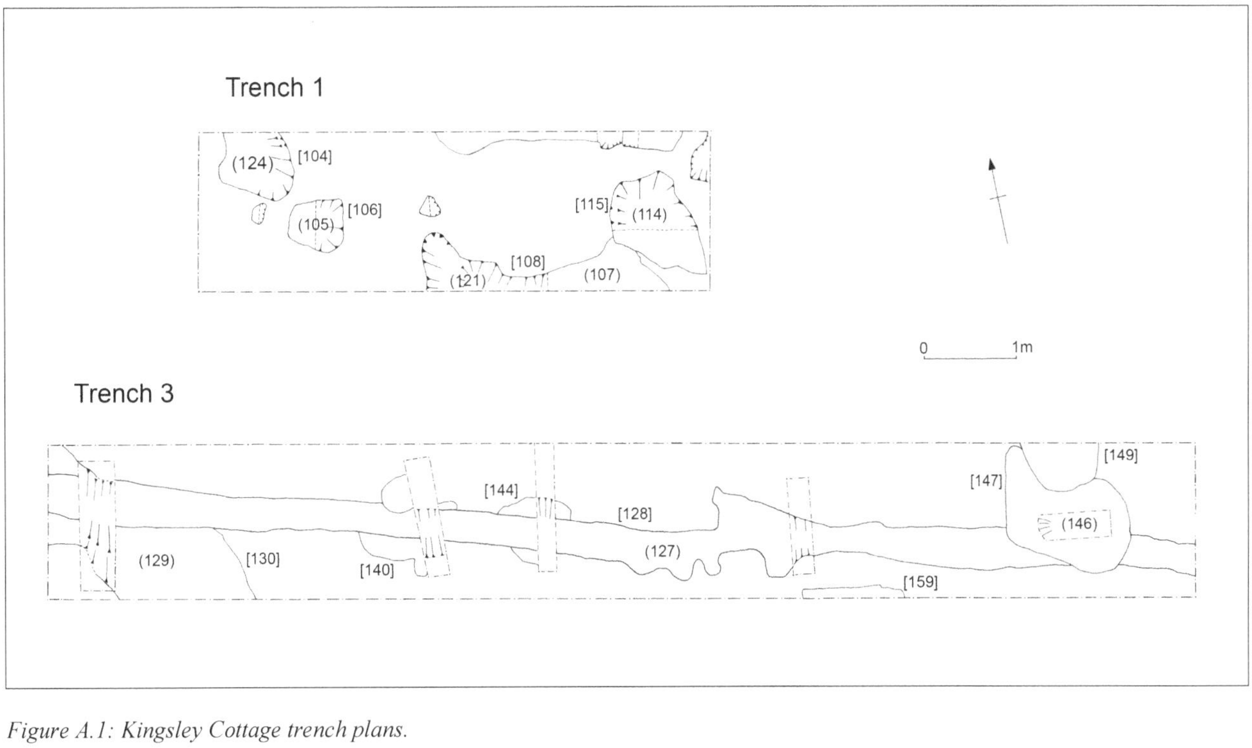

Figure A.1: Kingsley Cottage trench plans.

Post-Medieval Activity

In Trench 1 pit [115] contained a single fill (114) which contained 19th-century material. This feature cut pit [108]. Trench 2 produced only 17th- to 19th-century material, found within two pits, [135] and [155], and a linear cut [151]. Towards the east end of Trench 3, linear feature [128] was cut by pit [147]. Its fill (146) contained a mainly 19th-century assemblage, which also included a residual sherd of 15th-century date. Pits [149] and [159] in the same trench had fills consisting of shale, ash, charcoal, modern glass and pottery.

Discussion

The evaluation showed the continuation of Roman activity to the east of the Kingsley Fields site, although the limited extent of the trial trenching means that the full nature and the extent of activity on the Kingsley Cottage site are uncertain.

The north-west - south-east trending linear cut [130] in Trench 2 closely matches the projected alignment of the Roman road discovered at Kingsley Fields and may be a continuation of the road's northernmost, primary, ditch [5054]/[4153] (see Chapter 3).

Linear feature [128], which cuts [130], might represent a Roman feature postdating the disuse, probably in the late 2nd century, of the primary roadside ditch. However, the feature lay on an alignment which would have met the Roman road at an oblique angle and which corresponded more closely with the orientation of the Kingsley Cottage boundary. This may favour the feature being of a later origin, with the Roman finds being redeposited within its fill.

The post-medieval pits are likely to represent garden or domestic activity in the form of refuse pits. The linear feature [151] in Trench 2 appeared to be a drainage ditch.

Acknowledgements

The evaluation was carried out by Carolanne King, Graham Mottershead and Kieran Power, and was monitored by Mark Leah of Cheshire County Council. This note is based on a draft report by Kieran Power.

Appendix B. St Anne's Lane, Nantwich, A Note on the Excavation by Gifford in 2006

Michael Nevell

Introduction

In March and April 2006 Gifford carried out an excavation at the site of the former Burgess store, St Anne's Lane, Nantwich, on behalf of Muller Property Holdings Limited. The site was situated at NGR SJ 648 523, 250m to the south-east of the Kingsley Fields site and to the south of Welsh Row (Fig 1.2). Prior to the construction of the Burgess store in the 1960s, the site was occupied by the gardens of Whitehall, a large house first referred to in 1429 as belonging to John Kingsley. The house was set back from the street frontage and outhouses to the south crossed what was to become the eastern half of the site. Whitehall was demolished in the 1950s, deemed to be structurally unsound.

The area of the excavation occupied a raised terrace at the northern end of the Burgess compound. An evaluation by Gifford in December 2005 demonstrated that this terrace preserved the original ground level, with garden soils overlying a higher area of natural alluvial sand in the north-west corner in which pits and postholes yielding Romano-British pottery were identified. The natural ridge appeared to drop away to the south and east, and trenches excavated around a standing building contained deep waterlogged silts, sealing features perhaps associated with small-scale waterfront activities adjacent to a former watercourse, at a depth of around 1.2m below the existing ground level.

The excavation area was stripped of the overlying topsoil by machine, which was also used to remove a large concrete revetment along the southern edge of the area to facilitate access. All archaeological features were excavated by hand, although some soils were removed by machine after the hand-excavation of test pits through the deposit to ascertain its depth and nature.

The topography of the underlying alluvial sand was uneven, with the highest point in the north-western corner of the trench reaching 36m AOD. The surface dropped away to the south (35m AOD in the south-western corner) and east, reaching *c* 35m AOD in the middle of the trench and dropping steeply away to 33.5m AOD at the eastern end of the trench. In the lower lying areas (below *c* 34.5m AOD) the ground became waterlogged.

Roman Activity

No deposits predating the Romano-British period were identified in the trench, though a single sherd of grog, quartz and limestone tempered ware of Bronze Age date was recovered from one of the hand-dug test pits. Its position within spit 5 of Test Pit 2 places the piece within the lowest fill of a north-south aligned channel [205] in the middle of the excavation area (Fig B.1). This piece was almost certainly residual, as the same level within Test Pit 2 produced Gaulish samian ware dated to the 2nd century AD.

East of the channel the ground dropped away rapidly, possibly marking the former limit of the floodplain of the River Weaver. In the base of a section excavated by machine against the northern edge of the trench a narrow east-west aligned gully [203] was located which contained 2nd-century Roman pottery. The gully was bounded on its western side by a row of timber posts (202) aligned north-south, which probably marked a fence line.

Plate B.1: The St Anne's Lane site viewed from the south-east. The Roman channel [205] runs as a dark band across the site from left to right.

The channel itself was the most clearly visible feature within the trench [205] (Plate B.1). Cutting through the natural sand, it ran down the slope from north to south and was filled with dark waterlogged silt. The channel was *c* 7m wide with a flat base and appeared to begin turning westwards at its southern end. A section excavated across the southern end of the feature exposed what appeared to be the remains of a wattle panel (208), comprising interlacing hazel branches and running from east to west across the base of the channel cut. The panel lay within a deposit (207) from which pottery dating to the early 2nd century was recovered, all in poor condition, and a single rectangular off-cut of leather. A

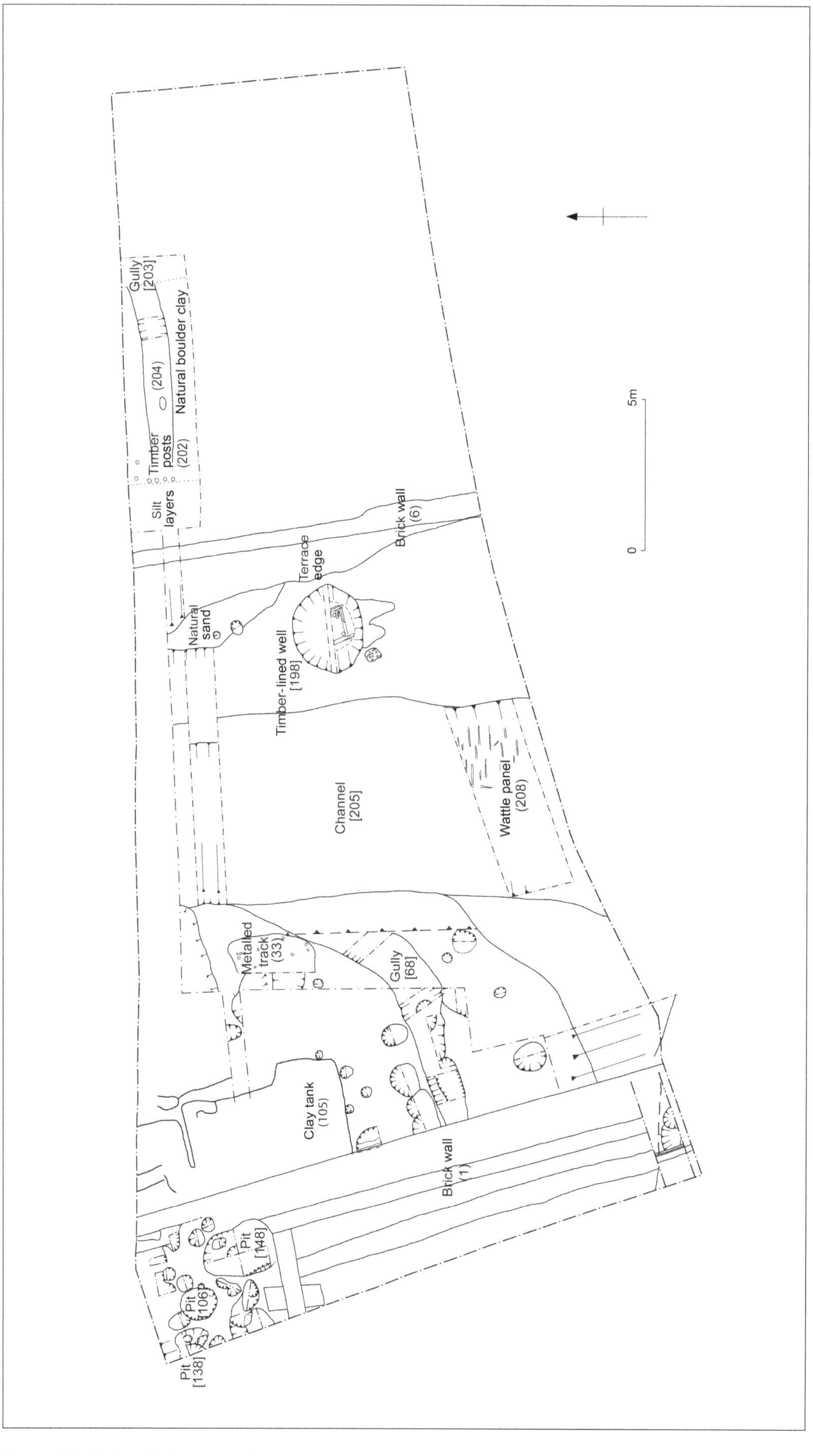

Figure B.1: St Anne's Lane, site plan.

section excavated towards the northern end of the trench recovered more pottery of this date, including southern Gaulish samian ware and Dressel 20 type amphora, alongside a quantity of sheep/goat lower leg bones. The function of the channel is unclear but it could possibly have been a man-made feature bringing water south from a watercourse which formerly flowed at the surface along Welsh Row (known as the Frog Channel).

Along the western edge of the channel, at the northern end of the site, was a short section of metalling forming a track surface (33). This surface was 2-3m wide and up to 0.3m thick and was constructed from small to medium-size pebbles, extending almost 7m into the trench from its northern edge. The surface contained 2nd-century pottery, including examples of Nene Valley colour-coated ware, Severn Valley ware and locally produced orange wares, as well as a lead weight, lead-casting waste and a possible *denarius* in very poor condition.

To the west of the track was a large rectangular feature, comprising a pad of clay oriented roughly north-south which measured at least 7.2m north-south (it extended beyond the northern limit of the excavation) and 3.5m east-west (105) (Figs B.1 & B.2). Two slots were excavated across the length and width of the feature to determine its construction. It was shown to be a layer of puddled orange clay 0.4m thick which had been laid in a flat-bottomed cut with sides sloping at an angle of roughly 45 degrees. There were no deposits underlying the clay, and no overlying features suggesting a superstructure. At the southern edge of the clay pad was a lip of clay 0.9m wide standing 0.4m above the level of the surface to the north, suggesting that perhaps the feature originally had edges which have been lost.

The surface of the clay feature was discoloured, appearing green/grey as if there had been some reduction due to mild heat or anaerobic conditions. A series of shallow deposits were noted pressed into the surface, apparently comprised of the same material. At the centre of the feature was a sub-oval spread of material (197) to the north of which was a curving linear feature, (186), (187) and (188), which doubled back on itself before extending beyond the northern limit of the trench. In addition there were a number of oval depressions 0.3-0.4m in diameter along the western side of the clay pad. These shallow deposits, typically no deeper than 50mm, were filled with pale grey mottled silty clay which contained many artefacts in fair condition.

The clay feature was sealed by a layer of dark brown silty loam 0.1m thick, (93), (94), (99), (100) and (101). This contained many artefacts dating to the mid to late 2nd century, including Mancetter-Hartshill mortarium, amphora (with the inscribed graffito 'AN'), sandy greyware, black-burnished ware, samian (including one base with a partial maker's stamp), white-slipped ware, Severn Valley ware, local orangeware (one base chipped away to within the diameter of the footring), local greyware, Wilderspool mortarium, possible kiln furniture, Romano-British tile and plenty of animal bone including sheep/goat and cattle. Some of these had butchery marks and the assemblage of bones is probably typical of domestic refuse, with many bones from meat cuts and some evidence of burning and cutting apparent. Dump context (94) produced a large quantity of coal and 2135g of iron slag, indicating iron smelting nearby.

This feature and these deposits were in turn sealed by a mottled soil (99), and contexts (87) and (91), a clayey soil possibly representing the sides pushed into the middle of the 'tank', sealing earlier fills. A series of 19th-century features were cut into the surface of (87) and (91). The clay pad was truncated on its eastern side by a gully [210] which appeared to be Roman in date. Its sandy fill (211) contained locally produced coarsewares in poor condition.

A single urned cremation burial was recovered from slot fill (188) and was found to sit directly above the clay pad. The contents comprised clayey silt with a high concentration of charcoal, interspersed with thin lenses of very finely rendered calcified bone with no recognisable fragments. The vessel was sealed by a layer of clean grey silty clay, possibly deposited after the cremation was placed in the tank. The vessel itself was an Oxfordshire colour-coated jar decorated with brown slip and rouletting.

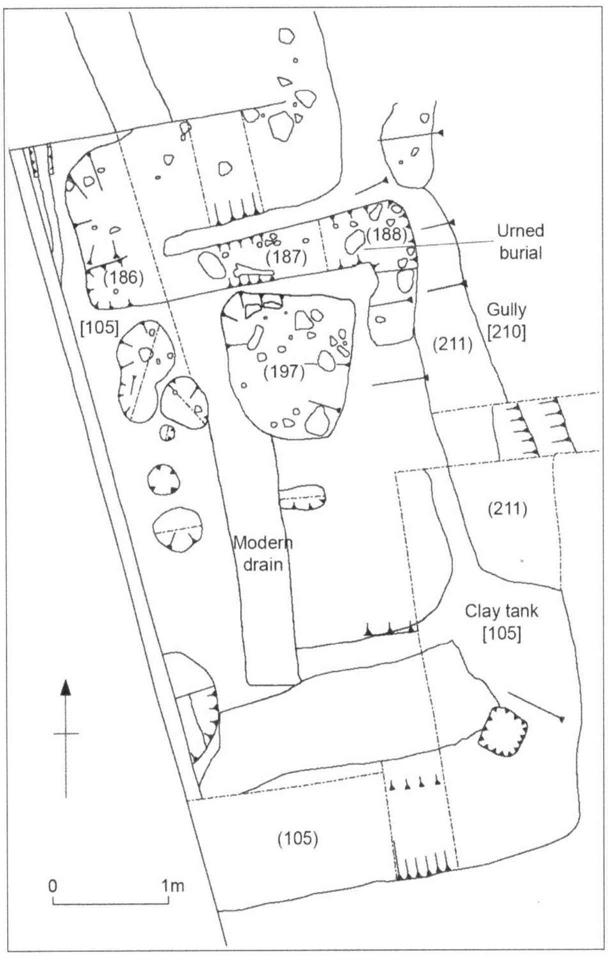

Figure B.2: St Anne's Lane, plan of clay tank [105].

Finally, six coins were recovered from unstratified deposits across the excavation trench by metal detecting. These were a *denarius* of Trajan, three further *denarii* of broadly 2nd-century date but too badly worn to be positively identified, and two badly worn *sestertii* of either the 1st or 2nd century AD.

Medieval and Post-Medieval Activity

In the medieval period the whole site, including the water channel, was sealed by a 0.4m deep layer of ploughsoil. There were a number of possible medieval pits in the north-west corner of the trench, [106], [138] and [148] (Fig B.1). West of the channel was a gully [68] running roughly east-west which produced 12th- to 14th-century pottery. In the middle of the trench and east of the Roman channel a timber-lined well [198] was discovered. This had been constructed from oak planks and square-section oak timbers in a round cut to form a square tank, braced with uprights in each corner. The fill of the well, (199) and (200), contained a large quantity of medieval and early post-medieval pottery dating from the 12th to the 16th century, bone and some fragments of lead-casting waste. At the eastern end of the trench up to 1m of clean silts lay above the Roman levels and these were capped by a thick layer of redeposited orange clay. Given the site's proximity to the floodplain of the River Weaver, it is likely that these deposits were laid down through successive flood episodes.

The upper levels of the trench were dominated by the footings for walls associated with the gardens and outbuildings of Whitehall. Two substantial brick walls with sandstone footings divided the trench and were aligned north-south. One (1) was 2.5m from the western trench edge and the other (6) 14.5m from the eastern trench edge. These walls marked the western and eastern limits of the Whitehall garden and between was a layer of garden soil. To the east was a network of walls representing the outhouses or stables of the main house.

Charred Plant Material

Five 10 litre samples were collected by hand from a range of mostly Roman features during the course of the excavation, these being slot fill (188), tank layer (194), spread (197), channel fill (207) and medieval well fill (199). The samples were processed using a Siraf tank fitted with a 500 micron mesh and flot sieve. The resulting flots were dried and examined using a microscope (x10) and the residues were sorted for artefacts and ecofacts.

Preservation was by both charring and waterlogging. Charred cereal was found in all contexts sampled, except the channel fill (207). Waterlogged wild/weed types were present in spread (197), well fill (199) and channel fill (207). The condition of all seeds and grains was good with a low level of surface abrasion and fragmentation. Charcoal was present in all samples and was mostly heavily fragmented, as were any examples of waterlogged wood. Cereal included hulled and naked barley (*Hordeum vulgare* var. nudum) from the Roman contexts. Two possible oat/rye (*Avena/Secale*) type grains were noted in the fill of the medieval well (199). The low numbers of cereal grains present suggested that activity was not directly connected with crop processing, and that their presence was more likely the result of rubbish burning.

Wild/weed types were restricted to fat hen (*Chenopodium album*), dock (*Rumex* sp.) and cleavers (*Galium aprine*), and were mostly from the medieval well. These are common ruderal weeds often associated with occupation sites. The method of preservation (ie waterlogging) could suggest that their presence is not the result of human intervention, the seeds being blown or washed into the contexts instead.

Discussion

The Roman activity at St Anne's Lane appears, from the pottery evidence, to date from the late 1st/early 2nd to the 3rd century AD, with an emphasis on the 2nd century, and seems to have been located on the north-western edge of the floodplain. This activity was focussed on a probable man-made channel with a metalled trackway running along the western edge and, further west, a rectangular shallow clay-lined feature with angled sides (105) interpreted as a tank, with an urned cremation in the fills. The cremation suggests that the site lay on the edge of the Roman settlement, with the tank possibly filled with domestic rubbish at the end of its life.

A parallel for the tank feature was excavated in 1999 by Gifford at Kinderton Hall Farm in Middlewich where it was interpreted as a shallow brine tank (Garner 2005, 22-3). At St Anne's Lane six fragments of possible briquetage were recovered from a single context (69), weighing a total of just 75g, although the group may include a small piece of firebar. This context was the fill of a medieval gully which ran west of the channel and which produced 12th- to 14th-century pottery, so these fragments were redeposited. There is no other evidence for salt production on the site and the tank may have had another use. The man-made channel, the adjacent clay-lined tank, and the presence of loom weights in the channel may suggest the processing of fibres in this area, with the channel supplying water for the tank which could have been used for soaking flax as part of the retting process in linen making.

Such an interpretation is, however, highly conjectural and is not supported by the palaeoenvironmental assessment (no flax pollen was found for instance). What is clear is that the character of the industrial acitvity at St Anne's Lane is far more domestic (witness the loom weights) or at least small-scale in nature compared to that at Kingsley Fields and appears to have been on the floodplain fringes of the Roman settlement.

Acknowledgements

The excavation team for Gifford were Tam Webster,

Laurence Hayes, George Lacey, Dave Stubbs and Dave Pelling. Metal detecting services were kindly provided by Colin Sharratt of the Crewe and Nantwich Metal Detecting Society. Tim Malim and Anthony Martin managed the project for Gifford. Mark Leah of Cheshire County Council monitored the excavation. This note draws on an unpublished post-excavation assessment report by Gifford.

Bibliography

Adams M H, in prep, *Court Farm, Halewood, Merseyside: Excavations of a Romano-British and Early Medieval Site*.

Allason-Jones L, 1985, '"Bell-shaped studs"?', in M C Bishop (ed), *The Production and Distribution of Roman Military Equipment: Proceedings of the Second Roman Military Equipment Research Seminar*, BAR International Series **275**, 95-108.

Allason-Jones L & Bishop M C, 1988, *Excavations at Roman Corbridge: The Hoard*, English Heritage Archaeological Report **7**.

Archaeological Services, 2004, *Kingsley Fields, Welsh Row, Nantwich, Cheshire; plant macrofossils, wood, diatoms and faunal remains assessment*, unpublished report **1122**, for University of Manchester Archaeological Unit, Archaeological Services Durham University.

Archaeological Services, 2007, *Kingsley Fields, Welsh Row, Nantwich, Cheshire; plant macrofossils, wood, diatoms, faunal remains and beetle analysis*, unpublished report **1485**, for University of Manchester Archaeological Unit, Archaeological Services Durham University.

Barber B & Bowsher D, 2000, *The Eastern Cemetery of Roman London: Excavations 1983-1990*, MoLAS Monograph Series **4**, Museum of London Archaeology Service, 61-7, 264-77, 360-5.

Barfield L, 2006, 'Bays Meadow villa, Droitwich', in Hurst, 2006, 78-242.

Barnett C, forthcoming, 'The Wood Charcoal (Romano-British)', in P Andrews, E Biddulph & A Hardy, *Settling the Ebbsfleet Valley. CTRL Excavations at Springhead and Northfleet, Kent: The Late Iron Age, Roman, Saxon and Medieval Landscape*, Oxford Wessex Archaeology.

Battarbee R W, Juggins S, Gasse F, Anderson N J, Bennion H & Cameron N G, 2000, 'European Diatom Database (EDDI), An Information System for Palaeoenvironmental Reconstruction', *European Climate Science Conference, Vienna City Hall, Vienna, Austria, 19.-23. October 1998*, 1-10.

Bayley J & Butcher S, 2004, *Roman Brooches in Britain: a technological and typological study based on the Richborough collection*, Report of the Research Committee of the Society of Antiquaries of London **68**.

Beagrie N, 1989, 'The Romano-British pewter industry', *Britannia* **20**, 169-91.

Beales P W, 1980, 'The Late Devensian and Flandrian vegetational history of Crose Mere, Shropshire', *New Phytologist* **85**, 131-61.

Behre K-E, 1986, *Anthropogenic indicators in pollen diagrams*, Rotterdam.

Bellhouse R L, 1955, 'Roman sites on the Cumberland coast, 1954', *Transactions of the Cumberland and Westmorland Archaeological and Antiquarian Society* **54**, 28-55.

Bestwick J D, 1975, 'Romano-British Inland Salting at Middlewich (Salinae), Cheshire', in K W de Brisay & K A Evans, *Salt: the Study of an Ancient Industry*, Colchester Archaeology Group, 66-70.

Birley A R, 2005, *The Roman Government of Britain*, Oxford University Press.

Blagg T F C, 2002, *Roman Architectural Ornament in Britain*, BAR British Series **329**.

Booth P, 2000, *The Oxford Archaeology Pottery Recording System*, unpublished manual, Oxford Archaeology.

Booth P M & Green S, 1989, 'The Nature and Distribution of Certain Pink, Grog Tempered Vessels', *Journal of Roman Pottery Studies* **2**, 77-84.

Bouby L & Marinval P, 2004, 'Fruits and seeds from Roman cremations in Limagne (Massif Central) and the spatial variability of plant offerings in France', *Journal of Archaeological Science* **31**, 77-86.

Braithwaite G, 2007, *Faces from the past. A study of Roman face pots from Italy and the western provinces of the Roman Empire*, BAR International Series **1651**.

Brewer R J, 1986, 'The beads and glass counters', in J D Zienkiewicz, *The Legionary Fortress Baths at Caerleon: II the Finds*, Cardiff, National Museum of Wales and Cadw, 146-56.

Brickstock R J, Cardwell P A, Busby P A, Cool H E M, Huntley J P, Evans J, Makey P & Wilson P R, 2007, 'Catterick Metal Detecting Project 1997-1999', *Yorkshire Archaeological Journal* **79**, 65-453.

Brown A, 1994, 'A Romano-British shell-gritted pottery and tile manufacturing site at Harrold, Bedfordshire', *Bedfordshire Archaeology* **21**, 19-107.

Buckland P C, 1980, 'Insect remains from the well', in I M Stead, *Rudston Roman Villa*, Yorkshire Archaeological Society, 162-7.

Buckland P C, Holdsworth P & Monk M, 1976, 'The interpretation of a group of Saxon pits in Southampton', *Journal of Archaeological Science* **3**, 61-9.

Buckland P I & Buckland P C, 2006, Bugs Coleopteran Ecology Package (Versions: BugsCEP v7.53; Bugsdata v7.09; BugsMCR v2.0; BugStats v1.2), available at http://www.bugscep.com.

Buikstra J E & Ubelaker D H, 1994, *Standards for data collection from human skeletal remains*, Arkansas Archaeological Survey Research Series **44**.

Bullock J A, 1993, 'Host Plants of British Beetles: A List of Recorded Associations', *Amateur Entomologist* **11a**, 1-24.

Butterfield B G & Meylan B A, 1980, *Three-Dimensional Structure of Wood. An Ultrastructural Approach*, London, Chapman and Hall.

Cameron N G, Birks H J B, Jones V J, Berge F, Catalan J, Flower R J, Garcia J, Kawecka B, Koinig K A, Marchetto A, Sánchez-Castillo P, Schmidt R, Šiško M, Solovieva N, Štefková E & Toro M, 1999, 'Surface-sediment and epilithic diatom pH calibration sets for remote European mountain lakes (AL:PE project) and their comparison with the Surface Waters Acidification Programme (SWAP) calibration set', *Journal of Paleolimnology* **22**, 291-317.

Carrott J & Kenward H, 2001, 'Species associations among insect remains from urban archaeological deposits and their significance in reconstructing the past human environment', *Journal of Archaeological Science* **28**, 887-905.

Caseldine A, 1990, *Environmental Archaeology in Wales*, Lampeter, Saint David's University College.

Clarke S, 1997, 'Abandonment, Rubbish Disposal and 'Special' Deposits at Newstead', in K Meadows, C Lemke & J Heron, *TRAC 96. Proceedings of the 6th Annual Theoretical Roman Archaeological Conference Sheffield 1996*, Oxford, Oxbow, 73-81.

Cleere H, 1974, 'The Roman iron industry of the Weald and its connexions with the Classis Britannica', *Archaeological Journal* **131**, 170-99.

Colledge S & Greig J, 1992, 'Environment', in Woodiwiss, 1992, 96-105.

Collingwood R G & Myres J N L, 1937, *Roman Britain and the English settlements*, Oxford, Clarendon Press.

Collis J, 1977, 'Owslebury (Hants) and the problem of Burials on Rural Sites', in R Reece (ed), *Burial in the Roman World,* CBA Research Report **22**, Council for British Archaeology, 26-35.

Connelly P & Power D, 2005, 'Salt Making in Roman Nantwich. Recent Discoveries at Kingsley Fields, Welsh Row', in Nevell & Fielding, 2005, 31-40.

Cool H E M, 2004, *The Roman Cemetery at Brougham Cumbria: Excavations 1966-67*, Britannia Monograph **21**.

Cool H E M, 2006, *Eating and Drinking in Roman Britain*, Cambridge University Press.

Cox M & Mays S (eds), 2000, *Human Osteology in Archaeology and Forensic Science*, London, Greenwich Medical.

Crone A, 1992, 'Wood', in Woodiwiss, 1992, 106-13.

Crowson A, 2001, 'Excavation of a Late Roman Saltern at Blackborough End, Middleton, Norfolk', in Lane & Morris, 2001, 162-82.

Cunliffe B, 1988, *The Temple of Sulis Minerva at Bath. Volume 2: The Finds from the Sacred Spring*, Oxford University Committee for Archaeology Monograph **16**.

Cunliffe B & Poole C, 1991, *Danebury: An Iron Age hillfort in Hampshire. Vol 4. The excavations 1979-1988. The finds*, CBA Research Report **73**, Council for British Archaeology.

Curle J, 1911, *A Roman Frontier Post and its People: the fort at Newstead in the parish of Melrose*, Glasgow, J Maclehose and Sons.

Darrah R, 2009, 'The Structural Wood', in Howard-Davis 2009a, 781-805.

Davis A with de Moulins S, 2000, 'The plant remains', in Barber & Bowsher, 2000, 368-9.

Davis A, 2006, *The Charred Plant Remains from Pepper Hill (ARC THB95)*, unpublished report for Museum of London Archaeology Service, Channel Tunnel Rail Link Phase 1.

Den Boesterd M H P, 1956, *The Bronze Vessels in the Rijksmuseum G.M. Kam at Nijmegen*, Nijmegen.

Denys L, 1992, *A check list of the diatoms in the Holocene deposits of the Western Belgian Coastal Plain with a survey of their apparent ecological requirements: I, Introduction, ecological code and complete list,* Service Géologique de Belgique, Professional Paper **246**.

Detienne M & Vernant J-P, 1989, *The Cuisine of Sacrifice among the Greeks*, University of Chicago Press.

Dickinson B, 1986, 'Potters' Stamps and Signatures on the Samian', in L Miller, J Schofield & M Rhodes, *The Roman Quay at St. Magnus House, London*, London and Middlesex Archaeological Society Special Paper **8**, 186-98.

Dinwiddy K Egging & Schuster J, 2009, 'Thanet's longest excavation: Archaeological investigations along the route of the Weartherless-Margate-Broadstairs wastewater pipeline', in P Andrews, K Egging Dinwiddy, C Ellis, A Hutcheson, C Philpotts, A B Powell & J Schuster, *Kentish sites and sites of Kent: a miscellany of four excavations*, Wessex Archaeology Report **24**, 57-174.

Dobney K M, Jaques S D & Irving B G, 1996, *Of Butchers and Breeds: Report on vertebrate remains from various sites in the City of Lincoln*, Lincoln Archaeological Studies **5**.

Dodd L J, 2005, 'Salt Making in Roman Middlewich: Part 1. Discovery and Rediscovery. Excavations along King Street', in Nevell & Fielding, 2005, 25-30.

Dodd L J, forthcoming, 'Excavation of Medieval and Early Post-Medieval Salt Works at Second Wood Street, Nantwich 2003-4', *Journal of the Chester Archaeological Society*.

Duthoy R, 1969, *The Taurobolium: Its Evolution and Terminology*, Leiden.

Earthworks, 1997, *Land Adjacent to Riverside, Nantwich, Cheshire: An Archaeological Evaluation*, unpublished report (R2217) held in the Cheshire Historic Environment Record.

Earthworks, 2001, *Proposed Redevelopment at Second Wood Street and Rear of the Former Cheshire Cat, Welsh Row, Nantwich, Cheshire: An Archaeological Evaluation*, unpublished report (R2377) held in the Cheshire Historic Environment Record.

Earthworks, 2003, *Proposed Residential Development on Land off Mill Street, Nantwich, Cheshire: An Archaeological Evaluation*, unpublished report (R2454) held in the Cheshire Historic Environment Record.

Earthworks, 2005, *Proposed Redevelopment at the Three Pigeons Inn, Welsh Row, Nantwich, Cheshire: An Archaeological Evaluation*, unpublished report (R2569) held in the Cheshire Historic Environment Record.

Earthworks, 2008, *Renewal of Gas Mains at High Street, Welsh Row & Wych House Bank, Nantwich, Cheshire: An Archaeological Watching Brief*, unpublished report (R2826) held in the Cheshire Historic Environment Record.

Earwood C, 1991, 'Objects of Wood', in N Holbrook & P T Bidwell, *Roman Finds from Exeter*, Exeter Archaeological Reports **4**, Exeter County Council and University of Exeter, 275-8.

Earwood C, 1993, *Domestic Wooden Artefacts in Britain and Ireland from Neolithic to Viking Times*, University of Exeter Press.

Earwood C, Cool H E M & Northover P, 2001, 'Two pewter bowls from a mire in South Wales', *Britannia* **32**, 279-85.

Edlin H L, 1949, *Woodland Crafts in Britain. An Account of the Traditional Uses of Trees and Timbers in the British Countryside,* London, B T Batsford.

Eggers H J, 1966, 'Römische Bronzegefässe in Britannien', *Jahrbuch des Römisch-Germanischen Zentralmuseums Mainz* **13**, 67-164.

Emons H-H, 1984, *Mit dem Salz durch die Jahrtausende*, Leipzig, Deutscher Verlag für Grundstoffindustrie.

English Heritage, 1998, *Dendrochronology: guidelines on producing and interpreting dendrochronological dates*, London.

Evans J, 1985, *Aspects of later Roman pottery assemblages in northern England*, PhD dissertation, University of Bradford.

Evans J, 1993, 'Function and finewares in the Roman north', *Journal of Roman Pottery Studies* **6**, 95-118.

Evans J, 1996, 'The Gas House Lane (AL23) Roman pottery', in S Cracknell, *Roman Alcester. Vol 2: defences and defended area*, CBA Research Report **106**, Council for British Archaeology, 58-97.

Evans J, 1997, 'The pottery from Brithdir Roman fort', in D Hopewell 'Archaeological survey and excavations at Brithdir', *Journal of the Merioneth Historical and Record Society* **12**, 310-33.

Evans J, 2001, 'Material approaches to the identification of different Romano-British site types', in S James & M Millett, *Britons and Romans: advancing an archaeological agenda*, CBA Research Report **125**, Council for British Archaeology, 26-35.

Evans J, 2002, *The Roman pottery from Middlewich: an assessment*, unpublished report for L-P Archaeology.

Evans J, 2003, 'The Later Iron Age and Roman pottery', in H Hinman & E Edges, *A Late Iron Age farmstead and Romano-British site at Haddon, Peterborough*, BAR British Series **358**, 68-107.

Evans J, forthcoming a, 'King Street and the Roman frontier in the North-West', *Journal of Roman Pottery Studies.*

Evans J, forthcoming b, '*The Plas Coch, Wrexham, Roman pottery'*, in G Wait, *Report on excavations at Plas Coch, Wrexham.*

Evans J & Mills P, 2009, 'The mortaria', in H E H Cool & D J

P Mason (eds), *Roman Piercebridge; excavations by D.W. Harding and Peter Scott 1969-1981*, Architectural and Archaeological Society of Durham and Northumberland Research Report **7**, 197-207.

Evans J & Rátkai S, forthcoming a, 'The Roman pottery from Walton-le-Dale', in Oxford Archaeology North monograph on excavations at Walton-le-Dale.

Evans J & Rátkai S, forthcoming b, 'Roman pottery from Mitchell's Brewery 1999 and 2000 (L99/072 and L99/109)', in A Lupton, *Report on excavations at Mitchell's Brewery Lancaster*, Oxford Archaeology North.

Evans J & Rátkai S, forthcoming c, 'The Pottery', in I Ferris, *Excavations at Binchester*.

Fairbairn N, 2002, 'Birch Heath, Tarporley: excavation of a rural Romano-British settlement', *Journal of the Chester Archaeological Society* **77**, 59-114.

Fielding A P, 2005, 'Practical salt making and the identification of early inland salt making sites in Cheshire', in Nevell & Fielding, 2005, 55-64.

Fitzpatrick A P, 1997, *Archaeological Excavations on the Route of the A27 Westhampnett Bypass, West Sussex, 1992, Volume 2*, Wessex Archaeology Report **12**.

Fogliazza D D & Pagani M, 1993, 'Insect pests in stored foodstuffs in Italy, Part 1: Coleoptera', *Tecnica Molitoria* **44**, 937-51.

Fowler W W, 1889, *The Coleoptera of the British Islands*, London.

Fox C, 1958, *Pattern and Purpose: a Survey of Early Celtic Art in Britain*, Cardiff, National Museum of Wales.

Fulford M, 2001, 'Links with the past: pervasive 'ritual' behaviour in Roman Britain', *Britannia* **32**, 199-218.

Funari P P A, 1996, *Dressel 20 inscriptions from Britain and the consumption of Spanish olive oil. With a catalogue of stamps*, BAR British Series **250**.

Gale R & Cutler D, 2000, *Plants in Archaeology*, Westbury and Royal Botanic Gardens, Kew.

Garner D, 2005, 'Salt Making in Roman Middlewich: Part 1. Discoveries before 2000', in Nevell & Fielding, 2005, 15-24.

Garner D & Reid M, forthcoming, 'Roman Middlewich: A Reassessment of its Form, Function and Chronology', *Journal of the Chester Archaeological Society*.

Gejvall N G, 1981, 'Determination of burned bones from prehistoric graves: Observations on the cremated bones from the graves at Horn', *OSSA Letters* **2**.

Gidney L J, 1999, 'The Animal Bones', in A Connor & R Buckley, *Roman and Medieval Occupation in Causeway Lane, Leicester*, Leicester Archaeology Monograph **5**.

Gifford & Partners, 1995, *Report on an Archaeological Desk-based Assessment of Land at Kingsley Farm, Welsh Row, Nantwich*, unpublished report (R2121) held in the Cheshire Historic Environment Record.

Gillam J P, 1951, 'Dales Ware: a distinctive Romano-British cooking pot', *Antiquaries Journal* **31**, 154-64.

Gillam J P, 1970, *Types of Roman coarse pottery vessels in Northern Britain*, 3rd edition, Newcastle upon Tyne, Oriel Press.

Gillam J P, 1976, 'Coarse fumed ware in northern Britain', *Glasgow Archaeological Journal* **4**, 57-80.

Giorgi J, 2000, 'The plant remains', in A Mackinder, *A Romano-British Cemetery on Watling Street*, MoLAS Archaeology Studies Series **4**, Museum of London Archaeology Service, 10-12, 65-6.

Goodall I H, 1990, 'Tanning, currying, and leather-working tools', in M Biddle, *Object and Economy in Medieval Winchester*, Winchester Studies **7.ii**, Oxford, Clarendon Press, 247-50.

Goodburn D, 1991, 'A Roman Timber Framed Building Tradition', *Archaeological Journal* **148**, 182-204.

Grant A, 1982, 'The use of tooth wear as a guide to the age of domestic ungulates', in B Wilson, C Grigson & S Payne (eds), *Ageing and Sexing Animal Bones from Archaeological Sites*, BAR British Series **109**, 91-108.

Gray L, 2008, 'Charred plant remains', in Powell *et al*, 2008, 173-6.

Greep S, 1995, 'Bone inlay', in W H Manning, J Price & J Webster, *Report on the excavations at Usk 1965-1976. The Roman Small Finds*, Cardiff, University of Wales Press, 272.

Greep S, 2004, 'Bone and antler veneer', in Cool, 2004, 273-82.

Häberlin C, 1934, *Die Nordfriesischen Salzsieder*, Hamburg.

Hall A R & Kenward H K, 1990, *Environmental evidence from the Colonia*, Archaeology of York **14/6**, Council for British Archaeology for York Archaeological Trust.

Hall A R, Kenward H K & Williams D, 1980, *Environmental evidence from Roman deposits in Skeldergate*, Archaeology of York **14/3**, Council for British Archaeology for York Archaeological Trust.

Hanson W S, 1982, 'Roman Military Timber Buildings: Construction and Reconstruction', in S McGrail (ed), *Woodworking Techniques before A.D. 1500*, National Maritime Museum, Greenwich, Archaeological Series **7**, BAR International Series **129**, 169-86.

Harcourt R A, 1974, 'The Dog in Prehistoric and Early Historic Britain', *Journal of Archaeological Science* **1**, 151-75.

Harris B E (ed), 1987, *The Victoria History of the County of Chester. Volume 1*, Oxford University Press.

Hartley B, Barber H G, Carter J R & Sims P A, 1996, *An Atlas of British Diatoms*, Bristol.

Hartley B R, 1972, 'The Samian Ware', in S Frere, *Verulamium Excavations: Volume I*, Report of the Research Committee of the Society of Antiquaries of London **28**, 216-62.

Hartley B R & Dickinson B M, 2008-, *Names on Terra Sigillata: an Index of Makers' Stamps and Signatures on Gallo-Roman Terra Sigillata (Samian Ware)*, London, Institute of Classical Studies.

Hartley K F & Webster P V, 1973, 'Romano-British Pottery Kilns near Wilderspool', *Archaeological Journal* **130**, 77-103.

Hassall M W C & Tomlin R S O, 1984, 'Roman Britain in 1983: Inscriptions', *Britannia* **15**, 333-56.

Hather J G, 2000, *The Identification of the Northern European Woods: A guide for archaeologists and conservators*, London, Archetype Publications.

Henig M, 2004, *Corpus Signorum Imperii Romani. Corpus of Sculpture of the Roman World. Great Britain, Volume I, Fascicule 9: Roman Sculpture from the North West Midlands*, Oxford University Press.

Heslop D H, 2008, *Patterns of Quern Production, Acquisition and Deposition: A Corpus of Beehive Querns from Northern Yorkshire and Southern Durham*, Yorkshire Archaeological Society Occasional Paper **5**.

Holden J L, Phakley P P & Clement J G, 1995a, 'Scanning electron microscope observations of incinerated human femoral bone: a case study', *Forensic Science International* **74**, 17-28.

Holden J L, Phakley P P & Clement J G, 1995b, 'Scanning electron microscope observations of heat-treated human bone', *Forensic Science International* **74**, 29-45.

Horion A, 1960, *Faunistik der Mitteleuropäischen Käfer, 7, Clavicornia, Sphaeritidae - Phalacridae*, Uberlingen-Bodensee.

Howard-Davis C (ed), 2009a, *The Carlisle Millennium Project. Excavations in Carlisle, 1998-2001. Volume 2: The Finds*, Lancaster Imprints **15**, Oxford Archaeology North.

Howard-Davis C, 2009b, 'The Wooden Artefacts', in Howard-Davis, 2009a, 805-16.

Howard-Davis C & Whitworth A, 2000, 'Structural Wood', in

K Buxton & C Howard-Davis, *Bremetenacum: Excavations at Roman Ribchester 1980, 1989-1990*, Lancaster Imprints **9**, Lancaster University Archaeological Unit, 323-35.

Howe M D, Perrin J R & Mackreth D F, 1980, *Roman pottery from the Nene Valley: a guide*, , Peterborough City Museum and Art Gallery Occasional Paper **2**.

Huntley J P, 1987, *Woodland management studies from Carlisle: Castle Street 1981*, Ancient Monuments Laboratory Report **119/87**, English Heritage.

Huntley J P, 1989, *Woodland management studies from Carlisle: Annetwell Street 1983-4*, Ancient Monuments Laboratory Report **51/89**, English Heritage.

Huntley J P, 1991, 'Woodland management studies', in M R McCarthy (ed), *The structural sequence and environmental remains from Castle Street, Carlisle*, Cumberland and Westmoreland Antiquarian and Archaeological Society Research Series **5**, 60-4.

Huntley J P & Stallibrass S, 1995, *Plant and vertebrate remains from archaeological sites in northern England: data reviews and future directions*, Architectural and Archaeological Society of Durham and Northumberland Research Report **4**.

Hurst J D, 1992, 'Shell Table', in Woodiwiss, 1992, fiche 3, E1.

Hurst J D (ed), 1997, *A Multi-Period Salt Production Site at Droitwich: Excavations at Upwich*, CBA Research Report **107,** Council for British Archaeology.

Hurst J D (ed), 2006, *Roman Droitwich: Dodderhill fort, Bays Meadow villa, and roadside settlement*, CBA Research Report **146,** Council for British Archaeology.

Hurst J D & Hemingway J A, 1997, 'The Excavation', in Hurst, 1997, 9-67.

Hustedt F, 1930-66, *Die Kieselalgen Deutschlands, Österreichs und der Schweiz unter Berücksichtigung der übrigen Länder Europas sowie der angrenzenden Meeresgebeite*, in L Rabenhorst (ed), *Kryptogamen-Flora von Deutschland, Österreich und der Schweiz* **7**, Parts 1-3, Leipzig, Akademische Verlagsgesellschaft.

Hustedt F, 1953, 'Die Systematik der Diatomeen in ihren Beziehungen zur Geologie und Ökologie nebst einer Revision des Halobien-systems', *Svensk Botanisk Tidskrift* **47**, 509-19.

Hustedt F, 1957, 'Die Diatomeenflora des Fluss-systems der Weser im Gebiet der Hansestadt Bremen', *Abhandlungen des Naturwissenschaftlichen Vereins zu Bremen* **34**, 181-440.

Hutchings P, 1983, 'The Wooden Artefacts', in McNeil, 1983, 82-4.

Jacomet S, Jones G, Charles M & Bittmann F (eds), 2002, *Archaeology of Plants, Current Research in Archaeobotany, Proceedings of the 12th IWGP Symposium, Sheffield 2001*, Vegetation History and Archaeobotany **11**.

Johnston D E, 1972, 'A Roman building at Chalk, near Gravesend', *Britannia* **3**, 112-48.

Johnstone C, 2008, 'Commodity of logistics? The role of equids in Roman supply networks', in S Stallibrass & R Thomas, *Feeding the Roman Army*, Oxford, Oxbow, 128-45.

Jones G B D & Webster P V, 1968, 'Mediolanum: Excavations at Whitchurch 1965-6', *Archaeological Journal* **125**, 193-254.

Jones L, 2000, 'Romano-British Coarse Pottery', in R W Cowell & R A Philpott, *Prehistoric, Romano-British and Medieval Settlement in Lowland North West England: Archaeological excavations along the A5300 road corridor in Merseyside*, National Museums and Galleries on Merseyside.

Kenward H K & Hall A R, 1995, *Biological Evidence from 16-22 Coppergate*, Archaeology of York **14/7**, Council for British Archaeology for York Archaeological Trust.

Kenward H K & Hall A R, 1997, 'Enhancing bioarchaeological interpretation using indicator groups: stable manure as a paradigm', *Journal of Archaeological Science* **24**, 663-73.

Kenward H K, Hall A R & Jones A K G, 1980, 'A tested set of techniques for the extraction of plant and animal macrofossils from waterlogged archaeological deposits', *Science and Archaeology* **22**, 3-15.

Kenward H K, Hall A R & Jones A K G, 1986, *Environmental Evidence from a Roman Well and Anglian Pits in the Legionary Fortress*, Archaeology of York **14/2**, Council for British Archaeology for York Archaeological Trust.

Keppie L J F, 1975, 'Stone to Clay', in A Robertson, M Scott & L Keppie, *Bar Hill: A Roman fort and its finds*, BAR British Series **16**, 31-125.

Koch K, 1989, *Die Käfer Mitteleuropas*, Ökologie **2**, Krefeld.

Koch K, 1992, *Die Käfer Mitteleuropas*, Ökologie **3**, Krefeld.

Koster A, 1997, *Description of the Collections in the Provincial Museum G.M. Kam at Nijmegen13: The Bronze Vessels 2. Acquisitions 1954-1996 (including vessels of pewter and iron)*, Nijmegen.

Krammer K & H Lange-Bertalot, 1986-91, *Bacillariophyceae*, Stuttgart.

Lageard J G A, Chambers F M & Thomas P A, 1999, 'Climatic significance of the Scots Pine (*Pinus sylvestris L.*) c. 2500BC at White Moss, south Cheshire, UK', *The Holocene* **9**, 321-31.

Lane T & Morris E L (eds), 2001, *A Millennium of Saltmaking: Prehistoric and Romano British Salt Production in the Fenland*, Lincolnshire Archaeology and Heritage Reports Series **4**.

Leary R, 1996, 'Roman Course Pottery', in A S Esmonde-Cleary & I M Ferris, 'Excavations at the New Cemetery, Rocester, Staffordshire, 1985-1987', *Staffordshire Archaeological and Historical Society Transactions* **35**, 40-60.

Leary R, 2004, 'Roman Coarse Pottery' & Appendix 2, in R Gregory, *Archaeological Evaluation at Castlefield Quay, Manchester*, unpublished report, University of Manchester Archaeological Unit, 18-23, 48-73.

Leary R, 2005a, *The Romano-British coarse pottery from excavations at Barton St, Manchester*, unpublished report for University of Manchester Archaeological Unit.

Leary R, 2005b, 'Mellor: The Romano-British Pottery 1998-2003', in M Nevell & N Redhead (eds), *Mellor: Living on the edge. A Regional Study of an Iron Age and Romano-British Upland Settlement*, Manchester Archaeological Monographs **1**, University of Manchester Archaeological Unit, Greater Manchester Archaeological Unit and Mellor Archaeological Trust, 44-8.

Leary R, 2008, 'Coarse Pottery', in Williams & Reid, 2008, 65-116.

Leney L & Casteel R W, 1975, 'Simplified Procedure for Examining Charcoal Specimens for Identification', *Journal of Archaeological Science* **2**, 53-159.

Locker A, 1992, 'Animal Bone', in Woodiwiss, 1992, 84-92, 172-81.

Loughlin N, 1977, 'Dales Ware: a contribution to the study of Roman coarse pottery', in D P S Peacock, *Pottery and early commerce: characterization and trade in Roman and later ceramics*, London, Academic Press, 85-162.

LRBC: Hill P V, Carson R A G & Kent J P C, 1960, *Late Roman Bronze Coinage*, London, Spink and Son.

Lucas G, 2007, 'Appendix 3: Pottery', in Rogers & Garner, 2007, 71-7.

Lucht W H, 1987, *Die Käfer Mitteleuropas, Katalog*, Krefeld.

MacGregor M, 1976, *Early Celtic Art in North Britain*, Leicester University Press.

Mackreth D, 1985, 'Brooches from Roman Derby', in J Dool & H Wheeler, 'Roman Derby: Excavations 1968-1983', *Derbyshire Archaeological Journal* **105**, 281-99.

McAvoy F, 2006, 'Dodderhill, Droitwich: excavations 1977-85', in Hurst, 2006, 3-45.

McErlean T & Crothers N, 2007, *Harnessing the tides: the early medieval tide mills at Nendrum monastery, Strangford Loch*, Belfast, The Stationery Office.

McKinley J I, 1993, 'Bone fragment size and weights of bone from modern British cremations and its implications for the interpretation of archaeological cremations', *International Journal of Osteoarchaeology* **3**, 283-7.

McKinley J I, 1994a, *The Anglo-Saxon cemetery at Spong Hill, North Elmham Part VIII: The cremations*, East Anglian Archaeology **69**.

McKinley J I, 1994b, 'Bone fragment size in British cremation burials and its implications for pyre technology and ritual', *Journal of Archaeological Science* **21**, 339-42.

McKinley J I, 1997, 'The cremated human bone from burial and cremation-related contexts', in Fitzpatrick, 1997, 55-72.

McKinley J I, 2000a, 'The Analysis of Cremated Bone', in Cox & Mays, 2000, 403-21.

McKinley J I, 2000b, 'Cremated human remains', 'Cremation burials', 'Cremated Remains', in Barber & Bowsher, 2000, 61-7, 264-77, 360-5.

McKinley J I, 2000c, 'Phoenix rising; aspects of cremation in Roman Britain', in M Millett, J Pearce & M Struck (eds), *Burial, Society and Context in the Roman World*, Oxford, Oxbow, 38-44.

McKinley J I, 2000d, 'Putting cremated human remains in context', in S Roskams (ed), *Interpreting Stratigraphy. Site evaluation, recording procedures and stratigraphic analysis*, BAR International Series **910**, 135-9.

McKinley J I, 2004a, 'Compiling a skeletal inventory: disarticulated and co-mingled remains', in M Brickley & J I McKinley (eds), *Guidelines to the Standards for Recording Human Remains*, British Association for Biological Anthropology and Osteoarchaeology and Institute for Field Archaeology, 13-16.

McKinley J I, 2004b, 'The cremated bone', in V Birbeck & C Moore, 'Preservation and investigation of Roman and medieval remains at Hyde Street, Winchester', *Proceedings of the Hampshire Field Club and Archaeological Society* **59**, 98-103.

McKinley J I, 2004c, 'The human remains and aspects of pyre technology and cremation rituals', in Cool, 2004, 283-309.

McKinley J I, 2004d, 'Aspects of the cremation ritual as evidenced by the animal bones', in Cool, 2004, 331-2.

McKinley J I, 2008a, 'In the heat of the pyre: efficiency of oxidation in Romano-British cremations – did it really matter?', in C W Schmidt & S Symes (eds), *Beyond recognition: the analysis of burned human remains*, London, Academic Press, 163-83.

McKinley J I, 2008b, 'Ryknield Street, Wall (Site 12)', in Powell *et al*, 2008, 87-190.

McKinley J I, 2009, *Channel Tunnel Rail Link Section 1: Human bone scheme-wide overview*, CTRL specialist report series, in ADS 2009.

McKinley J I, forthcoming, '"How did it go?"... putting the process back into cremation', in J Pearce (ed), *Death as a process*, Oxford, Oxbow.

McKinley J I & Smith P, 1997, 'Cremated animal bone from burials and other cremation-related contexts', in Fitzpatrick, 1997, 253.

McNeil Sale R, 1978, *Archaeology in Nantwich: Crown Car Park Excavations*, Chester, Bemrose Press.

McNeil Sale R, 1980, *Wood Street Salt Works*, Cheshire County Council and Liverpool University.

McNeil R, 1983, 'Two 12th-century Wich Houses in Nantwich, Cheshire', *Medieval Archaeology* **27**, 40-88.

McNeil R & Roberts A F, 1987, 'A Plank Tank from Nantwich, Cheshire', *Britannia* **18**, 295.

Major H, 2002, 'Roman decorated styli', *Lucerna, Roman Finds Group Newsletter* **23**, 2-6.

Manning W H, 1976, *Catalogue of Romano-British Ironwork in the Museum of Antiquities Newcastle upon Tyne*, Department of Archaeology, University of Newcastle upon Tyne.

Manning W H, 1985, *Catalogue of the Romano-British Iron Tools, Fittings and Weapons in the British Museum*, London, British Museum Press.

Marnival P, 1993, 'Étude carpologique d'offrandes alimentaires végétales dans les sépultures gallo-romaines: réflexions préliminaires', in A Ferdière (ed), *Monde des morts, monde des vivants en Gaule rurale. Actes du colloque ARCHEA/AGER (Orléans, Conseil Régional, 7-9 février 1992)*, 6éme supplément à la Revue Archéologique du Centre de la France, Tours, 45–65.

Martens M, 2004, 'Re-thinking sacred "rubbish": the ritual deposits of the temple of Mithras at Tienen (Belgium)', *Journal of Roman Archaeology* **17**, 333-53.

Martin T S, 2003a, 'The Late Iron Age and Roman Pottery', in S Forman & D Maynard, 'A Late Iron Age and Romano-British farmstead at Ship Lane, Aveley. Excavations on the line of the A13 Wennington to Mar Dyke road improvement 1994-5', *Essex Archaeology and History* **33**, 138-47.

Martin T S, 2003b, 'Roman Pottery', in M Germany, *Excavations at Great Holts Farm, Boreham, Essex, 1992-1994*, East Anglian Archaeology **105**, 96-155.

Martin T S, 2007, 'Techniques for exploring context, deposition and chronology', in R Hingley & S Willis, *Roman Finds: Context and Theory*, Oxford, Oxbow, 86-99.

Martin-Kilcher S, 1987, *Die römischen Amphoren aus Augst und Kaiseraugst. Ein Beitrag zur römischen Handels- und Kulturgeschichte. 1, Die südspanischen Ölamphoren, Gruppe 1*, Forschungen in Augst **7**, Römermuseum Augst.

Mason D J P, 2001, *Roman Chester: City of the Eagles*, Stroud, Tempus Publishing.

Mason D J P, 2003, 'The Heronbridge Archaeological Research Project: An Interim Report on the 2002 and 2003 Seasons of the Society's New Fieldworks Initiative', *Journal of the Chester Archaeological Society* **78**, 49-106.

Mattingly D, 2006, *An Imperial Possession: Britain in the Roman Empire*, Penguin, Allen Lane.

Meddens B, 1997, 'Animal Bone', in Hurst, 1997, 100-6.

Mees A W, 1995, *Modelsignierte Dekorationen auf südgallischer Terra Sigillata*, Forschungen und Berichte zur Vor- und Frühgeschichte in Baden-Württemberg **54**, Stuttgart.

Mikler H, 1997, *Die römischen Funde aus Bein im Landesmuseum Mainz*, Monographies Instrumentum **1**, Montagnac.

Miller S N, 1922, *The Roman Fort at Balmuildy*, Glasgow.

Mills P J E, forthcoming, 'The Iron Age and Roman pottery from Hayton', in M Millett, *Excavations at Hayton, East Yorkshire*.

Morgan R A, 1989, 'Tree-ring studies of Iron Age and Romano-British wood from Prestatyn', in K Blockley (ed), *Prestatyn 1984-5: An Iron Age Farmstead and Romano-British Industrial Settlement in North Wales*, BAR British Series **210**, 194-211.

Morris C A, 1980, 'A Group of Early Medieval Spades', *Medieval Archaeology* **24**, 205-10.

Morris C A, 1981, 'Early Medieval Separate-bladed Shovels from Ireland', *Journal of the Royal Society of Antiquaries of Ireland* **111**, 50-69.

Morris C A, 1990, 'Wooden Finds', in Wrathmell & Nicholson, 1990, 206-30.

Morris C A, 1998, 'The Wooden Artefacts', in H E M Cool & C Philo (eds), *Roman Castleford Excavations 1974-85. Volume 1: The Small Finds*, Yorkshire Archaeology **4**, West

Yorkshire Archaeology Service, 335-46.

Morris C A, 2000, *Craft, Industry and Everyday Life: Wood and Woodworking in Anglo-Scandinavian and Medieval York*, The Archaeology of York: The Small Finds **17/13**, Council for British Archaeology for York Archaeological Trust.

Morris E L, 2001, 'Briquetage', in Lane & Morris, 2001, 351-76.

Morris M G, 1991, *Weevils*, Slough, Richmond Publishing.

Mould Q, 1997a, 'Leather', in T Wilmott, *Birdoswald. Excavations of a Roman fort on Hadrian's Wall and its successor settlements: 1987-92*, English Heritage Archaeological Report **14**, 326-41.

Mould Q, 1997b, 'Leatherwork', in Hurst, 1997, 126-33.

Mould Q, 2004a, 'Hobnails and shoes', in Cool, 2004, 391-2.

Mould Q, 2004b, 'Leather shoe', in Reid, 2004, 32-3.

Mould Q, 2007, *Leather from Middlewich (E531)*, unpublished report for L-P Archaeology.

Mould Q, 2009, 'Roman waste leather', in Howard-Davis, 2009a, *Volume 3:DVD-ROM*, Appendix 10, 1416-22.

Munro J W, 1966, *Pests of Stored Products*, London, Hutchinson.

Nevell M & Fielding A F (eds), 2005, *Brine in Britannia: Recent Archaeological Work on the Roman Salt Industry in Cheshire*, Archaeology North West **7** (issue 17, for 2004-5), Council for British Archaeology North West, Lion Salt Works Trust and University of Manchester Archaeological Unit.

Newstead R, 1902, 'A descriptive account of Roman and other objects recovered from various sites in Chester and District, 1898-1901', *Journal of the Chester Archaeological Society* **8**, 81-106.

Nielsen-Marsh C, Gernaey A, Turner-Walker G, Hedges R, Pike A & Collins M, 2000, 'The chemical degradation of bone', in Cox & Mays, 2000, 439-54.

Orton C, 1989, 'An introduction to the quantification of assemblages of pottery', *Journal of Roman Pottery Studies* **2**, 94-7.

Osborne P J, 1975, 'The Coleoptera from the Roman Well (on the intervallum road, east of the gyrus)', in B Hobley, 'The Lunt Roman Fort and Training School for Cavalry, Bagington, Warwickshire', *Transactions of the Birmingham and Warwickshire Archaeological Society* **87**, 44-5.

Osborne P J, 1983, 'An Insect Fauna from a Modern Cesspit and its Comparison with Probable Cesspit Assemblages from Archaeological Sites', *Journal of Archaeological Science* **10**, 453-63.

Osborne P J, 1994, 'Insect remains from pit F and their environmental implications', in S Cracknell & C Mahany (eds), *Roman Alcester: Southern Extramural Area, 1964-1966 Excavations, Part 2: Finds and Discussion*, CBA Research Report **97**, Council for British Archaeology, 217-20.

Oswald F, 1936-7, *Index of Figure Types on Terra Sigillata*, University of Liverpool Annals of Archaeology and Anthropology **23**, Supplement.

Palm T, 1959, 'Die Holz und Rindenkäfer der sud- und mittelschwedischen Laubbaume', *Opuscula Entomologica Supplementum* **16**.

Pals J P & Hakbijl T, 1992, 'Weed and insect infestation of a grain cargo in a ship at the Roman fort of Laurium im Woerden (Province of Zuid-Holland)', *Review of Palaeobotany and Palynology* **73**, 287-300.

Parker A J, 1988, 'The Birds of Roman Britain', *Oxford Journal of Archaeology* **7**, 197-226.

Parkhouse J, 2000, 'Objects of Stone', in E Evans, *The Caerleon Canabae*, Britannia Monograph **16**, 453-8.

Penney S & Shotter D C A, 2000-1, 'Further Inscribed Roman Salt Pans from Shavington, Cheshire', *Journal of the Chester Archaeological Society* **76**, 53-61.

Percival S, 2001, 'Briquetage', in Lane & Morris, 2001, 182-202.

Petch D F, 1987, 'The Roman Period', in Harris, 1987, 115-236.

Philpott R, 1991, *Burial Practices in Roman Britain*, BAR British Series **219**.

Philpott R & Adams M H, 2010, *Irby, Wirral: Excavations on a Late Prehistoric, Romano-British and Medieval Site, 1987-96*, Trustees of National Museums Liverpool.

Philpott R & Brennand M, 2007, 'The Romano-British Period Research Agenda', in M Brennand (ed), 2007, *Research and Archaeology in North West England, An Archaeological Research Framework for North West England: Volume 2, Research Agenda and Strategy*, Archaeology North West **9** (issue 19), Association for Local Government Archaeological Officers North West and English Heritage, with Council for British Archaeology North West, 55-72.

Powell A B, Booth P, Fitzpatrick A P & Crockett A D, 2008, *The Archaeology of the M6 Toll 2000-2003*, Oxford Wessex Archaeology Monograph **2**.

Power D, Connelly P & Askew S, 2004, *Land to the rear of Welsh Row, Malbank, Nantwich, Cheshire (also known as Kingsley Fields), Post-Excavation Assessment Report*, University of Manchester Archaeological Unit, unpublished report (R2540) held in the Cheshire Historic Environment Record.

Preston C D, Pearman D A & Dines T D, 2002, *New Atlas of the British and Irish Flora*, Oxford University Press.

Price J & Cottam S, 1998, *Romano-British Glass Vessels: a Handbook*, CBA Practical Handbook in Archaeology **14**, Council for British Archaeology.

Priess S, Matterne V & Latron F, 2005, 'An approach to funerary rituals in the Roman provinces: plant remains from a Gallo-Roman cemetery at Faulquemont (Moselle, France)', *Vegetation History and Archaeobotany* **14**, 363-72.

Prummel W, 1989, 'Appendix to Atlas for identification of foetal skeletal elements of Cattle, Horse, Sheep and Pig', *Archaeozoologia* **3**, 71-8.

Pugsley P, 2003, *Roman Domestic Wood: Analysis of the morphology, manufacture and use of selected categories of domestic wooden artefacts with a particular reference to the material from Roman Britain*, BAR International Series **1118**.

Rackham D J, 1987, 'Assessing the relative frequencies of species by the application of a stochastic model to a computerised database of fossil or archaeological skeletal material', in L van Wijngaarden-Bakker (ed), *Data Management of Archaeological Skeletal Material*, Strasbourg, 185-92.

Rackham O, 2003, *Ancient Woodland its history, vegetation and uses in England*, Kirkcudbrightshire, Castlepoint Press.

Reid M with J Lageard, Q Mould & F Wild, 2004, 'Archaeological Observations at Snow Hill Car Park, Nantwich, Cheshire', *Journal of the Chester Archaeological Society* **79**, 25-36.

Rhodes M, 1980, 'Leather Footwear', in D M Jones, *Excavations at Billingsgate Buildings 'Triangle', Lower Thames Street, 1974*, London and Middlesex Archaeological Society Special Paper **4**, 99-128.

RIB: Collingwood R G & Wright R P, 1995, *The Roman Inscriptions of Britain 1. Inscriptions on Stone*, new edition, Stroud, Alan Sutton Publishing.

RIC: Mattingly H *et al*, 1923-2007, *The Roman Imperial Coinage*, London, Spink and Son.

Ricken H, 1934, 'Die Bilderschüsseln der Kastelle Saalburg und Zugmantel', *Saalburg Jahrbuch* **8**, 130-82.

Robinson M, 2002, 'Domestic burnt offerings and sacrifices at Roman and pre-Roman Pompeii, Italy', in Jacomet *et al*,

2002, 93-100.

Rogers G B, 1974, *Poteries Sigillées de la Gaule Centrale I: les motifs non figurés*, Gallia Supplement **28**.

Rogers G B, 1999, *Poteries Sigillées de la Gaule Centrale II: les potiers*, premier Cahier du Centre Archéologique de Lezoux.

Rogers I R & Garner D J, *Wilderspool and Holditch: Roman Boom Towns on the North Road*, Gifford Archaeological Monographs **2**, BAR British Series **449**.

Rogers J & Waldron T, 1995, *A Field Guide to Joint Disease in Archaeology*, Chichester, John Wiley.

Rutter J B, 1968, 'The Three Phases of the Taurobolium', *Phoenix* **22**, 226-49.

Salway P, 1997, *A History of Roman Britain*, Oxford University Press.

Sawyer P & Thacker A, 1987, 'The Cheshire Domesday', in Harris, 1987, 293-370.

Schenk A, 2008, *Regard sur la tabletterie antique: les objets en os, bois de cerf et ivoire du Musée Romain d'Avenches*, Documents du Musée Romain d'Avenches **15**.

Scheuer L & Black S, 2000, *Developmental Juvenile Osteology*, London, Academic Press.

Schmid E, 1972, *Atlas of Animal Bones*, Amsterdam.

Schweingruber F H, 1990, *Microscopic Wood Anatomy*, 3rd edition, Birmensdorf, Swiss Federal Institute for Forest, Snow and Landscape Research.

Scott S E, 2000, *The Animal Bones from Green Shiel, Lindisfarne*, MSc dissertation, Department of Archaeology, Durham University.

Shaw M & Clark J, 2003a, *Cheshire Historic Towns Survey: Nantwich Archaeological Assessment*, Cheshire County Council and English Heritage, unpublished report held in the Cheshire Historic Environment Record.

Shaw M & Clark J, 2003b, *Cheshire Historic Towns Survey: Northwich Archaeological Assessment*, Cheshire County Council and English Heritage, unpublished report held in the Cheshire Historic Environment Record.

Shotter D C A, 1990, *Roman Coins from North-West England*, Lancaster, Centre for North-West Regional Studies.

Shotter D C A, 1993, 'Coin-loss and the Roman Occupation of North-West England', *British Numismatic Journal* **63**, 1-19.

Shotter D C A, 2000, *Roman Coins from North-West England: Second Supplement*, Lancaster, Centre for North-West Regional Studies.

Shotter D C A, 2004, *Romans and Britons in North-West England*, 3rd edition, Lancaster, Centre for North-West Regional Studies, Occasional Paper **51**.

Shotter D C A, 2005, 'Salt Proprietors in Cheshire. Realities and Possibilities', in Nevell & Fielding, 2005, 41-6.

Shotter D C A, 2007, 'Appendix 2: Coins', in Rogers & Garner, 2007, 69-70.

Shotter D C A, 2008, 'The Coins', in Williams & Reid, 2008, 37-40.

Shotter D C A, 2011, *Roman Coins from North-West England: Third Supplement*, Lancaster, Centre for North-West Regional Studies.

Silver I A, 1969, 'The ageing of domestic animals', in D Brothwell & E S Higgs (eds), *Science in Archaeology*, 2nd edition, London, 283-302.

Simpson T, 2001, 'The Roman well at Piddington, Northamptonshire, England: an investigation of the Coleopterous fauna', *Environmental Archaeology* **6**, 91-6.

Skidmore P, 1991, *Insects of the British Cow-Dung Community*, Field Studies Council.

Šoštarić R, Dizdar M, Kušan D, Hršak V & Mareković S, 2006, 'Comparative analysis of plant finds from early Roman graves in Ilok (Cuccium) and Šćitarjevo (Andautonia), Croatia – a contribution to understanding burial rites in southern Pannonia', *Collegium Antropologicum* **30**, 429-36.

Stace C, 1997, *New Flora of the British Isles*, 2nd edition, Cambridge.

Stanfield J A & Simpson G, 1958, *Central Gaulish Potters*, London.

Stevens C, 2008, 'Environment and agricultural economy', in Powell *et al*, 2008, 457-9.

Straker V, 1979, *Macroscopic plant remains from Droitwich, Bays Meadow 1972-6*, Ancient Monuments Laboratory Report **2812**.

Strickland T, 2001, *Roman Middlewich. A story of Roman and Briton in Mid-Cheshire*, Middlewich, Roman Middlewich Project.

Sudell T L, 1990, 'The insects from Well 1', in Wrathmell & Nicholson, 1990, 267-71.

Sunter N & Brown D, 1988, 'Metal vessels', in Cunliffe, 1988, 9-21.

Swan V, 1984, *The Pottery Kilns of Roman Britain*, RCHM supplementary series **5**, HMSO.

Tassinari S, 1993, *Il Vasellame Bronzeo di Pompei*, Rome.

Taylor J, 2004, 'The distribution and exchange of pink, grog tempered pottery in the East Midlands: an Update', *Journal of Roman Pottery Studies* **11**, 60-6.

Terrisse J-R, 1968, *Les Céramiques Sigillées Gallo-Romaines des Martres-de-Veyre*, Gallia Supplement **19**.

Thompson F H, 1965, *Roman Cheshire*, Cheshire Community Council.

Tomber R & Dore S, 1998, *The National Roman Reference Collection*, MoLAS Monograph Series **2**, Museum of London Archaeology Service.

Tomlin R S O, 1988, 'The curse tablets', in Cunliffe, 1988, 59-265.

Tomlin R S O & Hassall M W C, 2003, 'Roman Britain in 2002: Inscriptions', *Britannia* **34**, 361-82.

Toynbee J M C, 1996, *Death and Burial in the Roman World*, Johns Hopkins University Press.

Tylecote R F, 1986, *The Prehistory of Metallurgy in the British Isles*, London, Institute of Metals.

Ulansey D, 1989, *The Origins of the Mithraic Mysteries*, Oxford University Press.

Ulbert C, Wulfmeier J-C & Huld-Zetsche I, 2004, 'Ritual deposits of Mithraic cult-vessels: new evidence from Sechtem and Mainz', *Journal of Roman Archaeology* **17**, 354-70.

Ulrich R B, 2007, *Roman Woodworking*, Yale University Press.

van Beek G C, 1983, *Dental Morphology: an illustrated guide*, Bristol, J Wright.

van Driel-Murray C, 2001, 'Footwear in the North-Western Provinces of the Roman Empire', in O Goubitz, C van Driel-Murray & W Groenman-van Waateringe, *Stepping through Time. Archaeological Footwear from Prehistoric Times until 1800*, Zwolle, Stichting Promotie Archeologie, 337-76.

van der Veen M, Livarda A & Hill A, 2008, 'New Plant Foods in Roman Britain – Dispersal and Social Access', *Environmental Archaeology* **13**, 11-36.

Wahl J, 2008, 'Investigations on pre-Roman and Roman cremation remains from southwestern Germany: results, potentialities and limits', in C W Schmit & S A Symes (eds), *The analysis of burnt human remains*, London, Academic Press, 145-61.

Watkin W T, 1886, *Roman Cheshire*, Liverpool.

Webster P V, 1975, 'Late Roman Occupation at Wilderspool', *Journal of the Chester Archaeological Society* **58**, 91-2.

Webster P V, 1976, 'Severn Valley Ware: a preliminary study', *Transactions of the Bristol and Gloucestershire Archaeological Society* **94**, 18-46.

Webster P, 2005, review of Strickland, 2001, *Britannia* **36**, 525-6.

Wells C, 1981, 'Human Remains', in C Partridge, *Skeleton*

Green: a Late Iron Age and Romano-British Site, Britannia Monograph **2**, 277-303.

Wessex Archaeology, 2009, *Kingsley Fields, Nantwich, Cheshire (WR02), Assessment of the Charred Plant Remains & Wood Charcoal from Cremation Graves and Related Features*, unpublished report for University of Manchester Archaeological Unit.

White R, 1997, 'Pottery Distribution', in P Barker, R White, K Pretty, H Bird & M Corbishley (eds), *The Baths Basilica Wroxeter: Excavations 1966-90*, English Heritage Archaeological Report **8**, 312-18.

Wild F, 2002, 'The Development of the Roman Road System in the North-West: The Evidence of the Samian Ware', *Britannia* **33**, 268-74.

Williams M & Reid M, 2008, *Salt: Life and Industry. Excavations at King Street, Middlewich, Cheshire, 2001-2002*, BAR British Series **456**.

Willis S H, 1998, 'Samian pottery in Britain: exploring its distribution and archaeological potential', *Archaeological Journal* **155**, 82-133.

Willis S H, 2005, 'Samian pottery, a resource for the study of Roman Britain and beyond: The results of the English Heritage funded Samian Project. An E-monograph', *Internet Archaeology* **17**, http://intarch.ac.uk/journal/issue17/willis_toc.html.

Wilmott T, Cool H & Evans J, 2009, 'Excavations at the Hadrian's Wall fort of Birdoswald (*Banna*), Cumbria: 1996-2000', in T Wilmott (ed), *Hadrian's Wall: Archaeological Research by English Heritage 1976-2000*, English Heritage, 203-395.

Woodiwiss S (ed), 1992, *Iron Age and Roman Salt Production and the Medieval Town of Droitwich. Excavations at the Old Bowling Green and Friar Street,* CBA Research Report **91,** Council for British Archaeology.

Worrell S, 2006 , 'Roman Britain in 2005: Finds reported under the Portable Antiquities Scheme', *Britannia* **37**, 429-66.

Wrathmell S & Nicholson A (eds), 1990, *Dalton Parlours: Iron Age Settlement and Roman Villa Site*, Yorkshire Archaeology **3**, West Yorkshire Archaeology Service.

Wright E, 2002, 'Querns and Quernstones', in Wilson P R (ed), *Caractonium: Roman Catterick and its Hinterland, Part II,* CBA Research Report **129**, Council for British Archaeology, 267-85.

Young C J, 1977, *Oxfordshire Roman Pottery,* BAR British Series **43**.

Zach B, 2002, 'Vegetable offerings on the Roman sacrificial site in Mainz, Germany – short report on the first results', in Jacomet *et al*, 2002, 101-6.

www.ingramcontent.com/pod-product-compliance
Lightning Source LLC
LaVergne TN
LVHW072255300926
849160LV00018B/1668
* 9 7 8 1 4 0 7 3 0 9 5 9 0 *